A Prometheus on a Human Scale – Ignacy Łukasiewicz

Piotr Franaszek / Paweł Grata /
Anna Kozicka-Kołaczkowska / Mariusz Ruszel /
Grzegorz Zamoyski

A Prometheus on a Human Scale – Ignacy Łukasiewicz

PETER LANG

Bibliographic Information published by the Deutsche Nationalbibliothek
The Deutsche Nationalbibliothek lists this publication in the Deutsche Nationalbibliografie; detailed bibliographic data is available online at http://dnb.d-nb.de.

Library of Congress Cataloging-in-Publication Data
A CIP catalog record for this book has been applied for at the Library of Congress.

Translation: Robin Gill, Narrator Translation Bureau
Technical corrections: Katarzyna Rozmus, Jakub Stawarz

Reviewers: Prof. Alison Frank Johnson - Harvard University (USA)
Prof. Krzysztof Broński - Cracow University of Economics (Poland)

Publication financed by The Ignacy Łukasiewicz Oil and Gas Industry Museum in Bóbrka, GAZ-SYSTEM, PERN and Bank Gospodarstwa Krajowego

Project coordinator: Ignacy Łukasiewicz Energy Policy Institute   Partner: Narodowe Archiwum Cyfrowe

Thank you for agreeing to use the photos by: The Ignacy Łukasiewicz Oil and Gas Industry Museum in Bóbrka, Subcarpathian Museum in Krosno, The National Digital Archives, Museum of Pharmacy, Jagiellonian University Medical College, The Museum of History of Photography in Cracow, National Ossoliński Institute, The Walery Rzewuski Museum of History of Photography in Krakow, Przemyśl State Archive and Anna Kozicka-Kołaczkowska.

978-3-631-79291-9 (Print) · 978-3-631-80106-2 (E-PDF) ·
978-3-631-80107-9 (EPUB) · 978-3-631-80108-6 (MOBI)
DOI 10.3726/b16091

Peter Lang – Berlin · Bern · Bruxelles · New York · Oxford · Warszawa · Wien

This publication has been peer reviewed.

www.peterlang.com

Contents

I. The One Who Said Goodbye to the Candle 11

In nostra terra 11

From the temple of eternal fire to the Zglenicki platform 14

Unwritten masterpieces 18

And this is what Poland is 19

From Kościuszko's homeland 26

Carnival in the colour of blood 35

The conspirators from the Hotel near "Luftmaszyna" 40

How can we doubt the good results... 44

A manuscript under a lucky star 51

Studies – the most urgent intention of all 56

There will be no white spirit 60

We distil! 65

I need lamps for this and that – one of these and one of those 69

II. The One Who Spurred the World 79

Pharmacy and dreams of kerosene 79

Entrepreneur and manager 91

An employer from another era 110

In the fight for the interests of the oil industry 114

For the common good – political activity 118

A good man – philanthropist and social activist 126

Chorkówka – the charm of the Polish court 132

The man in the stained apron 140

Honours and distinctions 143

The memory of Łukasiewicz 149

Łukasiewicz – known unknown 158

Father Ignacy is a model for modern generations 159
Teamwork 160
Responsibility in business 160
Energy policy and Łukasiewicz's legacy 161

III. Underwater Oil Pioneer – Witold Zglenicki 163
The Heir, Witold Zglenicki 163
2,400,000 dollars in gold 165
In the Mazovian manor house and Subcarpathian Bóbrka 166
In Warsaw and St. Petersburg 168
In the footsteps of Stanisław Staszic. Flights and falls 171
The Assayist of Riga 174
In Baku, where heaven and hell exist 175
The Assayist Engineer 178
Learning and working, not throwing a hat 178
The inventor 180
Desiderata – the key to the unnoticeable Aladdin's cave 181
And yet they will make the sea sleep 182
165 oil-producing fields, marine fields, platforms and breakwaters 183
Fields according to Zglenicki 185
The first maritime field, or when there are enough important matters on land 186
Technology, science and art 187
No happy ending 188
A legacy "for everlasting times" 188
Shafts in the Gulf of Ilyich Lenin 190
And the oil is still pouring 191

IV. The Power of Bóbrka 195
Political and social life of the Austrian Partition until 1914 195

The economy of Galicia until 1914 200
Galician industry 202
Banking 210
Transport 211
Subcarpathian petroleum mining 213
Drilling, or the breakthrough 216
From Canada and Pennsylvania 217
Borysław fever 218
Entering the 20th century. Innovation, patents and competition 220
Geology or divining rod 223
Eruptions, fires, landslides, groundwater 225
The Subcarpathian refinery industry 227
Factories of drilling machines and tools in Subcarpathia and Lesser Poland at the turn of the 20th century 234
Austrian authorities vis-à-vis the Subcarpathian oil industry 237
Societies and journals 240
Professional vocational education 244
MacGarvey and others 247
Łukasiewicz's heirs in the world 250
Congresses and exhibitions 252
Success in spite of obstacles 255

List of Figures 259

Bibliography 261

Index of Names 287

Geographical Index 293

I. The One Who Said Goodbye to the Candle

In nostra terra. From the temple of eternal fire to the Zglenicki platform. Unwritten masterpieces. And this is what Poland is. From Kościuszko's homeland. Carnival in the colour of blood. The conspirators from the "Hotel pod Luftmaszyną." How can we doubt the good results... A manuscript under a lucky star. Studies – the most urgent intention of all. There will be no white spirit. We distill! I need lamps for this and that – one of these and one of those.

II. The One Who Spurred the World

Pharmacy and dreams of oil. Entrepreneur and manager. An employer from another era. In the fight for the interests of the oil industry. For the common good – political activity. A good man – philanthropist and social activist. Chorkówka – the charm of the Polish court. Honours and distinctions. The memory of Łukasiewicz. Łukasiewicz – known unknown. Father Ignacy is a model for modern generations. Teamwork. Responsibility in business. Energy policy and Łukasiewicz's legacy.

III. Witold Zglenicki – A Pioneer of Underwater Oil

The Heir, Witold Zglenicki. 2,400,000 dollars in gold. In the Mazovian manor house and Subcarpathian Bóbrka. In Warsaw and St. Petersburg. In the footsteps of Stanisław Staszic. Flights and falls. The Assayist of Riga. In Baku, where heaven and hell exist. The Assayist Engineer. Learning and working, not throwing a hat. The inventor. Desiderata – the key to the unnoticeable Aladdin's cave. And yet they will make the sea sleep. 165 oil-producing fields, marine fields, platforms and breakwaters. Fields according to Zglenicki. The first maritime field, or when there are enough important matters on land. Technology, science and art. No happy ending. A legacy "for everlasting times." Shafts in the Gulf of Ilyich Lenin. And the oil is still pouring.

IV. The Power of Bóbrka

Political and social life of the Austrian Partition until 1914. The economy of Galicia until 1914. Galician industry. Banking. Transport. Subcarpathian petroleum mining. Drilling, or the breakthrough. From Canada and Pennsylvania. Borysław fever. Entering the 20th century. Innovation, patents and competition. Geology or divining rod. Eruptions, fires, landslides, groundwater. The Subcarpathian refinery industry. Factories of drilling machines and tools in Subcarpathia and Lesser Poland at the turn of the

20th century. Austrian authorities vis-à-vis the Subcarpathian oil industry. Societies and journals. Professional vocational education. MacGarvey and others. Łukasiewicz's heirs in the world. Congresses and exhibitions. Success in spite of obstacles

I. The One Who Said Goodbye to the Candle

In nostra terra. From the temple of eternal fire to the Zglenicki platform. Unwritten masterpieces. And this is what Poland is. From Kościuszko's homeland. Carnival in the colour of blood. The conspirators from the "Hotel pod Luftmaszyną." How can we doubt the good results… A manuscript under a lucky star. Studies – the most urgent intention of all. There will be no white spirit. We distill! I need lamps for this and that – one of these and one of those.

In nostra terra

"For there is no light for it to hide under a bushel.
Nor the salt of the earth for kitchen spices"[1]

Cyprian Kamil Norwid "Promethidion" (1851)
"In nostra terra, scilicet Polonia"[2]

Vitello (1230–1314)

Take two rulers and a map of Europe. With one ruler, link Norway's highest fjords with Greece's lowest shore. With the other, divide the half-vertical Spanish shore and the mountain range of Europe's border with Asia. At the point where the sticks cross, lies Cracow. The heart of Europe. A thousand years old, the first capital of Poland. It is dominated by Wawel Hill. The castle of former Polish kings, which is viewed in the mirror of the Vistula River. The queen of Polish rivers writhes in a gentle arch at its feet. A few streets further, at the Jagiellonian University, Ignacy Łukasiewicz – a man who, in the name of mankind, took the first few steps along the road that the whole world is now rushing – studied pharmacy.

Here, in the Museum of Pharmacy of the 650+-year-old Jagiellonian University, you can enter his pharmacy. Because it is Polish pharmacy that is considered to be the mother of petrochemistry and the oil industry. Łukasiewicz's pharmacy was its cradle.

"In terra nostra, silicet Poloniae, habitabili quae set circa latitudinem 50 graduum"[3] more than seven hundred years ago the astronomer and

1 Cyprian Norwid, *Promethidion* (Warsaw, 1989), p. 70.
2 Aleksander Birkenmajer, *Astrologia i astronomia w wiekach średnich* [Astrology and Astronomy in the Middle Ages], pp. 11–13.
3 Birkenmajer, *Astronomia..*, pp. 11–13, see Jerzy Burchardt, *List Witelona do Ludwika we Lvivku Śląskim: problematyka teoriopoznawcza, kosmologiczna*

mathematician Erazm Ciołek wrote about his birth within the radiant range of Cracow under the Latin pseudonym Vitello, whose astronomical writings and legendary work *Perspectiva* were still studied by Copernicus and Galileo. Indeed, it is an extraordinary fact that it is in Cracow that the 50th parallel and the 20th meridian, i.e. full-degree grid lines, intersect. There are only five such intersections of full grid points around the globe, including New Delhi, Mecca and Jerusalem.[4] In addition, the 20th meridian connects Cracow with the farthest end of Africa – Cape Agulhas – the cradle of human civilisation. It was recognised as the water boundary of two oceans – the Atlantic and the Pacific.

And if between the cobblestones of the royal courtyard of Wawel you put a branch with two threads attached, and one of them is 500 miles long and the other is 1000, then, like a pair of compasses, you will mark two special circles. The shorter thread will mark Central Europe. The other will embrace it all. The Renaissance genius Leonardo da Vinci closed the circle of his Vitruvian man in a similar way. The lines of the outstretched limbs of Leonardo's model intersect in his navel.[5] They designate the centre of the microcosm of man and the macrocosm of the world. [6]

Seven hundred years after Vitello, the historian Norman Davies simply called his book about the history of Poland *The Heart of Europe*. He explains this metaphor of Poland located in the centre of the continent in the following way:

> The title 'Heart of Europe' came during the writing of the last parts of the book (...).The image of Poland as one of the most vital organs of our continent, the traditional homeland of our most intimate feelings and emotions, seemed to be

---

i medyczna [Vitello's letter to Ludwik in Lvivek Śląski: theoriological, cosmological and medical issues] (Wroclaw, 1979), Vitelo, *Witelonis perspectivae liber (...) = books of perspectivae liber*, (Wroclaw: Ossolineum, 1983).

4 See Mieczysław Czuma, Leszek Mazan, *Pępek świata nazywa się Kraków* [The Centre of the World Is Called Cracow] (Cracow 2000), pp. 29–30.

5 Charles Nicholl, *Leonardo da Vinci: Flights of the Mind* (London: Penguin Books, 2005), pp. 265–266.

6 Daniele Barbaro, publisher of the architectural books of Vitruvius, wrote in the foreword to them(*I dieci libri dell' architettura di M. Witruvio*, 1556, p. 57): "Divine is the power of numbers(divina é la forza dei numeri) (...) In the structure of the cosmos and microcosm there is nothing more dignified than the property of weight, number and measure, from which (...) all divine and human things have arisen, increased and have reached perfection." quote from: Władysław Tatarkiewicz, *Wybór pism estetycznych* [Selected Aesthetic Writings] (Universitas, Cracow 2004), p. 140, cf. Paweł Maurycy Sobczak. *Leonardo da Vinci, Natura i wynalazek* [Leonardo da Vinci. Nature and Invention], 2017.

> particularly appropriate. (…) (…) After examination, it turned out that the title also indicated the geographical location of Poland in the very centre of Europe.[7]

The work of Ignacy Łukasiewicz is one of those phenomena of history which made Poland the heart of Europe and the "centre of the world" in a no less symbolic way. In the middle of the night of the world, precisely in the middle of the 19th century AD, Ignacy Łukasiewicz switched on the light which instantly conquered the whole globe. Thus, he initiated the history of crude oil in the history of human civilisation, the development of methods of its extraction, its distribution into individual fractions and its widespread use. The mass use of kerosene in the lamp of his construction, on the other hand, already at the end of the 19th century, transformed into the stage of turbulent development of the internal combustion engine.

It was Ignacy Łukasiewicz, a pharmacist from pharmacies in Gorlice, Jasło and Brzostek located in Lesser Poland – a geographical district within the realm of Cracow – the future world's first sheikh, not much more than 150 years ago, who commissioned a local craftsman to make a lamp of his own design for everyday use, in which the wick immersed in kerosene could burn and shine with an even, regulated, calm and safe light according to the owner's will. A lamp in which the fire of burning kerosene distilled from petroleum, which had hitherto been despised by science and chemistry as a wild mustang, for the first time in the history of human civilisation surrendered to man.[8]

Because it was Ignacy Łukasiewicz who not only immersed people in the element of open fire, which had not yet been tamed by man, but also connected it with the phenomenon of crude oil, which so far had seemingly been somewhat useless, rather tedious, dangerous and taking land away from people. Surprisingly shrouded in the fumes of poisonous, gaseous vapours above the bogs and puddles of stinking slurry. Fascinating with fiery fountains. An astonishing view of distant, mysterious night-time fireflies based on legends, myths full of ghosts and spells, mighty gods and titans. When this pharmacist from Polish Subcarpathia tamed this unpredictable monster, he himself experienced its paroxysms. During his experiments with oil distillation in his laboratory, explosions often took place. One such explosion permanently burned his brow.

And yet he promoted crude oil. He consciously and deliberately opened the first oil well in the history of the world, caused the excavation of its first shaft in the forest of Bóbrka, halfway from Dukla to Krosno. As a

7 Norman Davies, *Heart of Europe* (London 1995), p. 7.
8 Cf. Jerzy Sikora, *Ignacy Łukasiewicz* (Katowice: Wydawnictwo "Śląsk", 1978), p. 229.

celebration of this fact and a symbol that can be compared probably only to the metaphorical semantics of the Greenwich Meridian, there has since then been a stone obelisk with a date engraved on it – 1854.[9]

A sign of the year in which man made the world aware that oil is a valuable and useful mineral, just like coal and other treasures of the Earth. To commemorate the landmark date. The year in which a new history of the world began in Bóbrka near Krosno.

From the temple of eternal fire to the Zglenicki platform

> *22 June 1830: Second visit to the gas fires. Indians call this place Joal, the Persians Atashkadeh, from which the Russians made Tazhki. Joal is for Indians what Mecca is for Mohammedans. (....)From afar you can see four pillars of stationary fire standing in the air. The white wall surrounding this place of pilgrimage for the Indians was clearly visible on the dark background. Fires, scattered here and there, crawled at various distances. The entire Absheron Peninsula is their volcano. The main crater is in Joal. By the whirlpool all the air is infiltrated by the strong smell of coal gas. A wall of measurable height, cut into teeth on the top, encircles Joal in an irregular pentagon. The outer part serves pilgrims as a garden, the inner part as accommodation. (...) (...) Having entered, you can see in front of you in the middle of the courtyard a small bell tower with four arcades into a square space. On each of the four corners is a chimney from which a pillar of fire explodes. This belfry was built on the well that was the most abundant in the fiery gas. (...) Some Indians, naked to their waist, sat stationary, while others prayed or played dice.*[10]
>
> *(Aleksander Chodźko-Borejko 1804–1891)*

This description of the temple of the worshippers of the Atesgias in the village of Surakhani in Azerbaijan on the Absheron Peninsula, located about 18 km northeast of Baku, is an account of the last decades of tradition, which in

9 *Bogactwo Bóbrki* [The Riches of Bóbrka], Muzeum Przemysłu Naftowego i Gazowniczego im. Ignacego Łukasiewicza w Bóbrce, (Bóbrka, 2009), p. 21.

10 Andrzej Chodubski, *O świątyni czcicieli ognia "Ateszgiah."Z relacji polskich zesłańców i podróżników* [On the "Atesgias" Temple of the Worshippers of Fire. From the Relations of Polish Exiles and Travellers], (Elbląg 1991), p. 7.

Figure 1: Memorial obelisk in the Museum in Bóbrka. Source: Anna Kozicka-Kołaczkowska

this place dates back to ancient times and was considered the epicentre of Zoroastrianism.

The "Country of Fire" fascinated the civilisations of Eurasia and the Middle East for thousands of years. In the 4th century BC, this was the name of the Caucasian state. In Persian it sounded like – "Ader – badagan," in Armenian "Aterpatakan" or "Artpatakan", while in Arabic it was "Aderbaijan" and "Azerbaijan."[11]

In the imagination of the ancient Greeks, the rocks of the Caucasus Mountains – the mysterious and terrifying land of fire – were the most appropriate place for Prometheus' eternal execution. The story of the cruel punishment imposed on the Greek titan by Zeus aroused mercy and fear of the spectators of ancient theatre already two thousand five hundred years ago in the tragedy of Aeschylus.[12] For many centuries it was the topos of Mediterranean civilisation. People grateful for the gift of fire have recalled the history of Prometheus for centuries. Prometheism was one of the great ideas of the Romantic era, the times of the author of the above-mentioned memoirs – Aleksander Chodźka, a close friend of the genius Polish Romantic Adam Mickiewicz. It was understood as a readiness to fight for people's happiness even in a dispute with the gods. To steal its divine treasures from the heavens.

But even there, in the temple of fire in Azerbaijan, which for almost 150 years has been an un-drying Aladdin's cave of oil, the permanence of eternal fire has only been combined over the centuries with flammable gas. All the memoirs of travellers of old speak only of gas. The oil with which the land of the Southern Caucasus is saturated like a sponge goes unnoticed by the travellers and diarists. They are not convinced to light it, to play with this unknown, wild liquid capable of the only fireworks in nature.

> "A travelling Frenchman expressed it thus: If the fires in Baku and all the details relating to them had been visited by experienced physicists and experts in chemistry, the gas would have been used for lighting far earlier."[13]

The Polish exile Butowt-Andrzejkowicz, who had been sentenced there for ten years as punishment for political activity by the Russian authorities between 1844 and 1855, noted the remarks, of course only about the gas, while since 1853 Łukasiewicz's kerosene lamp had already been sent out into the world.

If Łukasiewicz had not insisted on lighting with kerosene, oil could have long remained a stinking, useless slurry by which the earth becomes useless and worthless. After centuries of thinking and generations of brainstorming about what it could be used for, and getting used to the idea that nothing special, apparently few people, except perhaps pharmacists, wanted to pay attention to it.

---

11 Chodubski, O *świątyni..* [Atesgias Temple..], p. 5.

12 Cf. Aeschylus, *Prometheus Bound,* trans. Joel Agee, (New York: NYRB Classics, 2002)

13 Chodubski, O *świątyni..,,* p. 13.

And unsuccessful attempts to make it useful had been going on since the earliest times. Oil was known to the Babylonians, Egyptians, Greeks, Romans, Arabs and Byzantines. The word "naphtha" is derived from the Assyrian *naptu*, which meant "earth oil." Mediaeval Europe noticed oil. The occurrence of oil in Poland was mentioned by the 15th-century history writer Jan Długosz. It was seen in old Polish encyclopaedias, herbaries and physiographic descriptions. Gabriel Rzączyński (*Historia naturalis curiosa Regni Poloniae, Magni Ducatis Lithuaniae XX divisa*, 1721, *Auctuarium historiae naturalis*, 1736) and Jan Krzysztof Kluk, a geologist and outstanding natural scientist, described its resources in the Subcarpathia region (*Exploration, cognition and use of peculiarly fit fossil objects* 1781, 1782). A particularly outstanding Polish geological paper from the turn of the 19th century was the work of Stanisław Staszic (*On the Upper-Sarmatian and later Polish landscaping*, 1806).

Samuel Bogumił Linde defined the word "kerosene" in his monumental "*Dictionary of the Polish Language*" of 1807–1815. We read in it that the word means "a species of natural oil, very liquid, easily igniting."[14]

However, still in the middle of the 19th century oil was ignored by great chemists of this world, although from today's perspective its first distillation and invention of common use for lighting was no less a discovery than electrification, and certainly more important than steam engines. On the other hand, even as a goodbye to the century the same dons, proud of the memorable successes of science, called their age just the age of "steam and electricity." All the more so, the technology of the second half of the 19th century – in the times of Łukasiewicz – was unable to see the extraordinary future of crude oil. It was Ignacy Łukasiewicz who made fuel and energy from the explosive, unpredictable "rock oil" uninteresting to the worlds of science, industry or business, which in the future would start engines, motors, cars and machines of successive decades.

At the end of the 19th century, travellers who visited the temple of Ateshgah did not see there yogis, who until recently had come to make a life by the sacred fire of burning gas. The place was guarded by the last, honorary custodian of the temple – the Pole Gabriel Wrzosek, who came to Surakhani in his youth as a recruit to the Russian army and stayed there.[15] At that time Azerbaijan was already a promised land for oil-hungry people.

It was not a coincidence, but the result of Łukasiewicz's example that every metre of its surface and adjacent bays of the Caspian Sea in search of this wealth was measured, researched, sketched and described by the Polish

---

14 M. Samuel Bogumił Linde, *Słownik języka polskiego* [Dictionary of the Polish Language] (Warsaw: PIW, 1951), vol.III.,p. 230.

15 Linde, *Słownik*.., p. 20.

engineer Witold Zglenicki at that time. He would fight with obstinacy worthy of his predecessor Łukasiewicz for permission to build the first ever offshore drilling platform, several decades before others matured to this project.[16]

Unwritten masterpieces

Thanks to Łukasiewicz's kerosene lamp, a civilisational change in everyday life habits took place. People made a mass, common, great step towards independence from the power of the rotation of their planet Earth, in liberation from the rhythm of day and night, darkness and light, life and death.

In 1853, Jane Austen had only been dead for 36 years. George Byron had been dead for 29 years. Goethe had died 21 years earlier. Only four years earlier Fryderyk Chopin. All of them, when they wanted to write, create, play from a score, have fun, dance, eat, just live, the darkness of evenings and nights, of living rooms, bedrooms, halls, nooks and crannies of home and corridors, were illuminated above all by the uncertain, flickering, expensive light of the open flame of burning candles.

In 1849 Johann Strauss, the father, died. If he had lived a bit longer, would he have matched his son in creating larger musical forms? How many miraculous operettas and maybe even operas could have been created earlier, if Strauss the Elder had had a chance to make better use of his time, thanks to the cheap, available, safe light of an oil lamp? There is no doubt that Johann Strauss the Younger was already writing his masterpieces by kerosene light.

Łukasiewicz's lamp advertised by the Pole in Vienna, sent in numerous prototypes around Europe, made a rapid career on all continents and changed the daily rhythm of people's lives. It appeared in flats, workshops, salons, ballrooms, and school, university and hospital rooms. It remained in them for a long time before Edison's light bulb and electricity became the daily bread of the next century for everyone. Unlike electrical installations, it was reliable during the wars that ruined the 20th century and during natural disasters. For the first time in the history of the world it saved man's life during the nighttime, urgent operation on 31 August 1853, in then Polish Lviv.[17]

Łukasiewicz's lamp project, based on Romantic ideas – for the good of mankind – unpatented by its originator and constructor, captured by a Viennese

---

16 Cf. A. Kozicka-Kołaczkowska, *Ropa z głębi morza* [Oil from the Depths of the Sea], www.instytutpe.pl/wp-content/uploads/2017/09/Historia-sektora-energii-nr-1-2018-Witold-Zglenicki.pdf, http://www.instytutpe.pl/wp-content/uploads/2017/09/Historia-sektora-energii-nr-2-2018-Witold-Zglenicki-cz.-II.pdf.

17 *75 lat temu pierwszą lampę naftową dla świata zapalił Polak, Był nim Ign. Łukasiewicz – pierwszy destylator nafty – twórca pierwszej kopalni nafty na świecie – twórca przemysłu naftowego i destylacyjnego* [75 years ago the first kerosene lamp for the world was lit by a Pole, Ign. Łukasiewicz – the first kerosene

company, multiplied in Europe and around the world. Thousands of designs for kerosene lamps were created, becoming a showcase for the artists' imagination. Luckily for them soon, at the end of the 19th century, came Art Nouveau with its capricious line, wealth of floral motifs, colours that turned the moods of twilight and night into a colourful fairy tale. Petroleum lamps were clad in delicious glass crinolines and stained glass, until they became works of art.[18] It is difficult for people to forget about them even in the era of electricity, which is why electric lamps which pretend to be oil are still lit today.

It was thus only with the first petroleum lamp that fire – the gift of Prometheus – began to bring to all mortals not only warmth, cooked food, the wobbly light of an easily extinguished candle, the pinching smoke of torches, the element of bonfires and fire disasters, but friendly, tamed, mobile, safe, home light.

Uncontrolled fire – the unpredictable dragon of burning kerosene, breathing with stinking breath and fire – turned in this lamp into a friendly, service-minded, domestic dog. Ignacy Łukasiewicz – this true Prometheus on a human scale – today gives the world his lamp from the portrait of the museum in the historic oil mine – the first in the world. In the unique open-air museum, The Ignacy Łukasiewicz Oil and Gas Industry Museum in Bóbrka onear Krosno you can also admire an exact replica of sheet metal and mica panes. Its first flame marked a new starting point in the history of the world in the race for the new, desirable treasure of oil. That was its power. That is its power.

And this is what Poland is

If Europe is a nymph – Naples
is the nymph's blue eye – Warsaw
Heart – thorns in the leg – Sevastopol,
Odessa, St. Petersburg, Jelgava –
Paris – its head – and London – its collar

Starched[19]

---

distiller – creator of the first kerosene mine in the world – creator of the oil and distillation industry], Ilustrowany Kuryer Codzienny, 1927, nr 342; http://www.instytutpe.pl/inne/165-lat-temu-zaplonela-po-raz-pierwszy-lampa-ignacego-lukasiewicza/.

18 Cf. Teresa Jabłońska, *Lampa naftowa* [The Kerosene Lamp], (Warsaw, 2012), *Blask lampy naftowej* [The Light of the Kerosene Lamp], Muzeum Piastów Śląskich, Opole 2010, *W kręgu światła lampy naftowej* [In the Light of the Kerosene Lamp], ed. Jan Gancarski, Krosno Muzeum Podkarpackie, Chorkówka, Muzeum Przemysłu Naftowego i Gazowniczego im Ignacego Łukasiewicza w Bóbrce, 2001.

19 Juliusz Słowacki, *Wyjazd z Neapolu* [Departure from Naples], in: *Wiersze i poematy* [Verse and Poems], (Wroclaw: PIW, 1959), p. 202.

Figure 2: Collection of kerosene lamps from The Ignacy Łukasiewicz Oil and Gas Industry Museum in Bóbrka. Source: Anna Kozicka-Kołaczkowska

When the 19th century entered its last two decades, an eighteen-year-old from the French provinces named Jarry bundled together a script for the school puppet theatre. After several years an unpredictable, but surprisingly frequent in the world of art, turn of things, his theatrical piece called "Ubu Roi"[20] won dramatic scenes. The young author did not even live to the end of the first decade of the 20th century, but his play became a precursor of the coming times of the theatre of the absurd and grotesque. Today, in the postmodern era, after more than a century and a half, it still appears on the posters for modern theatre, including in Poland, which can afford to distance itself from the grotesque, because it trains its taste on Witkacy, Gombrowicz and Mrożek. The composer Krzysztof Penderecki even used it as a canvas for the libretto of an opera buffa. [21]

As dizzying a career as it made itself, it made a phrase which the young author used to determine the place of action of his play. He is believed to have said during the premiere that the play was set "In Poland, meaning nowhere."[22] This phrase, as a typical "winged word" concerning a work that has nothing to do with realism, is still an object of various, fantasy interpretations, often absurd. Yet throughout the 19th century until the end of World War I, after which Europe's borders changed, it defined the real political reality. For 123 years, Poland did not function as a state, plundered by a company of the three neighbouring countries – Russia, Prussia and Austria.[23]

This does not mean, however, that it did not exist and the Polish nation dissolved into non-existence. Even the young Frenchman from the provinces knew that it existed, although tragic, it was a twist of fate for a country which two centuries earlier had been a major power. In previous centuries, Polish grain had been sent to the port of Gdansk and saved the peoples of Europe from hunger. It built the power of a kingdom 10 % of whose population, as in no other country, were free citizens – the nobility. It was King Jan Sobieski who was the only one who, with the genius of a strategist

20 Alfred Jarry, *Ubu Roi*, (Dover Publications Inc., 2003).
21 Cf. Krzysztof Penderecki, *Ubu rex: burleske suite*, development Hennig Brauer, (Schott Music International, 1996).
22 Cf. Tadeusz Żeleński-Boy, in: Alfred Jarry, *Ubu Król czyli Polacy* [Ubu Roi, or The Poles], (Bydgoszcz, 1993), p. 6.
23 Norman Davies, *The Legacy of Spiritual Power, Poland under the Partitions 1795–1918*, in: *Heart of Europe; A Short History of Poland*, pp. 154–185, Kieniewicz, *Historia Polski..*, p. 22.

Polish lands in the 19th century

Figure 3: Map of Polish lands after 1815. Source: Authors.

and Polish military, still at the end of the 17th century defeated and halted at Vienna the unhindered Turkish invasion of Europe, which terrified the France of Louis XIV and the deadly threatened Vatican.

Unfortunately, the entire 18th century, in the disintegration of the degenerated noble democracy incapable of yielding to the King's power, the Commonwealth of Poland, dismantled from within by domestic agents and surrounded by neighbouring powers, demoralised by confederations of foreign parties, the bribery of deputies breaking off the *Sejm*, the entanglement of the Polish king in Russian interests, was attacked from three sides by its neighbours. As late as 3 May 1791, in order to save the country, the patriotic communities led to the enactment of the Basic Law – the Constitution as the first in Europe and the second after the American Constitution – in the *Sejm*. All the more mobilised were the enemies who, among themselves, partitioned the Polish state struggling with internal disintegration, unable to withstand military resistance finally in 1795.

From the perspective of the next century, however, the non-existence of the Polish state and its simultaneous existence in the European iconosphere was, as can be seen, a living issue, obvious even for a young Frenchman from the provinces, and even more so for his parents and forbears. It has to be admitted that the articulation of this particular 19th-century absurdity of existence in the consciousness of Europe of a state without a state was this boy's success. Moreover, it was consistent with the feelings of the majority of Poles who disagreed with this absurdity. This absurdity was ended only by the weakening of the partitioning states with the war and the movement of the Bolshevik revolution in the 20th century. The diplomatic, organisational and armed efforts of the Polish political and military elite were included in the drawing of new maps after the World War I in 1918, and as a result Poles regained national sovereignty. [24]

However, in the middle of the 19th century, when Ignacy Łukasiewicz's life and activity culminated, the oppression of living and developing economic activity in a country colonised by foreign rule was at its peak and lasted until his death. The hopes of regaining freedom during the first decade of the 20th century seemed to be insignificant. Even the mighty 19th-century Europe, in their wildest dreams, did not expect their order to collapse soon either. Thus, twenty years after Łukasiewicz's death, the Polish artist Stanisław Wyspiański expressed this state of conviction about the permanence of things and the mystery of the Polish spirit in a beautiful, symbolic scene of the poet with the bride from a village near Cracow in a colourful, regional costume in the following dialogue from the play *The Wedding* written in the first year of the new, 20th century:

> POET
> Worldwide
> you can look for Poland, o bride,
> And nowhere shall ye see it.
> BRIDE
> Then maybe it is a waste of time to look for it.
> POET
> But there is one small cage –

24 Stefan Kieniewicz, *Historia Polski 1795–1918* [History of Poland], (Warsaw: Państwowe Wydawnictwo Naukowe, 1968), p. 22.

o, may Jagusia cover up
her hand under her breast.
BRIDE
This is a pleat
of the corset, sewn together a little tightly.
POET
- And there he knocks?
BRIDE
And what kind of learning is this?
Heart -/ -?
POET
And this is what Poland is.[25]

The symbolism of the heart, brought here by the poet, in which Poles kept the memory of the free Commonwealth of their ancestors for the age of their captivity, is characteristic and intrudes in the thinking about Poland and Poles in a natural way, not only to Norman Davies.

Finally, already at the beginning of the 19th century, the national bard of the Romantic era Adam Mickiewicz made a recommendation to his compatriots: "Have a heart and look to the heart,"[26] which survived as a cultural message over the next centuries.

In Polish culture, the whole 19th century, in contrast to the unfortunate political situation of the nation, was a wonderful period of the eruption of brilliant artists. It began with Romanticism, and after a solemn phase of the philosophical idea and pragmatism of positive work in the second half of the century, it ended with an accent of the phenomenon of the work of Stanisław Wyspiański – the main representative of the neo-Romanticism of Young Poland. Because Romanticism in its most severe waves of sober, stubborn Positivism until the end of the 19th century, which in other countries at that time had been floating for a long time in the depths of often extreme naturalism, actually never peeled away from the Polish heart.

Without any exaggeration, we can also say that Ignacy Łukasiewicz was, in his deepest essence and action, permeated by the spirit of the noblest ideas of both these great epochs of Polish culture. These were the motto of his charismatic behaviour and tireless efforts throughout his life. Today, without being aware of this cultural and historical background, many of his

25 Stanisław Wyspiański, *Wesele* [The Wedding], (Cracow: Społeczny Instytut Wydawniczy, 2016), pp. 148–149.

26 Adam Mickiewicz, *Romantyczność* [Romanticism], in: *Wybór poezyj, Ossolineum* [Selected Poems, Ossolineum], (Wroclaw – Warsaw – Cracow – Gdansk – Lodz, 1986), p. 103, – in the stanza "Dead you know the truths, unknown to the people// You see the world in powder, in every star spark// You don't know the truths of the living, you won't look at a miracle! Have a heart and look into the heart!"

deeds can look eccentric, incomprehensible or even absurd from the perspective of contemporary standards. This is how his charismatic, idealistic behaviour was assessed even by many of his contemporaries, albeit from other cultural circles.

For example, purely idealistic, Romantic, but also positivist in the Polish spirit was not patenting the invention of the kerosene lamp as a gift for humanity. After receiving Łukasiewicz's prototypes, however, the Viennese Dittmar did not hesitate to patent it, capturing unscrupulously someone else's invention and usurping the right to the fame of the constructor.[27] John Davison Rockefeller's envoys were unable to believe that Łukasiewicz did not accept any remuneration for allowing them to enter the secrets of the technological process of oil extraction and refining, which in their opinion was almost a mental illness.[28] There are many similar examples from his life.

However, apart from the inhabitants of the Polish land, did anyone go through such a course of Mickiewicz's Romanticism as the bard served to his compatriots in "Ode to Youth"?

> "En masse, young friends!
> In happiness our ends.
> Strong in unison, reasoned in rage
> Move on, young friends!"[29]

For centuries, the words of the "Ode to Youth" have been engraved in the hearts and minds of the poet's compatriots, who, as "reasoned in rage," did not endow such things for others. And they even shed blood "For our freedom and your freedom" according to the adage known to each of them. This is the eternal Polish background.

However, it was precisely this cultural, spiritual heritage, this "incomprehensible frenzy" and not naivety, as others thought, in order to strive for the happiness of all, that was also the main source of Łukasiewicz's personal strength and dynamics of action. His success in initiating, organising and developing a new industry in his own country and in the world grew out of the idea and deep sense of mission, the sense of working for the good of his fellow countrymen and all mankind. With a sense of absolute pragmatism, without false pathos, with a full adaptation of Promethean, Romantic ideas flowing straight from the sensitivity of heart and spirit, not distracting also

27 Ludwik Tomanek, Ilustrowany Kuryer Codzienny [Illustrated Daily Courier], 1927, No. 342.

28 Sikora, *Łukasiewicz a Rockefeller…*, pp. 259–270, *John Rockefeller did business on… Łukasiewicz*, Ilustrowany Kuryer Codzienny, 1927, No. 344.

29 Mickiewicz, *Oda do młodości* [Ode to Youth], in: *Dzieła zebrane* [Collected Works], p. 65.

from the fight with weapons and positivist, consistent action. Because it is "In happiness our ends."

From Kościuszko's homeland

"Every righteous American, every sincere believer in freedom loves you and worships you."[30]

Thomas Jefferson to Tadeusz Kościuszko, 1813

Jan Boży Józef Ignacy Łukasiewicz was born on 8 March 1822, so in the background of his biography a long string of key events of the 19th century for Poles developed. They determined both his personality and fate.

When at the end of his life he was already rich and famous, and his countrymen began to spread the news that as a sponsor and benefactor he had paved the streets around Krosno with his guilders, and when, in recognition and universal respect, a hail of honours, anniversaries, honorary fetes, diplomas and medals began to deluge him, to the end only modest and ascetic with regard to himself, he requested:

> "Gentlemen, I was born in a slipper, I've walked all my life in a slipper, and let me die in it."[31]

He said this at a time when the Polish nobility, although pauperised, thrown out of estates and pushed out to the cities, to intelligent occupations or to emigrate, still knew the steps of a sumptuous mazur[32] and cultivated the old Polish fashion of splendour and pride.[33] It was the Polish nobility that caused the European ball to begin with a polonaise for many decades of past centuries.[34] Contrary to the cosmopolitans and their Parisian costumes, it was they who, from the very beginning of a holiday, put on a Polish, rich robe lined with sobols, a kaftan with diamond buttons, daggers trimmed with genuine gold, a Slutsk belt – a real work of art – as well as a precious sabre at

---

30 Barbara Wachowicz, *Nazwę cię Kościuszko* [I Name You Kościuszko], (Warsaw: Oficyna Wydawnicza RYTM, 2000), p. 3.

31 Ludwik Tomanek, *Ignacy Łukasiewicz Twórca przemysłu naftowego w Polsce. Wielki inicjator – wielki jałmużnik* [Ignacy Łukasiewicz, Founder of the Oil Industry in Poland. A Great Initiator – A Great Almsgiver], (Miejsce Piastowe 1928), p. 72.

32 Cf. Mazur from the opera *"Halka"* Stanisław Moniuszko, https://youtu.be/kUMnvSnKnpo accessed 10/06/2019.

33 Magdalena Bartkiewicz, *Polski ubiór do 1864 roku* [Polish clothing until 1864], (Wroclaw 1979), pp. 126–133.

34 Cf. https://www.youtube.com/watch?v=f-3K2NYJl3E, https://www.youtube.com/watch?v=_PpfJUDGyBI.

the side. A substitute for this style can be admired today in the operas of the composer Stanisław Moniuszko and vigorous dances. This style, especially in the first half of the 19th century, was still natural and contemporary in Polish lands, and after enchanting previous epochs with the French fashion, it returned in the time of captivity.

> "Because wanting to be a good Pole
> It is not enough to have a Polish soul
> It's necessary to part with the frock coats,"[35]

said the words of a popular song.

Ignacy Łukasiewicz was a nobleman by birth. Tradition connected his family with the coat of arms of Łada[36]. The family of Ignacy's mother – Apolonia nee Świetlik – was even connected with the historical dynasty of the princes of the eastern Świętopełk dynasty. However, he brought out of his home a consistently professed conviction that neither birth nor great wealth constitute the most important values of man. In Łukasiewicz's childhood home, one earned one's bread solemnly, but apart from money, other values were sincerely respected.

Łukasiewicz was born at a time when the years of over a quarter of a century of captivity had already had an impact on the social moods of Poles. As Magdalena Bartkiewicz writes:

> "Poles longed for their lost Homeland, and the symbol of their fate became the motif of a wandering soldier, who had 'garments like rags, but they will pass among the ladies'. In this way, society paid homage to those fighting for Poland; departing from the old principles of caring for clothes, it was not the elegance of clothes that was valued now, but the patriotic attitude, courage and bravery – and they appreciated them most."[37]

This is how the Parisian emigrant and Professor of Slavic literature at the Sorbonne, the Polish, spiritual, romantic idol, poet Adam Mickiewicz, who died in the middle of this century, ostentatiously wore a *czamara*[38] – a coat of the lesser Polish nobility – and instead of the fashionable stick of the elegant man and star of salons, which he was, he clacked on the cobblestones of Paris with a walking stick cut from somewhere in the bushes of the Parisian forest. And it was not the prank of an eccentric, but a symbol of a pilgrim and forced exile whose homeland suffered.[39]

---

35 Bartkiewicz, *Polski ubiór...*, p. 127.
36 Łada, Ładicz, Mancz – polski herb szlachecki z Mazowsza according to T.Gaj, *Herbarz polski* [Polish Emblems], (Gdansk 2007).
37 Bartkiewicz, *Polski ubiór...*, p. 128.
38 Bartkiewicz, *Polski ubiór...*,pp. 128–129.
39 Kossak – Szczucka, *Dziedzictwo* [Heritage], T.I., p. 151.

The father of Ignacy Łukasiewicz – Józef – a man born in the 18th century in Mazovia, the central district of the former Republic of Poland and the soil of Chopin's mazurkas – distinguished himself in the Tadeusz Kościuszko Uprising[40], who in the last, great fight of that century against the Russian invader even swept the Polish peasants behind him, and after losing, by virtue of a deal with Russia, in exchange for abandoning repression against military volunteers, committed himself to emigrating from the country.

The end of the previous century was tragic for the Commonwealth of Poland, when neither the armed effort against the collusion of the looting neighbours nor the work of Polish political parties managed to prevent the loss of the state.

From the turn of the 19th century, every Polish biography had been a series of moments of hope for regaining freedom and discouragement, the euphoria of the uprisings and resignation or complete surrender. Subsequent generations born in the 19th century were confronted with the dilemma of choice – whether to accept the mediocre role of the conquered nation, forced to renounce its own tradition, language, dignified position, property and development opportunities, or to fight. Is it more effective to ensure physical vegetation, the silent cultivation of tradition under the guise of obsequiousness towards the foreign powers and wait for a convenient moment, or open, armed action as soon as possible?

The Polish uprisings, maintaining rebellious thoughts against foreign violence, hidden actions and secret associations, perfectly harmonised with the Romantic spirit of the age, the worship of individualism, heights of extraordinary achievements, original excesses, the mood of the turn of the century, which recalled the great heroes of Shakespeare, brought Byron and the great poets of Polish Romanticism, who became the spiritual leaders of Poles. It was Romanticism that resurrected the ancient myth of Prometheus in culture, which perfectly harmonised with the belief in the capabilities of the outstanding individual and at the same time with Polish dreams of great heroes and the struggle for the highest goals, among which freedom always came first.[41]

In Ignacy's family house the sabre of his father from the Kościuszko Uprising and a commemorative, iron ring with the inscription "Homeland to its defender" received, as one may guess, from the hands of Chief Kościuszko himself, were kept with reverence. The small brochure "*Will the Poles stand*

40 Wachowicz, *Nazwę cię Kościuszko,* (Warsaw 2000).

41 Cf. M. Janion, "Powstrzymać Prometeusza" [To Stop Prometheus], in: *Wobec zła* [In View of Evil], (Chotomów 1989), pp. 147–157.

up for independence"[42] by Kościuszko's adjutant, kept secret in a pharmacy cache, became one of the reasons for the criminal trial and imprisonment of young Łukasiewicz in the future.

Tadeusz Kościuszko lived out his days in Solura in Switzerland. He was not able to bear being across the ocean to the end, although in America he received awards for historic battle victories, a monument at the West Point military academy and the friendship of two American presidents. In the homeland of the Swiss hero Winkelried, who was worshipped by Polish Romantics, he died five years before the birth of Ignacy Łukasiewicz, but in Polish homes the forbidden memory of him, the great leader of the fight of the 18th century, lasted. Even in the 20th century, portraits of Kościuszko were still hanging on the walls, and serviettes decorated his figures on horseback and busts. In alabaster and bronze, and at market stalls in plaster and clay. The young Łukasiewicz received the cult of Kościuszko and his ideas with his mother's milk.

The family tradition indicated that the Łukasiewicz family belonged to the Łada coat of arms, but there are no sources that would bring the family's pedigree and its past financial condition closer to the family. On the birth day of Ignacy – the youngest of his siblings – his father Józef farmed in manor farms in Lesser Poland near Mielec, which at that time had been incorporated with the entire Lesser Poland West and East, including the surroundings of the Polish Carpathians and Subcarpathians, to the part occupied by the Austrian Empire under the rule of Vienna, called Galicia by the partitioner, for 27 years[43].

The unclear status of Ignacy's ancestors, impoverished landowners, was nothing unusual in a country ravaged by decades of war and the sequestrations of the invaders, in the area of inheritance and confiscation of family estates, which threatened the new rulers of the Polish lands. It is highly probable that the family did not pay with prison, nor with Siberia, but with impoverishment, for their participation in the fight under Kościuszko's command. After all, Ignacy's father came from the vicinity of Plock, which together with Warsaw came under the rule of Russia. His relocation to Galicia, the Austrian Partition, so already then, for half a century, abroad, was probably a consequence of that past, and a lack of family documentation and specific information about the past.

---

42 Cf. Józef Pawlikowski, *Czy Polacy wybić się mogą na niepodległość* [Can Poles Break Out to Independence?], (Warsaw: Wydawnictwo Ministerstwa Obrony Narodowej, 1967).

43 Kieniewicz, *Zabór austriack..i*, pp. 26–30.

After the Kościuszko Uprising, Józef Łukasiewicz was first a clerk at the courts of the Austrian Partition in Dynow and Chorzelow near Mielec, where his four children were born. Two sons – Aleksander and Franciszek – and two daughters – Emilia and Maria. The youngest, Ignacy, was born in the nearby Zaduszniki estate on the Vistula River. After the Łukasiewiczs had collected the appropriate funds, Zaduszniki was already a lease[44], just like two other manor farms in Czajkowa and Zachwiejow.[45]

The Łukasiewicz brothers and sisters owed their initial education, in accordance with the practice of the time, to home education. The teacher had a dignified atmosphere in the house with Kościuszkovian traditions. He was a reserve colonel of the former Polish army – Woysym Antoniewicz – a war veteran, who, because of his disability, together with his family, had the status of resident in the Łukasiewicz household. It was a common custom of Polish homes to accept individuals of meritorious service to the Polish cause who had either sacrificed their property or lost it because of this and were left without any means of subsistence under a foreign authority. There is no doubt that Colonel Woysym Antoniewicz also breathed the spirit of steadfastness and freedom into his pupil Ignacy.

Józef Łukasiewicz's entrepreneurship must have paid off quite quickly, since already in 1824 he could buy a tenement house in Rzeszow[46]. He was

---

44 The content of the document entitled Instrument Aukcyi located in the Rzeszow State Archive shows, among other things, that on 24 June 1828 Józef Łukasiewicz took two manor farms in Czajkowa and Zachwiejow on a three-year lease. The annual lease amount in Czajkowa was 864 guilders and in Zachwiejow 315 guilders. In addition, there was a 10 % relocation protection, i.e. the sum of 117 guilders 54 crowns. This gave a total of annual liabilities in the amount of 1296 guilders and 54 crowns, which were recorded on the mortgage of the house in Rzeszow under no. 368. Its value was "according to the attached tabular protocol court-taxed" at 12057 guilders 16 crowns. The document was signed by Józef Łukasiewicz possessor of Zadusznik and witnessed by Mikołaj Solmann and Władysław Szymanowski. The document became legally binding on 30 June 1828, after the session of the Rzeszow Magistrate and the entry in the book *Liber Onerum et obligationum.*

45 Preserved documents inform that at that time Józef's brother Jan Łukasiewicz and his wife Rozalia from the Ivanickis also lived in Zaduszniki. Probably it was their daughter Emilia who was one of Ignacy Łukasiewicz's godmothers. This branch of the Łukasiewicz family lived in Zaduszniki in 1830, because their son Marcin died there in the autumn of that year.

46 The Łukasiewicz's house was located at ulica Pańskiej no. 368. Its location can be determined on the basis of a map from 1841, preserved in the Rzeszow State Archive. Analysis of the map shows that the building with this number was located at the corner of the street defined on the map as Popijarska Street (*Expiaristen Gasse*) and present Fircowskiego Street, opposite the complex of former Piarist buildings,

undoubtedly a farmer as resourceful, fluent in management and management as he was brave during the war. Rzeszow was a big city[47], the seat of the *cirkuł*[48], and for the group of children it was important as it had a *gymnasium* school with many centuries of tradition. At the time when Ignacy Łukasiewicz was studying there, in accordance with the policy of denationalisation conducted by the Austrian partitioner, German was the language of instruction, and the curriculum did not include teaching the pupils' mother tongue or the history of Poland. After several decades of captivity, even reading literature in Polish was badly perceived by the authorities, while the most strictly forbidden readings, the distribution of which was treated as a criminal offence, included works by the most outstanding Polish artists of the Romantic era, who, as the bearers of the idea of freedom, were almost *en bloc* forced to live and create in exile.

Ignacy started attending the Rzeszow gymnasium in 1833[49], as recorded in the first class enrolment book (*Studiosi 1^{mae} Grammaticae Classis*).[50] Starting school at the age of about 10–11 years after home schooling was the rule throughout the 19th century. In the annals of the school, which today functions as a general secondary school, the final grades of Ignacy Łukasiewicz's completion of four grammar-school classes in 1837 have also

---

which housed the gymnasium (*Gymnasium*), the church (*Gymnasialkirche*) and the District Office (*Kreisamt*). At the back of the gymnasium church, the map shows the location of a primary school – the so-called *Normalschulen*.

47 *Słownik geograficzny Królestwa Polskiego i innych krajów słowiańskich* [Geographical dictionary of the Kingdom of Poland and other Slavic countries], Vol. 10, pp. 151–154, Vol. 15, pt.2, p. 568.

48 *Cirkuł* – old administrative area.

49 Earlier, in the years 1823–1828, Aleksander was a pupil of the Rzeszow *gymnasium* – the oldest son of Józef Łukasiewicz was a pupil of the Rzeszow *gymnasium* in the years 1823–1828. He graduated from the Rzeszow *gymnasium* in the 4th grade and one class of the humanities *gymnasium*.
Later on, he worked as a calculator at Wincenty Skrzyński's estate in Bachorz, which confirms the relationship between the Łukasiewicz family and the Skrzyński family. There is no accurate information for the next years of Aleksander's life. It is known for sure that he married Tekla nee Siekierski, to whom he left all his rights to his family home in Rzeszow. Probably this marriage did not have any children, which is evidenced by the fact that they were omitted from their will. Perhaps at the end of his life Aleksander Łukasiewicz lived in Ropczyce or its surroundings, as the preserved document from 1851 was drawn up in this very city. It is not known when he died.

50 In the first class pupils' enrolment book (*Studiosi 1^{mae} Grammaticae Classis*) it is written that he was aged 10, his place of birth was given as Zaduszniki in Galicia while the father's profession was given as a lease of a small land property. (*conductor praed.[iolum]*).

been preserved.[51] Unfortunately, despite very good grades in most of the subjects covered by the programme with Latin, Greek and German at the forefront, especially in mathematics, history and geography and behaviour, in the next school year fourteen-year-old Łukasiewicz no longer appeared at school. The family's good fortune was over.

At the end of 1836, the father of the family, Józef Łukasiewicz, died after a long period of illness. Probably his illness was one of the reasons for his earlier move to Rzeszow.[52] The surviving mother of Ignacy could not afford to support two students in the *gymnasium*, so in the list of students of the

---

51 See: APRz, I LO in Rzeszow, sign. 139, k. 23, 28, 32 and next, 42, 48, 61 and next, 59–60.
In the school year 1833–1834 he went to the first class (*I*[ma] *Grammaticae Classis*). His marks in the first semester (*Calculi, quos studiosii Gymnasii Ressoviensis priore sem. 1833 ex meritis retulerunt*) of the school year 1833–1834 were very good. Łukasiewicz, who was enrolled in the list of students under number 17, was marked on his progress in religious education (*Doctr. Relig.*), Latin language (*Lingua latina*), geography and history (*Geogr. Histor.*) and mathematics (*Arithmetica*). Also marked were his behaviour (*Mores*) and dedication (*Applicatio*). Excepting religious education for which he was awarded 2 – of the remaining he achieved the highest score – 1. In the second semester of that year (*altero sem. a.p. 1833*) he achieved top marks in every subject.
In the school year 1834–1835 Ignacy Łukasiewicz was a pupil in the second *gymnasium* class (*II*[da] *Grammaticae Classis*). The curriculum was the same as in the first class and he achieved top marks in the first and second semesters.
In the next school year 1835–1836 Łukasiewicz attended the third class of secondary school (*IIItia Grammaticae Classis*).The curriculum in third grade included Greek language (*lingua Græcia*). In the first semester he obtained a slightly lower grade in Latin and Greek, but the highest in history, geography and mathematics. Also highly marked were his behaviour, diligence and knowledge of religion. In the second semester of this school year he received the highest scores – only slightly less was his knowledge of religion.
In the school year 1836–1837, Ignacy Łukasiewicz was a student of the fourth year of *gymnasium* (*IV*[ta] *Grammaticae Classis*). In this class the curriculum remained the same as in the third class. In the first and second semester he had no problems with learning, gaining very high marks in mathematics, history and geography. His application and behaviour were marked very highly. His progress in Latin was rated one step lower.

52 Józef Łukasiewicz died on 6 November 1836 and was buried in the cemetery in Rzeszow. He was survived by his wife and five children – three sons – Aleksander, Franciszek and Ignacy, and two daughters – Emilia and Marianna. Marianna probably married in 1832 to Józef Folwarczny. This is confirmed by a document entered on 24 June 1840 from the year 1832. This is a pre-nuptial agreement of Marianna and Łukasiewicz and Józef Folwarczny – a reporter for the CK Noble Court in Tarnów.

gymnasium from the school year 1837/1838 there is only Ignacy's older brother – Franciszek.[53] The education of Franciszek was already further advanced, and this was probably the reason for the dramatic decision that it was he who had to complete his studies in order to study, and that the youngest son should start looking for a job and support his family financially. Apolonia Łukasiewicz was certainly not able to replace her husband as the manager of manor farms, especially since soon, in a letter to the Austrian authorities, she would describe herself as a disabled person.

From the modest facts that we know about the first years of Ignacy Łukasiewicz's life emerges the not very joyful time of his childhood, burdened by the illness and death of his father, successive impoverishment and everyday care of his sick mother for the material survival of the family. The first half of the 19th century was not a time when women of noble descent could obtain any kind of professional education on a larger scale, to support a family with gainful employment. Jane Austen's brothers, even in material troubles, thought to the end that demanding a decent royalty from the publisher for her and in her name offended them as obliged to keep their unmarried sister, even though the writer was already famous as a well-selling author during her lifetime.

This trend would continue in Europe until almost the end of the century. It was only in the penultimate decade of the 19th century that it became possible that the Polish woman Bronisława Skłodowska – the elder sister of Maria Curie-Skłodowska – as one of the first women could obtain a doctor's degree at the Sorbonne and run a gynaecological surgery in Paris. Maria Curie-Skłodowska would become the first female professor at that university at the end of the century. The Skłodowska sisters had such an opportunity to study and work only because in the Kingdom of Poland, from where they both came to study, women had the right to obtain the state baccalaureate. However, French women and women in many other European countries did not have the chance to obtain the baccalaureate in their own country at that time. In 1887, 114 women studied medicine in Paris, including 20 Polish women and only 17 French women.[54] Colleagues of the students were scornful at their sight, and often threw something ugly at them, not only verbally. Agatha Christie herself, whose formal education, like that of her

53 Franciszek – the middle son of Józef Łukasiewicz. He graduated from Rzeszow secondary school in 1837 and began to study law at the university in Lviv. It is known that after graduation he took a job in the Lviv city hall, but soon devoted himself to court and after moving to Brzezany became head of the local court. He was also a member of the local district council and chairman of its department. He left three sons and a daughter.

54 Cf. Natacha Henry, *Uczone siostry* [*Learned sisters*], Poznań, Wrocław 2016.

mother, was far from today's standards, testifies in her autobiography to the negative attitude towards women's education and professional work at the end of the 19th century. Both lost their status dramatically after the death of their husband and father.

Apolonia Łukasiewicz, therefore, struggled to stay on the surface during her life. Shortages at home, lack of funds to complete secondary education, apprenticeship in a pharmacy from the earliest youth were therefore not the perfect starting point for the professional career of young Ignacy. At that time, there was absolutely nothing to indicate that after a century and a half, the marks and entries in the registers, the name of this very modest student of Rzeszów gymnasium, whose completion he could not afford, would still be pored over by people of science. That archivists would scan and translate the Latin terms of the subjects that were taught in each class, give interviews to the radio and television during virtual walks on the streets he walked from school to home. That they would follow the long, hard-working, laborious path of this inconspicuous boy, which led him to new trails, without the slightest exaggeration, for the whole world. A similar assumption in his youth would have been a total absurdity.

On the subject of this boy from his youngest years survived only a family story about his rare instinct – sensitivity to the problems of others, financial help, for the family from the very beginning. As a child himself, this unfinished secondary school student spent the night in a pharmacy exchange office and worked in return for food and a penny. Empathy, sensitivity to the needs of others and reasonable, personal requirements were, apart from stubbornness and consistency in every action, the most important features of his character for the rest of his life. They would not always be appreciated, because some would consider them to be an exaggeration, bizarreness or naivety.

However, Ignacy's situation was successful in that as a fourteen-year-old and a graduate of a four-grade secondary school, he managed to acquire the right to practise in a pharmacy[55], and he managed to gain such a practice in nearby Łańcut.

There, in the pharmacy of Antoni Swoboda, whose wooden building survived until after World War II, Ignacy spent a good five years. Regardless of whether he was directed to it by blind fate or his own intuition, or maybe even dreams, it was certainly a very good choice. The years of apprenticeship as a pharmacy apprentice known in Latin as *puer*, or "boy," were not

55 W. Bonusiak (*Szejk z Galicji…*, p. 14) gives the date of commencing this practice as June 1836. However, there is information, discussed above, that Ignacy Łukasiewicz was still a fourth-grade student of Rzeszow gymnasium in 1836–1837.

a waste of time. Under the supervision of the Master of Pharmacy and his assistants and lab technicians, the young man, since childhood trained in reliable service by the old military veteran, managed to master an astonishing amount of knowledge and skills.

Four years of practice enabled him to take the so-called Tyrolean examination as a pharmacist's assistant. The *puer* passed the examination without problems. Before the Regional Pharmaceutical Committee in Rzeszow, he had to demonstrate knowledge of prescriptions, the ability to read prescriptions, translation of pharmacopoeias, knowledge of botany, pharmacognosy, pharmaceutical chemistry, chemical analysis and studies on practical pharmacy and the legislation in force in the monarchy[56].

After passing the exam in 1840 in Rzeszow, as evidenced by the documents preserved at the University of Vienna[57], the young pharmacist's assistant worked in Łańcut for 6 months, and from 6 October 1841, he took up the position of pharmacist's assistant in the Rzeszow regional pharmacy "Under the Black Eagle" owned by Edward Hübl. More or less in this place, where today a pharmacy operates at the intersection of 3 Maja and Jagiellońska streets, opposite the secondary school building – his former *gymnasium*.[58]

Carnival in the colour of blood

The scenario of this charity fantasy ball was intricate. Emilia Stacherska, nee Łukasiewiczówna, the older sister of pharmacist's assistant Ignacy, married for eleven years and wife of a tenant near Łańcut in Żołyń did a great job by visiting the surrounding manors. The ladies from the Society of Charity Ladies in Rzeszow were also good actresses. Amateur performances were shown regularly. Polish ladies, in love with Mickiewicz's delicious heroines – the brave Grażyna and heroic Emilia Plater, who was killed for the cause of her homeland in disguise in a soldier's uniform – were also able to organise themselves when it came to the highest goals.

56 Władysław Roeske, *Ignacy Łukasiewicz 1822-1882*, (Warsaw: Państwowy Zakład Wydawnictw Lekarskich, 1974), p. 18.

57 University of Vienna Archive, Vienna, Protocoll pharmacopei 1751–1854.

58 Opposite the secondary school building there were brick buildings numbered 365, 365b and 365c, described on the map as the seat of the pharmacy (*Apotheke*). They were owned by the local pharmacist Edward Hübl. Edward Hübl married Klara, a widow nee Knobloch Cynarska, who brought considerable wealth to the marriage. Since 1820 Edward Hübl was a widower. Two qualified pharmacists – Hübl and Jan Royss from Bochnia, two laboratory workers Jakub Wietecha from Staroniwa and Ignacy Sewernik from Boguchwała and a laboratory worker Rosenberg from Leżajsk – worked in the pharmacy.

Figure 4: Ignacy Łukasiewicz. Source: The Ignacy Łukasiewicz Oil and Gas Industry Museum in Bóbrka

In this first half of the century, Poles, as if in the momentum of old customs, not yet so maltreated and slaughtered with slavery and oppression without perspectives, still commonly adhered to the tradition of the Old Polish carnival. So they were dancing against all the odds, and the organisation of games and masquerades was nothing unusual.

The date of the ball in Rzeszow was set for Tuesday, 17 February 1846.[59] A few days earlier, the local nobility started to come to Rzeszow to meet before the end of the week. The custom of the landed spheres was for the ladies for elegance and noble manners to ask to the gentlemen to remove their weapons before dancing. After all, carabels and sabres at the side traditionally crowned the formal, noble clothing. For now, only in the other Partitions could you lose your head for a too ostentatiously worn Polish costume.

> "I well remember that in Warsaw we stood at Gerlach's and that I, with my head beaten with the fact that he ordered the Prince to arrest me for white hats *a la* Bolivar W. I was afraid that this fate would meet me, because I had a white hat."[60]

Recalled from his childhood in his journey with his grandmother to the capital, the writer Józef Ignacy Kraszewski, brought up in the province of the Russian Partition. In the conquered country, even teenagers of several years old experienced similar dilemmas, feeling like personal enemies of the brightest prince.

The weapons assembled by the men at the ball in Rzeszow were to be immediately delivered to the insurgents' units in the forest. The date of the Uprising was set for February 20–21, i.e. from Saturday to Sunday. The first point of the action was to be arson of his own house by the tailor Jan Czarnecki, and then the local priest Jan Tałasiewicz – one of the most active conspirators – was to hand over the keys from the bell tower, as usual during the fire, in order to give a signal to the insurgents to enter the city by

---

59 In mid-January 1846 Ludwik Mierosławski, designated as the leader of the future uprising, arrived in Cracow from Poznań. At a meeting of representatives of all Partitions, a national government was formed. The aim of the insurgents was to rebuild the state within the borders of 1772, and the date of the uprising's outbreak was set for the night from 21 to 22 February 1846. Franciszek Wiesiołowski came to Rzeszow on February 13th to check the preparations for the uprising. Łukasiewicz was to present him with a plan for conquering the city with the help of noble troops and hitting the garrison headquarters. However, probably the state of preparations did not satisfy Wiesiolowski, who ordered to convene a conspiratorial meeting in Rzeszow on 18 February. The pretext for it was to be a ball, planned for 17 February.
After Wiesiołowski's departure, the situation in Rzeszow changed. Two pieces of information reached the city – the first one about the dismantling of the conspiratorial network in Lviv on February 12–13. The arrests of several dozen people recruited by Edward Dembowski certainly worsened the mood in Rzeszow. In addition, the news came from Tarnów that the local Uprising commander – Leon Czachowski – arbitrarily sent orders to the area, accelerating its outbreak.

60 Kraszewski, *Pamiętniki...*, p. 331.

ringing the alarm.[61] Unfortunately, in the Habsburg state, for 25 years, it was the ruthless Chancellor Metternich, on behalf of the mentally and physically handicapped ruler, who had been in charge of police terror and a large bureau of agents, and it was his local mercenaries who thwarted the plan.

The mayor of Rzeszow, Edward Koschina, engaged a large number of secret confidantes to track down Polish conspirators in advance. Exactly on 17 February, on the day of the planned ball, Austrian troops under the command of General Legedisch encircled Rzeszow and seized the city, issuing a ban on organising the ball. At the corners of the town, landowners drawn from the surrounding manor houses were captured. Weapons were turned back, weapons were seized and suspects were captured in prisons.[62]

Two days later, on 19 February 1846, in the pharmacy of Hübel "Under the Black Eagle," in the presence of the mayor of Rzeszow, the pharmacist's assistant Ignacy Łukasiewicz was arrested as the chief organiser of the rebellion against the Austrian authorities. It turned out that the mayor's secret agents had been observing Edward Hübl's pharmacy for a long time, and it was indeed the logistics centre of the anti-Austrian conspiracy in Rzeszow. In suspecting Łukasiewicz of playing a key role in the conspiracy, the Austrians were not mistaken.

Metternich's machine of political terror raised doubts throughout the Habsburg Empire.

"Metternich was a fanatic of absolutism, an advocate of serfdom relations, a spokesman for the police state,"[63]we read in a concise description of the

---

61 The insurgent units of Pieniążek from Kielnarowa, Skrzyński from Zwięczyca, Tański from Krasny, Reja from Widełka and Wojnarowicz from Przybyszówka were to enter the city.

62 At the time, five divisions were stationed in Galicia and each of them was divided into two brigades.The area comprising Rzeszow, Leżajsk Przeworsk, Jasło, Gorlice and Dukla was controlled by a brigade under Major General Ignaz von Legeditsch. This brigade was part of the division headed by Lieutenant Field Marshal Albert von Csollich. The brigade of Von Legeditsch entered the garrison stationed in Rzeszow. It probably numbered about 440 infantry and 930 cavalry. They were stationed in the barracks at the then Mostowa St. (*Brücken Gasse*). In von Legeditsch's opinion, who predicted the uprising, the army was not prepared for its suppression. The commander of the brigade stated that the deployment of the Austrian units was ineffective in the case of the destruction of bridges, among others, in Wojnicz, Pilsen and Rzeszow communications with the capital would be interrupted and the various garrisons would be isolated.
Probably for this reason, since the beginning of 1846, the number of garrisons of Galicia increased, incorporating reservists – trained soldiers who had completed their military service.

63 Mieczysław Czuma, Leszek Mazan, *Austriackie gadanie czyli encyklopedia galicyjska* [Austrian Chats or the Galician Encyclopaedia], (Cracow 1998), p. 534.

rule of this creator of Austrian politics in the first half of the 19th century. In two years' time, in 1848, due to the total armed riots of the oppressed nations of the monarchy, it was to end with the retreat of this tyrant from office.

In the 1840s, the Austrian spying was at the height of the propaganda campaign among the peasants. The aim was to incite peasants to rebellion against Polish landowners, able to carry out the uprising effectively for both material and conscious reasons. The Austrians distributed rumours that they intended to abolish serfdom as a compulsory tribute of the peasants to the nobles. The paradox of this situation was the fact that the serfdom, contrary to the Polish demands of the Kościuszko group from the last century and the paragraphs of the Constitution of 3 May 1791, was still maintained in the Polish lands by the partitioners. The conspirators, in preparing the uprising at the beginning of the 1840s, were also going to officially announce the abolition of serfdom. In Cracow, in mid-January, the National Government was elected, which determined its actions as soon as it took power after the outbreak of the uprising. All this collapsed, however.

The provocation among the peasants in Galicia was successful. Within a dozen or so days from the scheduled date of the outbreak of the uprising on 21/22 February, peasants led by the self-proclaimed Herszta Jakub Szela, who was in collusion with the Austrian authorities, threw themselves into the extermination of local landowners, murdering landowners, court clerks, priests and even servants loyal to the nobility. The authorities wanted not only to get rid of the most patriotic citizens, i.e. the landowners, but also to destroy the resources of weapons and food accumulated in manor houses and forests for the purposes of the uprising.

Peasants, many of whom had previously reported to the authorities about the preparations, stores, meetings and movements of the nobility, murdered landowners with impunity for almost a dozen or so days in March 1846, annihilated manor buildings, robbed and killed manor livestock. The bloody carnival covered most of western Lesser Poland. Szela's agreement with the authorities, despite the existence of the military and police apparatus, gave free hand to peasant gangs, who were constantly fed lying propaganda. For example, the mayor of the town of Pilzno recommended that "the peasants should help the emperor to catch the insurgents who kill the peasants with saws and cut off the women's breasts."[64] The agitated peasants tracked down to the last one in the woods of the refugees from the manors turned into ruins, they killed the wounded and tortured their recent employers. The

64 Mieczysław Czuma, Leszek Mazan, *Austriackie gadanie..*, pp. 399–400.

massacred bodies were taken to the Austrian authorities in exchange for a fee for the corpse.[65]

On one of such carts carrying wounded prisoners for execution was a landowner from a manor house in the south, at the foot of the Carpathian Mountains in the district of Gorlice, Tytus Trzecieski, the owner of the Polanka estate. Much indicates that if a peasant whom he knew had not recognised him and spared him his life, the fate of the world oil industry could have taken a somewhat different course. His future partner Ignacy Łukasiewicz – the future leader of the world's oil industry – in the meantime was resident in Rzeszow prison for a longer time. The fates of the best Poles were typical.

The conspirators from the Hotel near "Luftmaszyna"

During less than five years of work at the pharmacy in Rzeszow, Ignacy Łukasiewicz, as a regional leader, created and developed a regional network for the resistance movement, becoming the personal nomination of Edward Dembowski[66], an activist in exile in Rzeszow and the surrounding area.

The 19th-century Polish Great Emigration abounded in free initiatives aimed at the military recovery of Poland within the borders of the pre-Partition period. Since 1832, the Polish Democratic Society based in Poitiers and later in Versailles had been operating in France. In 1835 the organisation *Lud Polski* (People of Poland) was established in London. In the district of Greater Poland, in the Prussian Partition, since 1842 there had been a Union of Plebians. These organisations sent emissaries to the territory of all three Partitions in order to attract coordinators of works in the

65 Cf. Tomasz Szubert, *Jak(ó)b Szela 14 (15) lipca 1787 – 21 kwietnia 1860* [Jak(ó)b Szela 14 (15) July 1787–21 April 1860], (Warsaw 2014), pp. 102–108.

66 The Galician conspiracy was revived by Edward Dembowski, a contemporary of Ignacy Łukasiewicz. Linked with the Association of the Polish Nation and conspiracy in the Prussian Partition, he represented more radical views in these organisations. Threatened with arrest by the police in the Prussian and Russian Partitions in the spring of 1845, he came as an emissary of the Polish Democratic Society to Galicia. Thanks to this work resumed on the preparation for the Uprising in this Partition. Dembowski had talks with Wiesiołowski, Wiśniowski and Viktor Heltman. They ended up deciding that in the eastern part of Galicia the insurgent action would be organised by Wiśniowski, and in the west Dembowski and Wiesiołowski. Franciszek Wiesiołowski was to conduct agitation among the nobility and Dembowski among townspeople and peasants. Dembowski also created a network of agents who were to gather information about military forces on their territory, draw up plans for an attack on military units and direct conspirators in the field.

regions for self-organisation and to cover the whole territory of the former Commonwealth. The Polish leaders of these groups, in accordance with the moods of the epoch, mostly leaned towards democratic views, but in their ideological work and manifestos they decided that they had to cover the entire Polish society, from the rural people through craftsmen and burghers to the Polish nobility with their independence propaganda. Regional agents were to collect data on the military forces of the partitioners, ways of diversion and their liquidation, and organise the arming of volunteers.

In July 1845, Edward Dembowski – the philosopher, chief organiser of the nationwide uprising under preparation – personally met with a seemingly modest but already outstanding conspirator, the pharmacist's assistant from Rzeszow.[67] Preparations for the uprising were in full swing in Rzeszow. In this work, Łukasiewicz, despite his young age, had not been a novice for many years. The mercenaries of the mayor of Rzeszow were not such intelligence geniuses. The police label *politisch Verdächtig*, i.e. a political suspect, was already following Łukasiewicz from Łańcut.

Łukasiewicz had already come into contact with the principles and methods of conspiracy at school, and after installing himself in Łańcut on his first pharmacy practice, he quickly realised how ideal such a place as a pharmacy, which is visited by a lot of people, was for developing a structure of secret contacts. It turned out that even in small Łańcut patriots do not sleep. Aleksander Tarłowski, an activist from the structures of the Democratic Chamber, made contact with the clever boy from the pharmacy. The friendship between the pharmacy apprentice and the much older, former participant of the 1830 Uprising could develop without arousing much suspicion, because both combined the joint production and research on the use of spirit-based liquid for lighting lamps.

It seems, therefore, that at this point in Łukasiewicz's biography one can also see the first impulses of his interest in and acquaintanceship with the construction of lamps.[68]

Łańcut was not only the seat of the magnificent palace residence of the Potocki family, but also the vodka factory founded by Count Alfred Potocki, where Tarłowski worked as a professional. The young pharmaceutical *puer*

67 Dembowski probably met Łukasiewicz through Wiktor Solman, who was, with Julian Goslar, one of Dembowski's closest associates. Solman had known the Łukasiewicz family since childhood because his parents were the godparents of Ignacy Łukasiewicz and friends with his father. However, membership of the conspiracy was definitely not decided by acquaintance, but Łukasiewicz's past membership of independence movements and probably his educational and self-education work.

68 Jerzy Sikora, *Łukasiewicz a* ..., pp. 27–28.

initiated his secondary school friend, who due to lack of experience or perhaps naivety appeared before the subject of a pharmacy in the nearby town of Leżajsk, in matters of conspiracy. Metternich's agents also understood quite well how an excellent contact box a pharmacy could be. The subject from Leżajsk turned out to be a police confidante and the conspiracy in Łańcut was revealed from top to bottom both in the pharmacy and in the vodka factory. Tarłowski was arrested and imprisoned. For Łukasiewicz, after several interrogations, apart from contacts with Tarłowski, nothing criminal was proven and the case ended only with his inclusion in the police files as a suspect.

It cannot be ruled out that the Łańcut defeat was the main reason for moving to Rzeszow, but the return to this city, anyway, was beneficial for the young Łukasiewicz. The pharmacy "Under the Black Eagle" was located on Pańska Street in Rzeszow in the immediate vicinity of his family home, and in addition, it served as a regional pharmacy, so it gave the pharmacist somewhat more opportunities to develop his professional career.

The Łukasiewicz family apparently had friendly relations with the owner of the pharmacy and close neighbour from Dukla, Edward Hübl. According to the law, Ignacy was still a minor. Edward Hübl was the legal guardian of the young assistant and provided him with food and a room at the pharmacy. His stay in Rzeszow also gave Ignacy a better opportunity to help his sick mother Apolonia and provide material support for the rest of the family. The young pharmacist's assistant, at his modest salary, which was the not too high for those days amount of 12 florins, was extremely concerned about the possibility of supporting the family. The term "man of good heart" would from this moment of his life be a constant *leitmotif* of the memories of witnesses. From Rzeszow he was also able to send money to his older brother Franciszek, who was studying law in Lviv. Conscientious in his pharmacy work, he gained greater respect over the course of time.[69]

In the pharmacy "Under the Black Eagle," not counting Łukasiewicz, there were two qualified pharmacists – including the principal, and two laboratory workers. As the political stumbling in Łańcut by no means extinguished Łukasiewicz's ideological spirit, he spread his political wings in Rzeszow. He operated in many ways, engaging both the immediate community, as well as going outside to the city and the surrounding area.[70]

---

69 Władysław Roeske, *Ignacy Łukasiewicz...*, p. 17.

70 As is clear from later testimony – Łukasiewicz managed to contact supporters of the Uprising, coming from different social strata in Rzeszow. Among them were Franciszek Jaekl (Secretary of the Pent magistrate), Feliks Zabierzewski (writer of the *cirkul* fund), court apprentices Tytus Lewandowski, Franciszek Gruszczyński and Stanisław Holzer. Łukasiewicz's contacts also included youth and secondary

Apolonia Łukasiewicz – his mother – rented a part of the house to an educational institute for girls founded by a teacher who came from the Russian Partition, Karolina Krynicka. The pupils of the *pension* came mostly from the noble community of Rzeszow, hence the parents of the girls managed to carry out propaganda, information and organisational work. These contacts were priceless. Sister Emilia, with whom he was connected throughout his life by probably the closest ties of all his siblings, helped her brother in this job "in the field" with great commitment.[71] Emilia, who wanted to help her infirm mother, was stationed with her at the time, together with her children Jan and Honorata. Years later, the tiny Honorata would become the subject of his feelings and be his wife. For several decades of the 19th century still, unlike today's medical and moral reasons, marriages between close family were not an unacceptable practice.

Łukasiewicz was able to involve a wide range of foreign ladies in his activities. He carried out organisational work during the preparation of amateur theatrical performances in which members of the Society of Charity Ladies took part. However, the official repertoire of these performances was certainly different from the confidentially transmitted stanzas created outside the censorship, in exile and in secret printing houses. Poems about freedom, about hardening the spirit, about the history of the former Commonwealth, smuggled from behind the border cordons, were written one from another, they were learnt by heart and read on important anniversaries. Officially, most often light comedies were played. Perhaps the ones that were assimilated by the great creator of the theatre at the request of the last Polish king, Wojciech Bogusławski.[72] Perhaps art of the Napoleonic veteran and

---

school graduates from Rzeszow – Jan Madejski, brothers Karol and Henryk Bieniaszewski, Konstanty Michlewski and Adolf Sławik. Outside Rzeszow Karol Chodziński from Łańcut (already known for participating in the Conspiracy), trustee of Straszydło Feliks Ziemba, and tenant of the Leżajsk estate Wladyslaw Okólski were also linked to the conspiracy.

71 Emilia Łukasiewiczówna married Ernest Stacherski – an emigrant after the November Uprising, who leased an estate in Żołynia near Łańcut – in about 1835. Ernest Stacherski died in 1873 and after his death, his wife Emilia moved to Chorkówka. The couple had three children – Jan, Honorata and Maria. Honorata was later the wife of her uncle Ignacy Łukasiewicz. Jan Stacherski had no luck with the conduct of his business. He was in debt with the Mokowo estate he owned and in 1874 eventually went bankrupt. Ignacy Łukasiewicz brought his wife – and his niece – Alojza Stacherska nee Folwarczna with their young children, to Chorkówka.

72 Wojciech Bogusławski (1757–1829) writer, actor, opera singer, director.

the genius of the Polish comedy of Count Aleksander Fredro[73], who, already locked away into his estate of Benkowa Wisznia in the Subcarpathian region, gave his countrymen the fantastic "Revenge," taking place in the castle in Odrzykon, not far from Rzeszow. These could have been singing performances, and under these conditions each song in Polish was succour for the spirit.

Łukasiewicz himself tried his hand at the art of acting, but unfortunately the stage exposed the paucity of his talent. Instead, it was the mainspring of theatrical initiatives, which were an excellent cover for the promotion of secret initiatives. Those involved in theatrical activity could freely visit Master Hübl's helper in the back room of his pharmacy regarding the organisation of shows. An invaluable pillar of the Rzeszow underground was Father Jan Tałasiewicz, who also held meetings at the presbytery under various pretexts, including preparations for performances.[74]

Conspiratorial meetings also took place in the city. One of the last, immortalised in reports from the Austrian investigation, went down in history. On the night of February 13th, a week before the planned uprising in the "Hotel pod Luftmaszyną," on the present Słowackiego Street, Łukasiewicz was catching up with Count Wiesiołowski – an emissary from outside on behalf of the Confederacy of Polish Democrats. After the meeting, the two conspirators set off on a horse wagon to Łańcut in order to persuade the local activists, including Tarłowski's former accomplice in underground work, to attack the local garrison, but they met with a refusal. In the week when Łukasiewicz was arrested, the tailor Czarnecki, priest Tałasiewicz and Tarłowski from Łańcut were also seized.

How can we doubt the good results...

"Say, eagle! My eagle!
White-winged, unattainable (?)
Whence came this swarm of black thoughts?
They grow – where the handcuffs be!

73 Aleksander Fredro (1793–1876) – Count, Polish comedy writer, diarist, romantic poet, veteran of the Napoleonic Wars including the Moscow Campaign, officer in the imperial staff, Knight of the Order of Military Virtue and the Legion of Honour.

74 In autumn 1845 Chaplain Jan Tałasiewicz, associated with the activities for independence ever since his studies at the seminary in Przemyśl, settled in the rectory of the Rzeszow church. At the end of 1845, the vicarage served as the meeting place for Wiesiołowski and Ignacy Łukasiewicz with Franciszek Wolański – an agent from the vicinity of Jaslo.

Ah! Wilfulness, dry order,
Which decomposes the Spirits store!
Like Siberia – like boars
And the king of fleshly torture!
But a nation's spirit poisoned
That's just pain!"[75]

(Zygmunt Krasiński, Psalm of Love)

The apparatus of the Austrian state machine, famous for its sluggishness, in this case showed the dynamics of a petard. In the case of Ignacy Łukasiewicz, who was arrested on February 17, the Regional Office sent a full letter to the national authorities in Lviv as the capital of Galicia – the Austrian Partition – which reported that:

> "Whenever the famous revolutionary leader Franciszek Wiesiołowski played in Rzeszow, he would come together with the pharmacist's assistant Ignacy Łukasiewicz and Father Jan Tałasiewicz each time."[76]

A special Criminal Commission appeared in Rzeszow, followed by a revue of the testimonies of the arrested, numerous confidantes, who had already sniffed around the activities of the would-be insurgents. There were also, as happens everywhere in similar circumstances, voluntary informers who testified against Łukasiewicz.

From the evidence of material violations against the authorities, in addition to the testimonies of spies following Łukasiewicz's earlier actions, there was a letter from Franciszek's brother instead of Ignacy, delivered directly to the Commission by a helpful acquaintance, who seemed to be a trustworthy person.[77] But the most serious evidence against Łukasiewicz turned out to be "subversive" books (because they were Polish) found in a pharmacy drawer by a new pharmacist accepted in the place of the arrestee.

In Łukasiewicz's drawer there were found philosophical and historical treatises, brochures published under pseudonyms, most often in exile, poems whose reading or recitation was a frequent item on the agenda of social meetings, and inevitably also on many evenings of the Society of Charity Ladies under the guise of rehearsals for some kind of farce. Among these were:

75 Zygmunt Krasiński, *Psalmy przyszłości* [Psalms of the Future], (Cracow: Universitas, 2002), pp. 24–25.

76 Roeske, *Ignacy Łukasiewicz...*, p. 21.

77 It was betrayed by one of the participants in the conspiracy – court apprentice Tytus Lewandowski, who took the letter to his superior.

> "Adam Mickiewicz's 'Books of the Polish Nation and Pilgrimage', 'Old Tales about Polish Customs', 'A Course on the Military Art', 'The Current State of Galicia', 'Memoir of the Polish Democratic Society', 'Memoir for the rule of August III'."[78]

A special item in this set was a thin booklet, a hit amongst illegal readings, secret printing houses in Poland and packages smuggled from foreign publishers under the title *Can Poles Break Out for Independence*. The insiders knew perfectly well that this book, often reprinted under a pseudonym or anonymously, was written by Józef Pawlikowski – Tadeusz Kościuszko's own adjutant, and the thoughts expressed there maybe expressed the words of the Chief himself to his compatriots.

There were also "Psalms of Love" by Zygmunt Krasiński, not without reason considered one of the three greatest leaders of the spirit of Polish freedom. Every invader understood perfectly well that the "a nation's spirit poisoned," which the poet lamented, was a warranty of his reign. Therefore, Edward Hübl the owner of the "Under the Black Eagle" Pharmacy, terrified by the vision of his involvement in "revolutionary" matters, saw no other way than delivering this criminal package to the authorities immediately. And could a complacent citizen of Metternich's state read the words of Kosciuszko's officer without the utmost horror? In order to gain independence, he commanded Poles:

> "We should consider Poland in its first, our borders."[79]
> He asked again:
> "Is it possible to doubt the good results if the whole nation moved unanimously?"[80]
> And he rightly noticed a no less terrifying perspective:
> "We have one more advantage, that there are Poles in the armies of the three powers surrounding us."[81]
> – extracting a comforting positive from the national tragedy.

Meanwhile, already in March, several days after the arrest of the organisers of the would-be uprising, behind the walls of the prison, an Austrian provocation of the Polish courts took place. It is not known to what extent the sounds of the tragic carnival of the slaughter of the Polish nobility by the hands of its own people – accidents so far unprecedented and difficult to imagine in the history of the Commonwealth – reached cell number 12 of the Rzeszow prison. Courts, military authorities, police and administration really had a lot of work apart from catching and punishing the would-be

---

78 Roeske, *Ignacy Łukasiewicz...*, p. 19.
79 This means from the First Polish Partition in 1772, J. Pawlikowski, *Czy Polacy ...*, p. 87.
80 Roeske, *Ignacy Łukasiewicz...*, p. 102.
81 Roeske, *Ignacy Łukasiewicz...*, p. 110.

insurgents. Despite the efforts of Łukasiewicz's family to get to see him, he was not granted this right. Even a touching, begging letter from his blind mother did not receive any reaction.

In the Lviv documentation of his trial, we can find this letter of Apolonia Łukasiewicz to the starost of Rzeszow von Lederer, dated 4 May 1846. This is the mother's dramatic request for a lenient sentence.

> "Such a conviction encourages me, the most unhappy of mothers, to approach you Sir, and to throw at your feet with bitter tears a sprinkling request, through which I beg mercy to myself for the sake of my imprisoned son [...]. If my son, so young in his age, has gone astray, oh – do not judge him, Sir, from his inexperience, but look with a merciful eye on his misfortune, his disability, and his aging mother, whose beloved son is the only support, the whole source of her existence...."[82]

Similarly to the harrowing pleas to spare her minor son a penalty, Mrs. Rollinson the blind lady in the ballroom scene at the Russian senator's home in the drama *Forefathers' Eve* written by the poet Adam Mickiewicz a decade or so earlier in Paris.[83]

The first document from Łukasiewicz's interrogations before the Criminal Commission is dated from the end of June 1846, so the interrogation took place only after four months of the suspect's detention. Łukasiewicz adopted a tactic of denying his participation in the conspiracy and any plotting against the authorities. He claimed that his meetings with the emissary Wiesiołowski were purely social, he did not discuss political issues, although the preparations for the actions were an open secret. The books were not his, but those of his deceased predecessor. Against these testimonies, reports of the Rzeszow pharmacy lab technician and the police informer were registered, who had serious suspicions that during evenings "over schnapps" in the back of the pharmacy, Łukasiewicz's guests spoke German especially so that he would not understand anything. In pure Polish, he also quoted Łukasiewicz's cry: "Exile a German from the country, that the Pole himself would stay!"[84]

On 16 June 1846, Ignacy Łukasiewicz's sister Emilia Stacherska was also questioned. Her testimony brought nothing new to the proceedings because she claimed that she knew nothing, she noticed nothing, and her brother Ignacy was focused primarily on the fair performance of his duties in the Hübl pharmacy and assisting his sick mother. In her testimony, Emilia Stacherska expressed herself very positively about her youngest brother, pointing also

---

82 CDIAU, Fond 152, description 2, case 5655, k. 47.

83 Adam Mickiewicz, *Dziady, część III* [Forefathers' Eve, Part III], (Wroclaw-Warsaw -Cracow -Gdansk –Lodz: Ossolineum 1984), p. 120.

84 Sikora, *Łukasiewicz a ...*, pp. 161–172.

to his care for his orphaned niece Alojza, who remained without any means of subsistence.

Antoni Swoboda, the pharmacist from Łańcut, turned out to be extremely loyal. He claimed that at that time, just before 17 February, Łukasiewicz came to return to him the silver teaspoon he once borrowed, which was probably supposed to be some kind of a tool for making pharmaceuticals.

However, the finale of the many months' investigation was draped in black. The partitioners' routine repression against the Polish was the confiscation of assets. Luckily, the authorities agreed to a written request from the prisoner for permission to rewrite to his sister Emilia a part of the Rzeszow family house due to him from the division after his father's death. Undoubtedly, the official efforts of the family were supported by bribery, all the more so as the signing of the act required bringing to Łukasiewicz's cell notary officials accompanied by appropriate guards.[85]

There was a never-ending carousel of twisted testimonies, including serious accusations of Łukasiewicz's reports on the state of armaments and the number of troops in the region, and his own armaments. He was even incriminated by the testimonies of the national emissaries – Dembowski and Wiesiołowski.[86] The end of 1846 ended with a verdict of the court in Lviv and recognition that the accused was guilty of treason by with an order to the Rzeszow court to reconsider the case.[87] The new trial started with the new year and lasted five months, but one would be wrong to think that

---

85 On 4 August 1846, a so-called concession document was concluded, whereby Ignacy Łukasiewicz – owner of 1/5 of the real estate No. 113/368 in Rzeszow after the recognition of the inheritance of Józef Łukasiewicz of 9 November 1837, withdrew his shares in the house on behalf of his sister Emilia nee Łukasiewicz Stacherska for ownership "having received from her full satisfaction, this is an agreed sum 1000 guilders." The sale was conducted in Rzeszow on 4th August and entered in the city ledger on 24 August 1846. Again, it is worth recalling that at this time, Ignacy Łukasiewicz was in custody and there the document was signed and the members of the Criminal Committee participated in this act as witnesses.

86 Łukasiewicz's situation worsened in August 1846, when he was charged by the testimonies of Franciszek Wolański, a commissioner for the Jasło district, appointed by Wiesiołowski. Wolański testified in the investigation about his stay in Rzeszow in December 1845, when in the presbytery of Fr. Tałasiewicz he met with Łukasiewicz and Wiesiołowski. The leader of the conspiracy was to commission Łukasiewicz to prepare a report on the state of military preparations for the outbreak of the Uprising and weapons, located in Rzeszow.

87 At the end of August 1846, the Criminal Commission completed its investigation and sent the case of Ignacy Łukasiewicz to the judicial authorities in Lviv. He was accused of "actively supporting the crime of state treason and an open partner in the deed of Franciszek Wiesiołowski and his companions."

the spirit of the prisoner fed by the food of the Polish Romantics suffered a collapse. The foreign authorities knew perfectly well that their biggest enemy was the idea of freedom.

Łukasiewicz did not hesitate to start his work again from his prison cell. Tadeusz Kosciuszko's recommendations that the universal movement and unwavering faith in the meaning of action must bring good results and be crowned with success, gave strength to his spirit.

Through a guard, he made contact with an emissary from France and began to organise a network of the freedom movement, also involving his brother outside. The thing soon appeared to have been denounced by a fellow prisoner, and Łukasiewicz gained special status as a difficult prisoner. In the middle of the year he was transported to the capital of the Austrian Partition, to Lviv, and imprisoned in the building of the liquidated Carmelite monastery. Another investigation started. Due to the variable testimony of the co-accused, doubtful circumstantial evidence and the unbroken tactic of not admitting anything, without much result for the investigation.

It is worth noting that in view of the extremely wide range of his organisational activities during the preparations for the Uprising, the penetration of many noble, clerical, craftsmen and social organisations, there could be enough of both arguments and people who would actually incriminate Łukasiewicz the most severely for him to receive the most draconian penalties. However, the efforts of the authorities' informers alone were apparently not effective against the extraordinary, as one may guess, solidarity of the wide range of people who, after all, had seen Łukasiewicz in action perfectly well. After all, these were the most various of activities: forbidden, Polish readings and declamations in social groups, singing, discussions, organising a network of intelligence officers in weapon depots, institutions, offices, the logistics of armaments and assigning roles to individual conspirators, deadlines, arranging communication scenarios, diversions, fighting, mapping and marking routes, storehouses for weapons, food, dressings and medicines.

Typical in this respect is the extremely clever, misleading testimony of his sister Emilia, kept in the archive of the case. The accusations of dragging the local nobility into the underground, when, by the way, the matter of agitation led by the superior of the Krynicka's *pension* for girls came up, she answered as follows:

> Since my present stay in Rzeszow, I have not had any fun, especially at the balls, nor have I been at all in the houses mentioned here, i.e. at the manor house in Zwięczyca and Zaczerń. I know the heiress Kajetanowska Skrzyńska, because four years ago she lived here in Rzeszow in our tenement house, and secondly during my father's lifetime we kept the village of Rudki from Father Boczkowski in possession, so we

> were neighbours of Mr. and Mrs. Skrzyński, and who lives in Zaczerń in the manor house, I do not even know.[88]

The accused Łukasiewicz did not break down, despite the fact that violence was also one of the investigative methods. The old veteran Tarłowski, who had gone through more than one thing during the war, initially broke down in this investigation. After a year he revoked his testimony, claiming that he was forced to them by beatings. The absence of any comments in the file on this assertion confirms the version of his assertion.

At the end of August 1847 it was decided to discontinue Łukasiewicz's case, which was confirmed by the Court of Appeal two months later, and then the consent of the empire authorities was awaited. There was no hurry here either. The Vienna decree came in December. After almost two years in prison, this decree confirmed Łukasiewicz's positive sentence on condition that he took the oath that he would remain in Lviv[89] under the watchful eye of the capital city authorities and would report on every call. The oath was taken by him on December 27, 1847, just after Christmas, with the sluggishness and malice typical of any apparatus of violence. The wasting of life, depriving the victim of at least one more, even a small, human joy, like a Christmas family meeting, is its timeless feature.

This stage of Łukasiewicz's life had definitively closed. Łukasiewicz never came back to the Independence Conspiracy or Rzeszów. The years of conspiracy in the pharmacy in Łańcut, and especially his work as the Rzeszów conspiracy coordinator, unexpectedly showed quite an unexpected direction of Łukasiewicz's development as a man. A young man, who at school age seemed destined to be docile, struggling with deprivation, crushed by family problems, a man with a kind heart, turned out to be a courageous, steadfast, not insignificantly cunning individualist, who was able to lead the Austrian investigators by the nose and, despite very serious accusations, to wait for the investigation to be discontinued.

It seems quite interesting that apparently this modest pharmacist not only turned out to be an enthusiast of lighter music – comedies, theatrical entertainment and social balls – but also did well in salon contacts. He was able to co-organise social events, including the preparation of this main scenario with the active participation of ladies. It is possible that the numerous presence of schoolgirls in his mother's house gave him a sense of familiarity

---

88 Central State Historical Archive of Ukraine in Lviv (Центральний Державний Історичний Архів м. Львова – dalej: CDIAL), Fond 152, description 2, case 5678, p. 7–7v.

89 *Słownik geograficzny*... [*Geographic dictionary*], p. vol. 5, pp. 497–554, Vol. 11, p. 438.

with strange women. It is a pity that so little is known about how credible the mentions are of the feelings he was supposed to have had about Karolina Krynicka, the superior of the girls' *pension* in Mrs. Lukasiewicz's house. Certainly she was a maiden with excellent upbringing, manners and appropriate education. Why did not their union survive the test of time? Maybe because, although the time for this marriage was appropriate, Ignacy was released from prison as an unemployed, poor, and what is worse, politically stigmatised man without higher education and clearer prospects for the future. No one would have given a penny for his future fame and millions. Including Miss Krynicka and most probably himself.

A manuscript under a lucky star

The year 1848 was supposed to bring almost all of Europe a war, a general revolution called by historians the Springtime of Nations, and Austria a relative political imbalance.

For Łukasiewicz, after his release from prison, the year 1848 began seemingly successfully. However, the machinery of the Austrian state did not forgive its victims easily. Łukasiewicz had to wait for the verdict of the Supreme Court of Justice in Vienna to become final, and although this decision came on 12 January, the Pole was still kept in uncertainty for more than a month. In this state, he could both expect the prospect of a reinvestigation and could not plan for a concrete professional future. The fact that his brother Franciszek lived in Lviv, after studying law, with whom he was able to stay in the capital of the Austrian Partition, because he was forced to remain there, was a good thing. Franciszek took up a job in the Lviv magistrate. Soon he was to move to the provinces, where he became the head of the court.

Ignacy remained under police supervision for a long time. A secret letter sent from the criminal court to the governor's office in Lviv concluded:

> "Due to his obduracy and stubborn denial, as a dangerous individual he should be subject to special police supervision in the future."[90]

This opinion was long the cause of Łukasiewicz's special treatment and various administrative difficulties. After eight months of hospitality with his brother, the "dangerous individual" finally managed, in mid-August 1848, to obtain a post of pharmacist's assistant in the leading Lviv pharmacy "Under the Golden Star," whose star turned out to be happy for him also.

Here he worked reliably for the next two years under the care of the owner Piotr Mikolasch, who was the honourable president of the Board of Lviv Pharmacists. It must be admitted that this pharmacist's student showed

90 CDIAU, Fond 152, description 2, case 5699; Orlewski, *Spisek w aptece*... p. 169.

Figure 5: Pharmacy. Source: Museum of Pharmacy, Jagiellonian University Medical College

exceptional intuition, which in the future also brought him both material benefits and historical fame. He appreciated the acquired knowledge and possibilities of the new employee and personally trained him in the pharmacy laboratory.

The pharmacy "Under the Golden Star" turned out to be the best school for the young pharmacist thirsty for professional development. Anczyc, a diarist, wrote about Mikolasch:

> "After learning about his skills and professional education, he liked him very much and put him to various analytical tests under his supervision."[91]

The role of the laboratories maintained in previous centuries by pharmacists is difficult to overestimate at a time when the various types of scientific and industrial institutions were incomparably fewer than today. They formed

91 Anczyc, *Ignacy Łukasiewicz*, "Kłosy", 1882, No. 874.

an invaluable contribution to the development of the biological and physical sciences. They were used for the production of medicines, various parapharmaceuticals, drugstore products and sanitary technical chemicals, which had to be produced on site. The old pharmacopoeias, as official pharmacy codes, apart from medicines authorised for marketing, descriptions of their operation also included the conditions, methods of obtaining, purification and storage of raw materials for the production of articles sold in the pharmacy.

By fortunate coincidence, of this two-year period of Łukasiewicz's work in Mikolasch's pharmacy between 1848 and 1850 there has been preserved a 274-page, grey canvas-framed notebook, mostly written with his own hand, with this very handwritten signature – "Manuscript – Ignac Łukasiewicz – 1850."[92] It is, as is the custom of many pharmacists of old date, a private almanac of pharmacy prescriptions and experiences gained during work. This *silva rerum* seems to have passed into the possession of later owners, as evidenced by the presence in it of several other, unidentified pieces of writing. Of the 300 prescriptions written in this manuscript, 175 are made with the hand of Łukasiewicz, that is, of course, considered useful and worthy of notation by him. Łukasiewicz uses both traditional Latin for medical terminology and the official German language, often in Gothic transcription.

One hundred years later, professor of pharmacy, former director of the Museum of Pharmacy in Cracow, Wojciech Roeske, in a book about the great ancestor of Polish pharmacy, draws attention to the specificity of Łukasiewicz's notepad.[93]

Łukasiewicz's manuscript contains herbal medicines:

> aromatic waters *(Aqua aromaticae)* such as lavender, orange, cresotta, eye waters (*Aquae ophtalamicae – briliana, lutea, cerulea*), plasters (*anglicanum, nigrum Moscoviticum, Christi, ad rupturas*)
> elixir (*Elixir longae vitae*)
> electuary (*Electuaria*)
> plant extract for treatment of syphilis (*Rooh antisiphillitis*)
> tinctures (*Acoria aetherea, Chinae Comp. Pyrthti Comp. Anispasmatica, Rhei Darelli, Spiritus lavendulae*)
> antimony wine (*Vinum Stibiatum*)
> ointments (*Uguenta ad pedum, Pediculorum*)
> chocolate *(chocolatae et morsuli)* and sugars

92 The manuscript is now in the collections of the Museum of Pharmacy of the Jagiellonian University in Cracow, *Rkpp. Kp.inw. 3361:Manuskrypt I. Łukasiewicza z 1850 r.*

93 Wojciech Roeske, *Ignacy Łukasiewicz 1822–1882*, (Warsaw 1974), pp. 26–31.

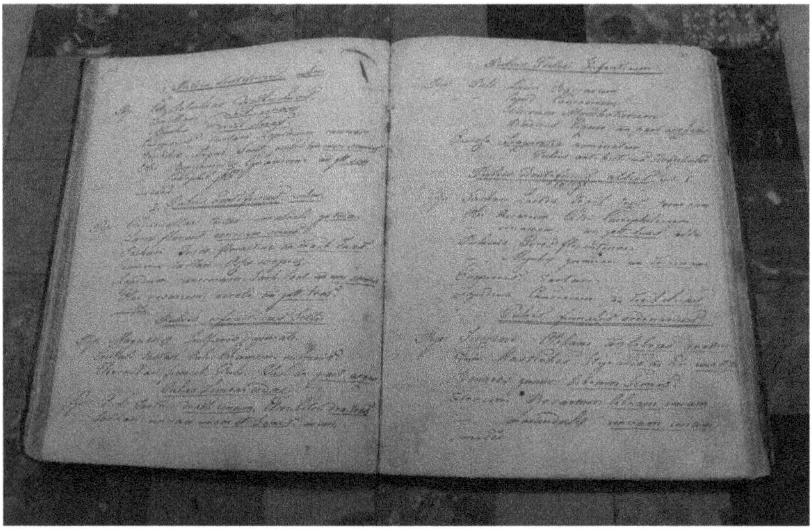

Figure 6: Łukasiewicz's manuscript. Source: Museum of Pharmacy, Jagiellonian University Medical College

Wojciech Roeske particularly distinguishes the complex drugs manufactured by Łukasiewicz, especially chemical preparations, which required chemical knowledge and laboratory work. He gives examples of chemical drugs which, according to his knowledge, were only introduced into the official list of drugs five years later.[94] It would thus appear that the Lviv pharmacy "Under the Golden Star" remained at the forefront of modern pharmacy at that time, as evidenced, among other things, by:

Oxidium stibium hydrosulfur, rubrum (minium), Ferrum citricum, Ferrum Magnesio Citricum, Hydrargyri Depuratio, Merucius Niger Hanemani, Mercurius praecipit albus, Acidum aceticum, Acidum Hydrocianicum, Aecetum Commun.

Professor Roeske also drew attention to an important feature of the prescriptions written in Łukasiewicz's hand. While other authors of the entries use signs used by mediaeval alchemists to determine the quantity of weight components of mixtures, Łukasiewicz gives their measurements in the then modern nomenclature – in pounds, ounces and drams.[95] It is probably a generational signal connected with keeping up to date with the new global, trends in his field. This means that Łukasiewicz's notebook was taken

94 Roeske, *Ignacy...*, p. 30.

95 Pharmacy measures introduced in Europe around the middle of the 19th century instead of the Nuremberg system.

over by his senior colleagues, and perhaps Mikolasch's entry is also among the entries. However it was, Łukasiewicz's notes were certainly also considered by his senior colleagues to be valuable, and worthy of preservation and continuation. But why did not Łukasiewicz take his notebook with him when it would certainly be useful in his further work?

From the cosmetics routinely produced in Łukasiewicz's times, his notebook contains recipes for creams, eau do colognes (*Eau de Paris, Eau de Cologne*), hair care waters, as well as lipsticks (*Ceratum ad labia*).

It was also obligatory for the pharmacy at that time to serve as a storehouse for various, nowadays unused chemicals. Łukasiewicz skilfully wrote down recipes for varnishes, inks, mastics and even fireworks in his copybook.

Łukasiewicz's notebook, to the satisfaction of his heirs, allows for a direct, fascinating insight into the world of pharmacy in the mid-19th century and proves the surprisingly broad, compared to the present day, practical competence of the former pharmacist. On the other hand, there are saddening reflections on how helpless medicine was in the face of illnesses which, long before the great discoveries of medicine at the end of the 19th century and the 20th century, decimated many societies. In Łukasiewicz's manuscript there are also recipes which could not be effective remedies for infectious diseases, but were at most preventive measures or so-called parapharmaceuticals.

As examples of such recipes from Łukasiewicz's "Golden Book," Professor Roeske mentions recipes for specifics especially needed in times of commonly occurring epidemics of tuberculosis, scarlet fever and other infectious diseases, such as: [96]

- A recipe for carrot starch – *Amillum Daucii* – as a remedy for lung diseases
- Nettle extract (*Belladonna*) dissolved in cinnamon water – a measure to prevent scarlet fever
- Vegetable oil tincture with camphor, ammonia and aromatic vinegar – *Yris fluidum* – for the disinfection of rooms during an epidemic with smoke from burning tissue paper soaked in this preparation.

No wonder, because only three months after Łukasiewicz's death, in 1882, Robert Koch announced the discovery of tuberculosis mycobacteria, and its effective treatment became possible only in the 20th century, thanks to the discovery of penicillin by Alexander Fleming. The recipes of Łukasiewicz's manuscript were fully consistent with the state of medicine and pharmacy of his time prevailing in all pharmacies in Europe.

96 Roeske, *Ignacy…*, p. 29.

Studies – the most urgent intention of all

Meanwhile, the Austrian Empire went through an enormous political upheaval in 1848 and, if it had not been for Russia's help in suppressing the Hungarian Uprising, it would have been little short of complete disintegration. Chancellor Metternich's career as a symbol of the widespread hatred of rebellious peoples collapsed, and the poor Ferdinand Habsburg, generously endowed with the nickname The Beneficent, abdicated. From December 1849, Austria already had a new Emperor Franz Joseph. The nations of the empire were promised the extension of autonomous freedoms.

Ignacy Łukasiewicz focused all these two extraordinary years of political change on conscientious work, but his ambitions and growing competences exceeded the modest professional position he still held. He dreamt of studying pharmacy at the Jagiellonian University in Cracow. Unfortunately, it was difficult to change anything right away in his situation, which was under police supervision and forbidden to leave Lviv.[97] The announcements that the corset of restrictions on the people of the Galician province had been relaxed did not entirely correspond to the truth.

Łukasiewicz wrote unsuccessful requests for a long time. He felt like a full-fledged pharmacist, but the authorities refused to allow him to go to Cracow. His letter to the ministerial authorities, which has survived, is full of regret. It was difficult for Łukasiewicz to come to terms with the fact that he would have to be "forced to stay in a subordinate position of a pharmacist's assistant for the rest of his life."[98] In his letter he assured that for ten years of his work to date he had raised funds for his dream studies and these were now "the most urgent of all intentions." After two years, however, his efforts, especially those of his principal Mikolasch, who enjoyed a high position in Lviv, finally brought results. After submitting a warranty, the authorities issued Łukasiewicz with a passport to Cracow.

According to the decree of the Ministry in Vienna of 10 May 1850, to take pharmaceutical university studies they required two years in public practice at a pharmacy as a pharmacist's assistant and passing a professional exam. Łukasiewicz had eight years of practice after the professional examination, so he had met the required conditions long ago. At the same time, in mid-September 1850, he was released from the pharmacy "Under the Golden

97 As Władysław Anczyc stated, "The Lviv Gubernium did not allow Łukasiewicz to complete his studies in Cracow and only after two years of the principal making every effort to ensure that Łukasiewicz was issued with a passport to Cracow, and to handle his warranty.", "Kłosy", 1882, No. 874.

98 Roeske, *Ignacy...*, p. 31.

Star," and then immediately went to Cracow. As is written in the documents in the Archives of the Jagiellonian University:

> "Ignacy Łukasiewicz (...) residing in Cracow Grodzka Street 103 (...) joined in pharmaceutical studies at the Faculty of Philosophy of the Jagiellonian University in the autumn of the academic year 1850/1851."[99]

The archives of the University also contain his study programme. This included lectures and exercises in general chemistry, inorganic chemistry, pharmaceutical and medical-forensic chemistry, botany, mineralogy, heat science, magnetism and electricity, as well as experimental physics and zoology.

The rank of his lecturers, most of whom rated his results as very good or good, was impressive.

Professor Ignacy Czerwiakowski[100], son of the court physician of the last Polish king Stanisław Poniatowski, the founder of the first Polish faculty of surgery, doctor and director of a military hospital during the defence of Warsaw during the November Uprising, lectured on Botany at the Cracow pharmacy. Under the direction of Czerwiakowski, the Warsaw Botanical Garden became one of the leading gardens in Europe.

Professor of mineralogy, geology and zoology Ludwik Zejszner, in turn, was the son of the pharmacist of King Poniatowski, an outstanding cartographer, whose many geological maps were used in their publications, regardless of copyright, by numerous European cartographers. Zejszner published the first geological map of the Tatra Mountains in history, explored deposits of the Kingdom of Poland and also contributed to the oil industry. He was fascinated by the presence of crude oil in the Subcarpathian region. Łukasiewicz perfectly remembered the student trip with Professor Zejszner to Wieliczka near Cracow and its famous salt mine, which had been in use for many centuries, in order to observe the salt firing under the influence of inflammable hydrocarbon gases. During this trip, the professor interrogated the students about the area they came from, especially interested in whether oil could be found in their homeland. From Professor Zejszner's suggestion it appeared that oil, in his opinion, could be used for lighting.[101] However, these were purely theoretical suggestions, not used anywhere in wider practice. These observations were remembered by Łukasiewicz as a student in order to soon come to life.

Władysław Ludwik Anczyc – a friend of student years and one of the greatest admirers of Łukasiewicz's person and work – at the time of his

---

99 Roeske, *Ignacy...*, p. 30.
100 Ignacy Rafał Czerwiakowski (1808–1882).
101 Ludwik Zejszner (1805–1871).

arrival in Cracow, was from 1848 an assistant professor of the Faculty of Chemistry and Pharmacy at the Jagiellonian University. It was he, just a year younger than Łukasiewicz, who gave him the greatest possible testimony, writing:

> "We met him as one of the most talented and diligent students and a best friend. He worked hard, and made it easier for less talented colleagues to study with private, selfless lectures. At that time, there were not so many scholarships (...), most of them went forward in thorny poverty, and in the last need, a friend saved a friend. Łukasiewicz was one of the first in this respect and he shared his last penny with the needy...."[102]

Ludwik Anczyc, who came from a famous noble acting family, in later years was above all known as the author and publisher of many magazines in both Partitions, both in Warsaw and in Cracow, as well as the author of many most kind memoirs about Łukasiewicz, all the more valuable as they were written from experience and heard from eyewitnesses.

There was not much exaggeration in the praiseworthy opinion about Łukasiewicz. A good nature and comradeship was always the hallmark of his personality, while during his first year of studies he was able to pass almost all the lectures and exercises provided for in his two-year study programme. This diligence, however, went hand in hand with quite down-to-earth problems, because it was after a year in Cracow that Łukasiewicz faced a lack of means of subsistence. By the end of the first year he had already started to apply for admission to a "strict examination," which could become the basis for awarding him a master's degree.

In his retained letter to the Ministry of Religions and Education, he declared that he had already mastered the subjects that he had not had time to complete in his first year during his Lviv practice:

> "as the enclosed certificate issued by the highly respected Lviv pharmacist and president of the Lviv Pharmacists' Committee Piotr Mikolasch shows, during my pharmacy work I devoted myself constantly to chemistry and pharmacognosy studies and also in this field I received education from that pharmacist."[103]

From the applications sent, his sense of delay in relation to the normal course of things, which was caused both by the lack of material means and lost time spent in prison, appears. This is probably the reason for his increased dynamism in his studies, as well as a clear certainty as to the level of his knowledge. Undoubtedly, the solemnly completed years of practice in the

102 Władysław Anczyc, *Ignacy Łukasiewicz...*

103 Archive of the Jagiellonian University, Cracow, WL 216, Letter from I. Łukasiewicz to Minister of Religion and Education Count Leon Thun of 17 May 1851.

pharmacy laboratory under the supervision of the principals and systematic self-education had to have an effect.

As a result of his letter, the Jagiellonian University Medical and Surgical Faculty authorities asked Professor Florian Sawiczewski, the rector of the University at that year, for an opinion on the obligation of Łukasiewicz to meet the formal requirements for studies. [104] Professor Sawiczewski referred to Łukasiewicz's unprecedented request with full fundamentality. He answered the question of the Department firmly:

> "Mr. Ignacy Łukasiewicz does not attach to his certificate proving that he attended a course in pharmacognosy at any university of the Empire, and I consider this subject to be very important and necessary for pharmacists, so the submission for strict exams cannot be accepted, especially that according to the existing regulations, pharmacists were obliged to attend the University for two years of study."[105]

However, Professor Sawiczewski did not leave Łukasiewicz without help. He recommended him to a friendly owner of an alumina factory in the village of Dąbrowa, 50 kilometres from Cracow, Waholz, who asked him to recommend him a suitable specialist for his plant. Thanks to this protection, Łukasiewicz gained a flat with board and a salary, which, although small, allowed him to survive the next eight months and complete the next semester of his studies.

The last, fourth semester of his studies was, however, completed by Łukasiewicz at the University of Vienna. We do not know how he earned his living there. In Vienna, he attended lectures with practicals by professor of analytical chemistry Josef Redtenbacher, with whom he later continued his correspondence. On 30 July 1852, he passed a post-master's examination[106] with a "good" grade on the basis of his diploma thesis "*Baryta et Anilinum*," i.e. a subject in the field of knowledge about hydrocarbons.[107]

---

104 The Ministry of Religion and Education sent his request to the Jagiellonian University on 20 May 1851 with the comment "Medical and Surgical Faculty in Cracow to be officially settled, in accordance with the applicable regulations and only in the case of particularly favourable circumstances."

105 Archive of the Jagiellonian University, Cracow, WL 231, Letter from F. Sawiczewski to the Dean of the Medical Faculty, UJ of 16 June 1851.

106 Rygorosum (łac. *rigorosus – rigorous*) – former name of the doctor's oral examination. The exam covered other areas of science and together with the dissertation was part of the requirements for gaining the title; See *Słownik wyrazów obcych* [Dictionary of Foreign Words], ed. Jan Tokarski, (Warsaw: Państwowe Wydawnictwo Naukowe, 1980), p. 658.

107 *"I. Łukasiewicz Ignatius, Galicianus e Zaduszniki, natus 8 Marti 1822, absolvit gymnasia et tirocinium Resovio, cursum pharmac. Cracoviae et Vienno, rigorose examinat pro magistr. Pharmac. Die 30 Juli 1852 c.c. Bene", Protocoll Pharmacopei 1751 1854*, Archive of the University of Vienna.

The second half of the 19th century, gaining more and more speed, seemed to be a new beginning. Industry developed intensively, with steam machines and technology forcing an avalanche of new inventions entering the industry on a large scale. The symbol of this new era was the First World Industrial Exhibition in London in 1851, which started the parade of exhibitions from the largest capitals to the ambitious provinces. Newspapers of the world wrote about the Parisian exhibition. Also fascinated by this event were the inhabitants of the Polish provinces under partition, living between the beginning and the end of this extraordinary half-century, triumphantly ending with the accent of the Paris Exhibition and the erection of the Eiffel Tower as a symbol of modernity, hard science, medicine, technology and inventiveness, which gave people new opportunities.

The turn of the 19th century also seems to be a period of Romantic, rebellious attitude on the part of Łukasiewicz, which had such a strong influence on all his youth. In March 1852 he reached the age of 30. Still not forgotten by the police authorities, once again during his studies he was arrested and questioned in connection with suspicions of poisoning a well in order to assassinate the lives of the Germans. This suspicion was probably born while, as a professional in chemical analysis, Łukasiewicz undertook a commission to study well water in Cracow. "I don't know of such a poison that would only serve the Germans," he was supposed to have told the judge quite ironically.[108]He was released from custody without consequences.

Since then, his life was the biography of the truest positivist, whose strength was in work and realistic pragmatics in gaining even the most romantic goals.

There will be no white spirit

> *"An Orthodox Jew in traditional costume went to the Mikolasch pharmacy in Lviv with a flask of rock oil under his arm. When asked what he wished, he stated that he had an idea. 'What?' he was asked. Could this black oil not be used for distillation to produce... white spirit?*
> *The pharmacy worker, to which Schreiner communicated this idea, reached for a flask with a yellow liquid and gave it to the Jew to smell.*
> *– Friend, do you want to have a drink?*
> *– Well, why should I have such a nasty drink?*
> *– Because this is your idea, vodka from oil.*

108 Roeske, *Ignacy..*, p. 32.

> *The pharmacy worker who showed Schreiner their finished product extracted from natural oil, was the young master in pharmacy, Ignacy Łukasiewicz."*

Such a scene from the Lviv pharmacy "Under the Golden Star" was quoted by Ludwik Tomanek, biographer and journalist in the Cracow issue of the *Illustrated Daily Courier* in 1927 as one of the still vivid stories and memories of people who remembered Łukasiewicz's times 45 years after his death.[109]

The newly graduated Viennese master in pharmacy came back to Piotr Mikolasch's company. Here, among other activities, he was particularly interested in the phenomenon of crude oil taken into account by the principal as a possible base for pharmaceutical products. The extraction of raw materials for preparations made in nature was normal in pharmaceutical practice at that time.

In the areas of natural leaks of the Subcarpathia region, especially on the surfaces of water reservoirs and bottoms, oil had always been known. For centuries, the surrounding population had tried to use it in various ways, especially for treating people and animals. Cottage oil production in the 19th century even developed. In 1835 there were 30 active pits with a capacity of 16 litres of oil per day in Borysław. It must be admitted, however, that this was a small yield. On an annual basis, these points gave 15,000 litres. In 1840, 75 pits were already being exploited in the Stanisławów area, giving a total of 24,000 litres.[110]

Peasants from the vicinity of Krosno and Gorlice in Lesser Poland, i.e. from the Western Subcarpathian region, similarly to Drohobycz or Borysław in the Eastern Subcarpathian region, i.e. centres located near Lviv, used oil to smear the body against rheumatism, frostbite, scabies, and even used it as a medicine against worms. They tried to produce greases for wheels and tools, for boat sealing and wood impregnation. They used it for primitive cressets instead of oil.

The natural consequence of this situation was, therefore, that crude oil was sent to the pharmacy, as the only place at that time where one could meet experts who knew about chemical analysis and who had laboratories and the necessary equipment. The peasants, who could probably already be called the first miners, long before Łukasiewicz's return to Mikolasch's pharmacy, tried to get to know the language of oil, and learn more about the way to distil it.

109 Ludwik Tomanek, *Ilustrowany Kuryer Codzienny*, No. 342, 12.12.1927.
110 Stanisław Brzozowski, *Ignacy Łukasiewicz*, (Warsaw: Interpress, 1974), p. 47.

For years, Subcarpathian peasants had been cleaning oil by means of heating. However, the Mikolasch pharmacy was the best pharmacy in the city and one of the leading pharmacies in the whole Austrian Partition. It had excellent facilities and a well-equipped laboratory used, as we know, not only for pharmaceutical purposes, but also for the needs of the economy, crafts and industry.

Piotr Mikolasch[111] was a graduate of pharmacy in Vienna. As a master he found himself in the pharmacy "Under the Golden Star," which he took over after getting married to the widow of the owner Ressig. This pharmacy was run by subsequent generations of Mikolaschs, and despite the turbulent and complicated history of almost two hundred years of the history of Poland and Lviv, the pharmacy is still operating in its place today.

Since 1848, the year in which Łukasiewicz happily left the Rzeszow prison, Jan Zeh, a Master of Pharmacy, worked in Mikolasch's pharmacy.[112] He mentioned that already in his previous workplace, in Sambor, he had met a peasant named Babtuła, who brought a smelly, colourless distillate of oil to the local pharmacy, but the pharmacy forwent longer deliveries. This fact was recorded by Jan Zeh, and in this way he immortalised in his memoirs the names of two traders Abraham Schreiner and Lejba Stierman, who in the suburb of Drohobycz bought oil from local peasants and carried out primitive "distillation" or unprofessional cleaning by economic means.[113]

Schreiner's promotional campaign in the Mikolasch pharmacy was crowned with success. The principal was tempted by the vision of producing not so much white spirit as a substitute for the expensive *Oleum Petrae album* imported from Italy, which was produced from oil then called rock oil. Mikolasch bought foreign rock oil at high prices and sold it in a pharmacy as a valuable specific. Łukasiewicz himself said in his memoirs that he shared this initial enthusiasm and immediately saw with his imagination the prospect of exporting his own *Oleum Petrae album*. He never lacked imagination and the ability to reach unlimited viewpoints.

Therefore, the offer of the Borysław crude oil bidders was accepted with interest by the pharmacy "Under the Golden Star," while the task of "fragmentation" of crude oil on a wholesale scale and the use of the separated components was entrusted to two pharmacy workers – masters Łukasiewicz and Zeh. It was probably Ignacy Łukasiewicz who convinced the boss that it was worth undertaking this task on a larger scale. He himself presented the course of events as follows, and this will be read by contemporaries,

111 Piotr Mikolasch (1805–1873).
112 Jan Zeh, b. Łańcut 2 September 1817, d. 25 January 1897 in Borysław.
113 Brzozowski, *Ignacy ...*, p. 59.

including living witnesses together with traffickers and his colleague Zeh and Principal Mikolasch, who knew very well what it was like:

> "And there was a pharmacist there, my good friend, a certain Zeh. I show him oil and advise him to cleanse it further. We added acids, passed it through an alembic and immediately got clean kerosene."[114]

These words are known from the only "interview" with Łukasiewicz – a rather innovative form for those times – by Szczęsny Morawski, published under the title "Own Testimonies." And even though mankind still had to wait for dictaphones, it is hard not to believe in the sense of this statement. Unfortunately, Jan Zeh himself denied this version of the initially joint work on obtaining pure kerosene, which Łukasiewicz honestly presents. Zeh emerged from oblivion in oil circles, suddenly, after several decades, publishing memoirs. In his version of the events he was even inclined to immortalise the names of some of the many traders of raw materials, rather than to mention even a word about Łukasiewicz's existence.

Łukasiewicz was never, in his whole life, guided by low motives, ambition nor self-interest, evidence of which he gave in the second half of his life, which cost him millions. The sense of Łukasiewicz's public relationship is clear testimony to the fact that Zeh, when Łukasiewicz was giving his "interview" to public knowledge, was a completely anonymous figure in the world of oil workers in the 1860s. And yet he was alive. Until the year of Łukasiewicz's death, he had enough time to recall his merits. He could also then, for example, have published such a text:

> "It was I who was the first to show Łukasiewicz oil and we decided to cleanse it further."

However, he did not do so at the time. He waited. He survived Łukasiewicz by fifteen years.

In 1852, as Jan Zeh claimed in his memoirs, the traders from Drohobycz brought several barrels of "distillate" of oil, offering it to various potential buyers, of which 2 Austrian cetnars, i.e. about 112 kg, they managed to sell to the pharmacy "Under the Golden Star."[115]

It was a difficult and risky task to develop a method of oil distillation in the absence of a technology for processing the explosive, flammable liquid. Despite the dynamic development of industry and work on technologies for the processing of various minerals, oil had not been taken seriously by industry so far, mainly due to its inapplicability. The only emerging property – flammability – was not an attractive feature, either because of its

---

114 Brzozowski, *Ignacy..*, p. 59.
115 *Brzozowski, Ignacy..*, p. 50.

explosiveness or because of its reeky, stinking smell, and the uneven flame emitted during burning. Earlier, on a more serious scale, such unsuccessful attempts had taken place in France and Russia, but they were abandoned.

In the Subcarpathia region, as assessed by Professor Roeske,

> "The closest success among these accidental empirics was in the years 1810–1817 an official of the Truskawiec salt mine Józef Hecker (...) however, it is known that the venture was unsuccessful, Hecker came to financial and mental ruin."[116]

As a pharmacist, through an analysis of sources, Professor Roeske stated that:

> "Hecker's distillate obtained by primitive distillation without the deliberate use of physico-chemical methods was a mixture of light naphtha fractions. Hecker did not illuminate with kerosene, but with gasoline fractions isolated at a temperature of 70 to 120°C, hence the short duration of his invention and the dramatic end of his venture."[117]

Hecker's story of an extremely spectacular, loud beginning and equally famous, tragic end echoed all over the Subcarpathia region and effectively discouraged experiments with oil distillate lighting.

In the middle of the 19th century, however, there was already a large-scale industry of light oils extracted from lignite or bituminous shale, and it seemed that there was no alternative to this direction. For example: Saxony already in 1830 had 40 plants with 2,200 retorta, the USA in 1860 had 50 huge plants. Lamp oil was produced in France and Russia, rich in bituminous shale. Justus Liebig, a chemist, was active in Estonia and still in the 1860s he postulated the construction of lamp oil factories based on *kir* – asphalt shale from the Southern Caucasus.[118] He did not give up the industry for a long time before he realised that he would not be able to cope with the uneven fight against the better. Regardless, the hard coal gas industry was in full swing, and the lighting of the city streets with gas lamps at that time was an irresistible hit and a symbol of the momentum towards modernity.

The distillation of the large amount of oil purchased by Mikolasch for lighting purposes was not the main reason why the boss commissioned his two masters to "fragment" it, because there were many other, better, as it seemed, oils to choose from. Some form, however risky and generic, not very different in its art from wild distillations in peasant fields, was recommended by an Austrian pharmacopoeia from the end of the 18th century, from 1790, suggesting how to deal with the lack of original *Oleum Petrae album*:

116 Roeske, *Ignacy...*, p. 43.
117 Roeske, *Ignacy...*, p. 44.
118 Brzozowski, *Ignacy...*, p. 47.

"Any quantity of yellow or red rock oil is extracted and fills a large glass retort; it is then poured in about ten times the amount of water, the receptacle is extended, the holes are closed and distillation on the sand bath begins. Initially, there is a light and colourless oil which does not bind to water and which is stored in well-sealed receptacles."[119]

We distil!

The first formula for extracting *Oleum Petrae album* used by the two employees of the Mikolasch pharmacy in 1852 is unknown, but it is known that the distillation effect turned out to be extremely satisfactory.

Years later, in a constantly enthusiastic tone, Łukasiewicz reported on his immediate reaction to this success:

"I am asking for the establishment of a company. Mikolasch agreed, so as a threesome with Zeh we gave 800 guilders each and we set up a company. They are sending me on a tour to Sambor – to Borysław, and Kołomyja – to Sołotwina and Bohorodczan. I bought several thousand pots of this, sent to Lviv and we distill!"[120]

This tour allowed Łukasiewicz to find out where the oil occurred. Not without initial disappointment, because it turned out that there was less oil in the field than it would seem. These first conclusions about the low abundance of deposits, in fact extremely oil-bearing, were misleading, because the primitive extraction of crude oil depended on the season. Łukasiewicz, on the other hand, travelled in search of oil in the autumn of 1852. He was lucky only in Pechenizhyn, near Kolomyja, where he found a well dug in the previous century. This first, wholesale purchase of crude oil by Łukasiewicz was reported by Anczyc thusly:

"According to old Polish mining laws, the well was a royal property, so after Galicia was taken away, it became the property of the imperial estates, the so-called camera. However, the Management Board did not exploit the oil on its own account, but let it lease to an old religious speculator who, unable to dispose of the product, was in arrears in payment. Therefore, the entire supply, amounting to tens of thousands of pots, was seized in court, and Łukasiewicz bought it very cheaply for the company and took it to Lviv. Immediately, purification and distillation on a larger scale were undertaken. Samples of the obtained product were sent to pharmacists, pharmacy materials factories and manufacturers of compressor products (elastic rubber, which dissolves in oil),"[121]

A man, who already in his youth was able to organise a logistic network of conspiracy for the whole area, immediately revealed a talent for business.

---

119 Brzozowski, *Ignacy..*, p. 51.
120 Brzozowski, *Ignacy..*, p. 52.
121 Brzozowski, *Ignacy.*, p. 53.

Łukasiewicz's own account of these first steps in the oil trade on a larger scale also exists.

> "Mikolasch's, an old and wealthy company, was well known in the trading world. We sent samples to Vienna, Prague, Trieste, and we dropped the price to 80 guilders per cetnar. Orders come, but small, here for the pound, there for 3, for 8."[122]

Therefore, even more he started to think that it was worth developing a way to use the oil for lighting.

> "Zejszner reminded me and the lighting of Sambor with earthly fat,"[123] he stated, as it was already known at the time that it made no sense to purchase larger quantities of oil for pharmaceutical purposes alone. At the same time *Oleum Petrae album* lost its popularity. Surely, simply, patients were becoming less and less convinced of its magical properties. However, since Mikolasch himself lost his interest in oil and left the company, the two of his master's students remained on the battleground with the oil reserves. Łukasiewicz and Zeh.

In a rented apartment in the Lviv district of Lychakiv, they continued their work on improving the distillation of crude oil to obtain excellent, odourless kerosene. Fractional distillation[124] required a gradual separation of individual fractions as the temperature increased. Łukasiewicz and Zeh, having brought the liquid to temperatures of 200–250 degrees Celsius, deprived it of light fractions, such as petrol. Heavy hydrocarbons such as bitumen and technical oils remained in the apparatus. The obtained chemical distillate was further refined with concentrated sulphuric acid and then with soda. This principle was applied at the basis of the technology of kerosene refining in the 20th century.

The fact that Łukasiewicz and Zeh worked together at that time on the crude oil distillation method is sometimes presented as a special Gordian knot to be solved by biographers and historians in order to clearly determine which of the two partners contributed more to the development of the first, professional, non-intuitive, conscious, chemical knowledge and laboratory practice based on the scientific and technical method of fractionated distillation. There are sources and many understatements about this. Above all, however, it is possible to recreate a logical sequence of events which unquestionably and in every aspect dispel doubts in favour of Ignacy Łukasiewicz.

The analysis of known facts as well as the professional path and personality of Jan Zeh is of particular importance in this decision. It is no coincidence that since the 1840s, since his first job in Sambor, Zeh was not interested in the possibility of exploring or processing oil, despite the imposition of

122 Brzozowski, *Ignacy..*, p. 54.
123 Brzozowski, *Ignacy..*, p. 54.
124 Roeske, *Ignacy..*, p. 44

willing suppliers and the need to import expensive Italian distillate. This idea was realised in the pharmacy "Under the Golden Star" only after the arrival of Ignacy Łukasiewicz with a fresh supply of the most modern knowledge acquired during his studies and after the university practice of chemical analysis, despite the fact that Zeh, being five years older than Łukasiewicz, had already worked for Mikolasch before.

Apart from this, it does not seem possible that the merchants brought from Drohobycz to Lviv several barrels of crude oil, which had not been particularly useful so far, as a "shot in the dark," spontaneously. Rather, it is obvious that they had already satisfied the demand for crude oil that was cottage-cleaned, either by themselves or by peasants, much earlier. Otherwise, as a matter of commercial prudence, they would not immediately opt for a large risky transport. It is probable that it was not the first time they had visited the best pharmacy in Lviv, known almost all over Galicia. However, they managed to make some money from their goods only when they found Ignacy Łukasiewicz in the pharmacy. The scene described in IKC[125] with one of them with a bottle of oil under the arm as a sample is simply the most probable scenario for the whole action.

The fact that Zeh graduated from the *gymnasium* in Drohobycz is very significant. It is not possible that merchants from this tiny town did not recognise their compatriot in the famous pharmacy in the capital. Moreover, Jan Zeh had been studying and working in the centre of activity of Czech Józef Hecker, a salinary official in Kossów and Drohobych, a decade or so earlier. It was Hecker who first tried to distil oil and use it for lighting, and even for 8 months illuminated the salina and barracks in Sambor and the suburbs of Drohobycz with oil. Already in 1816 he promoted oil lighting in Vienna and Prague. Although Hecker's attempts failed because the distillate he produced was dangerous, often exploded and there were no proper lamps to use this fuel, Zeh undoubtedly heard of Hecker's achievements and had many years before 1852 continued and improved on his work.

Jan Zeh was never before his memorable first, bigger oil purchase for Mikolasch's workshop interested in oil or even mentioned Hecker.[126] Many times he gave more sufficient evidence that oil was not in fact his fascination or passion, which would release particular inspirations for action, even when not only Łukasiewicz, but also the whole world, was

---

125 IkaC – popular abbreviation of the name *Ilustrowany Kuryer Codzienny* [Illustrated Daily Courier], a national political and informational daily published in 1910–1939 in Cracow.

126 Jan Zeh, *Pierwsze objawy przemysłu naftowego w Galicji* [First Appearances of the Oil Industry in Galicia], in: *Przemysł naftowy* [The Oil Industry] of 10 July 1935, pp. 391–393.

completely captured by oil. In his further life, Zeh consistently focused on the production and trade of petroleum products for pharmacy use. In the case of a creative personality with the dynamism of the discoverer and researcher, it is psychologically impossible to move around one's small field. Unlike Zeh, Ignacy Łukasiewicz, even on his deathbed, delirious and in agony, did not stop talking about new initiatives, an infinite number of projects and intentions, which were created in his creative mind, and which he would undoubtedly have pursued had it not been for his premature death.

Jan Zeh's lack of enthusiasm for perfecting his skills in the field of oil according to some interpreters is explained by the trauma after the death of his wife and sister-in-law, who died in an oil barrel fire. Facts do not seem to confirm this thesis particularly. The personal tragedy did not discourage Zeh either from marriage or from the oil trade. Only three years later, Zeh married again and enjoyed a happy family. He moved from Lviv to Borysław, which, just like Drohobych, was the centre of oil in Eastern Subcarpathia, which at that time was already under the fever of accelerating industrial extraction and processing of oil. Even in Borysław, the epicentre of the formation of oil castes, Zeh did not go beyond the horizon of his own pharmacy-wide trade, although he continued to produce and sell oil products.

Doubts of a slightly lighter form are without number. There is no certainty as to what really caused the big explosion in his pharmacy oil store. It is hard to believe that the perpetrator, who allegedly threw a cigarette butt into it, would admit it. Or rather, Jan Zeh found out very quickly that without the cooperation of someone like Łukasiewicz on a larger scale, he was not able to cope with this unpredictable liquid.

Zeh's lack of broader horizons are combined with an unpleasant, low envy, which would be given embarrassing expression in the future on the pages of his memoirs, seven years after Łukasiewicz's death, when he would no longer have the opportunity to look into his eyes. This happened only 37 years from the moment of joint work on oil refining.

During these almost four decades, Jan Zeh did not appear at jubilee celebrations, nor did he appear at the fetes of numerous initiatives for Łukasiewicz's person, of which many were organised. He never claimed his role at that time, he did not appear in the numerous companies of Łukasiewicz's, who quickly became a popular and at the end of his life a widely worshipped person, a man to whom oil workers even came from across the ocean. According to Łukasiewicz's accounts, Zeh never wanted to meet him or remind him of himself, and yet so many of Łukasiewicz's true friendships lasted for a long time. Zeh never sent a letter to a newspaper or praised his acquaintance with the famous colleague. Too many personal witnesses probably lived to deny Łukasiewicz's pioneering and decisive role in the development of this first starting point of the oil industry, which was the approach to the scientific method of oil refining.

I need lamps for this and that – one of these and one of those

The hopes for a wider use of kerosene in medicine failed, Mikolasch's pharmacy was satisfied with its mediocre sales for traditional purposes, but Łukasiewicz did not give up the idea of using it for lamps, because it burned evenly, without explosions. He started his experiments:

> "I'm trying to light up, of course in an oil lamp. The tank ignites from the inside, explodes and almost burned me. Then they proclaim new lighting: pinoline and camphine. I can't sleep from envy, I go to the famous tinsmith Bratkowski," he says[127].

As with the unfortunate tests with kerosene for an oil lamp, so were the earlier tests with photogenic lamps. Here, a completely new idea was needed.

Adam Bratkowski, a tinsmith specialising in the construction and production of lamps, was an extraordinary figure. A nobleman of origin, since childhood making his way through life alone after the death of his parents, he took up craftsmanship training and as an apprentice travelled across half of Europe, perfecting his profession in workshops in Vienna, Berlin, Hamburg, Switzerland and France, in Paris itself. Having returned to Poland, to Lviv, after only five years, in 1845 he became famous for a commission from Archduke Ferdinand d'Este, Governor of Galicia, for oil street lamps. Bratkowski's lamps shone beautifully and illuminated more and more Lviv streets, making the name of their creator famous.[128]

The appearance of lamps with camphine and pinoline sowed doubt in Łukasiewicz. He did not fail to test them:

> "I bring them in – I light them up, it's a miracle! Not light."[129]

And yet the thought of a kerosene lamp was already too tempting. Both quinoline from vegetable raw materials and camphine, i.e. a mixture of turpentine and alcohol, were incomparably more expensive and more difficult to obtain than kerosene.

---

127 Brzozowski, *Ignacy..*, p. 59.

128 Brzozowski, *Ignacy..*, p. 60.

129 *Ignacy Łukasiewicz i ropa galicyjska (Zeznanie własne)* [Ignacy Łukasiewicz and Galician oil (own testimony)] in: *1971–1974 Ku czci Ignacego Łukasiewicza* [1971–1974 In honour of Ignacy Łukasiewicz], Organizing Committee of the Ignacy Łukasiewicz Jubilee Year celebrations at the Krosno Land Association, ed. Zbigniew Ringier, Zdzisław Sroka.

Figure 7: Łukasiewicz's lamp – a replica of the prototype. Source: Subcarpathian Museum in Krosno

"I need lamps for this and that – one of these and one of those."[130] This is how he reported his dialogue during his first trip to Bratkowski. Where "for this and that – one of these and one of those" certainly meant for kerosene, for interiors and for wind and rain, that is, that they would be so certain as they already stand on the streets, safe. Numerous consultations and trials finally yielded results.

The first kerosene lamp was similar in appearance to a mug, because it had the shape of a cylinder with a large handle. The kerosene reservoir made of sheet metal was strengthened with a metal fitting. Behind a glass of mica, a porous suction wick was placed in an openwork burner for air supply burned.

With this lamp Łukasiewicz went first to his principal.

"I say to Mikolasch

– Sir! Let's buy more of this oil!

130 Roeske, *Ignacy..*, p. 45.

– What for? Will we drink it?
– We will shine!

I show him the lamp (...) yet he does not believe,"[131] says Łukasiewicz in his interview with Morawski.

Mikolasch, however, believed. Not only did he allow the pharmacy's laboratory to be illuminated at night, but he also won the tender for lighting the hospital. "For a penny less than everyone else," thus spoke Łukasiewicz, to whom the boss entrusted the work. The hospital purchased 500 kg of kerosene from the newly revived Mikolasch-Łukasiewicz-Zeh company.

"Still remembering Zejszner and the city of Sambor – I was the first in Galicia to illuminate the main hospital in Lviv, known as the Piarists,"[132]we read in Łukasiewicz's interview and no one dared to question it during all his years of life. The loyal and painfully earnest Łukasiewicz does not deny others a leaf on the wreath of fame to Zejszner, to whom he owes his inspiration for this success.

The work was completed on 31 July 1853, and the lamps from the Bratkowski plan designed by Łukasiewicz, shining since March in the window of the Mikolasch pharmacy, lit up the hospital for the first time, and on the same night a man's life was saved. This unquestionable fact was recorded in hospital books and newspapers. An operation on the appendix of a man named Władysław Cholecki was performed by the surgeon Zaorski.

When you get to know Łukasiewicz's personality, you can imagine his satisfaction from this successful event, which was perfect as a symbol of an act for the good and happiness of people. As for the fact that it was a personal success for Łukasiewicz, none of his contemporaries who, on the twentieth anniversary of the lighting of kerosene lamps at the Łychakiv hospital, on exactly 31 July 1873, organised a jubilee for this event and a great fete in honour of Łukasiewicz, had any doubt. Jan Zeh did not dare to come to this jubilee. Perhaps Piotr Mikolasch was still alive that July, but he died just this year, and the date of his death is not known. However, the young sons of their former principal lived and certainly knew the truth.

Łukasiewicz systematically improved the design of his lamp over the next six months of 1853. Its prototypes and various variants remained in the attic of the Mikolasch pharmacy until the end of the 19th century. The heirless death of its inventor and the unprecedented lack of imagination of his descendants, which Łukasiewicz's lamp became, the role it played in the promotion of oil, even when the world was already at the height of its exploration meant that no examples of it have survived to this day. The unfortunate

131 Brzozowski, *Ignacy..*, p. 61.
132 Brzozowski, *Ignacy..*, p. 62.

Figure 8: Łukasiewicz's lamp. Source: Subcarpathian Museum in Krosno

fate of Lviv and the whole of Poland, and especially the two world wars, resulted in desolation in which the memorabilia of Łukasiewicz and his property were lost. A replica of the prototype of the historical lamp was reconstructed for The Ignacy Łukasiewicz Oil and Gas Industry Museum in Bóbrka on the basis of a photograph found in the collection of the deceased relative of Łukaszewicz Franciszek Pik-Mirandola.[133]

And Łukasiewicz's invention opened great prospects for kerosene, which, considering the reality, and especially its cheapness, could only supplant other lighting products. *Hydrocarbure* imported from Hamburg for lighting purposes to Vienna and used in Stobwasser lamps cost 43 guilders per cetnar, for example, and a cetnar of petroleum 29. In addition, the camphine and pinoline lamps were not as safe as they seemed to be.[134] From the very beginning, Łukasiewicz was aware of the great future of kerosene and acted quickly.

Already in 1853 he, no one else, sent samples of kerosene to the Viennese pharmacist J. Heindl with prototypes of his lamps. In time, Heindl became the

133 Brzozowski, *Ignacy..*, p. 60.
134 Roeske, *Ignacy…*, p. 42.

owner of a chemical factory. The very good relations between Łukasiewicz and the former professor of the University of Vienna, Redtenbacher led him to send a similar package to him. Through the Lviv Chamber of Commerce Łukaszewicz promoted kerosene and his lamps in Paris.

His project was captured by the Viennese lamp manufacturer Rudolf Dittmar, who registered a patent, which quickly allowed him to gain a dominant position in the European market, where Łukasiewicz's lamp was named "Viennese." Dittmar lamps also displaced the lamps from the Bratkowski manufactory from the market. Although the Lviv-based manufacturer tried to spread his wings by differentiating the range of lamps from luxury to popular, for the poorer population, students, for utility rooms, but, operating in the province of the monarchy, he certainly had incomparably less chance than the Viennese company.

Łukasiewicz, without applying for a patent which he had had the chance to register as the first, observed the march of his invention with satisfaction and joy. He expected the development of mining and the domestic oil industry. He was typically charismatic and idealistic.

This is how Tomanek, his biographer from the beginning of the 20th century, sums up the matter:

> "The Viennese Dittmar lamps are famous and have even been adopted by America. As we know, Dittmar only improved the lamp, originally made by Łukasiewicz. However, he appropriated the whole invention completely, not taking care to give back the due part of the profits to Łukasiewicz or his heirs. Naturally, Łukasiewicz, who was broadly generous, did not care at all about such a detail as the reserved patent. The factories of machines, pipes and all the equipment needed by the oil industry developed enormously. The production of sulphuric acid and other chemicals needed to refine kerosene increased. Mechanical workshops were created everywhere, and employed a significant number of workers. In a word, the oil industry caused the expansion of many other industries and gave a whole wave of people the opportunity to earn."

In this respect, he led a revolution, taking into account the use of benzines to run motors similar to the revolution caused by the introduction of railways.[135]

As we know, where new sources of profit are discovered, the instincts of small businesses start working the most efficiently. Both the senior shareholders of Mikolasch's firm, with Zeh, went on their particular paths. Each of them started to trade in kerosene on a larger scale on their own. Zeh opened his own distillery in Borysław, while in Lviv he traded in kerosene. With samples of kerosene to Vienna, Schreiner went to the railway

135 L. Tomanek, Ignacy Łukasiewicz, *Twórca przemysłu naftowego w Polsce. Wielki inicjator - wielki jałmużnik* [The Creator of the Oil Industry in Poland. A Great Initiator – A Great Almsgiver], (Miejsce Piastowe 1928), p. 100.

management immediately. The news of a successful wholesale petroleum dealership with the Lviv hospital shocked the whole trading community. Mendel Sachs, a grain merchant, even ran a race with Schreiner for the same goal from Drohobycz to Vienna.[136]

Perhaps the names of these or other traders would be lost if it were not for the fact that, as a joke of history, they sometimes even consumed the glory of the first oil distillers. Schreiner, a would-be white spirit enthusiast, liked to be regarded as such.

To some extent, the arrival of the heralds of the fantastic civilisational coup in Vienna, the capital of the empire itself, must have really impressed. It is obvious, however, that the conscious development of the method of distillation of individual fractions could not be done by simple people without a university chemical education and a specialised laboratory. This opinion is strongly promoted as non-alternative by professor of pharmacy Roeske, and it is impossible to argue in this respect with an expert on the subject.[137] He also firmly raises in his work the issue of the unauthorised and unprofessional use of the term "distillate" in relation to amateur products, the "naïve," empirical purification of crude oil.

There were many attempts to take priority away from Łukasiewicz, both as a shareholder in the first fractionated distillation and the originator and constructor of lamp prototypes using kerosene, including the pioneering excavation of the first oil wells from the period of his later activity, hence the unauthorised freedom in this respect still exists today. There are many absurd cases in this respect, or deliberately lies, contrary to documents and facts. Among them is the aforementioned practice of Rudolf Dittmar, who took up the production of Łukasiewicz lamps, proclaiming himself to be their inventor, although he only improved them over time.

In turn, Yale College chemistry professor Benjamin Silliman junior is completely misrepresented as the designer of oil lamps, whereas Łukasiewicz's discoveries and the immediate transfer of the sensational news about his oil production to America only encouraged him to research crude oil. Silliman is indeed the author of the first professional scientific paper from 1855 on the subject of Pennsylvanian oil, but his theoretical work without Łukasiewicz's example of success would certainly not have induced business people to invest in an untested thing, and undoubtedly for a long time would not go beyond the walls of the university library. Silliman's distillation method,

---

136 Brzozowski, *Ignacy..*, p. 68.

137 See Wojciech Roeske, *Magister farmacji w pracy nad destylacją ropy naftowej* [Master of Pharmacy in the Work on the Distillation of Crude Oil], in: *Ignacy Łukasiewicz 1822–1882*, pp. 37–52.

analogous to that of Łukasiewicz, i.e. using sulphuric acid, did not obtain kerosene until two years later. In time, he became a shareholder in an oil company, but he was never involved in the construction or production of oil lamps.[138]

Crude oil, however, at the very beginning did not make much of an impression on America. Edwin T. Drake did not drill his first well in order to explore for oil on American soil until 27 August 1859, i.e. 5 years after Łukasiewicz opened his oil well in Bóbrka, despite the harassment, jokes and disbelief of his fellow countrymen. He repeated a similar feat only once, but from the very beginning the land of Pennsylvania offered unusual and richer deposits than those discovered by Łukasiewicz, and the already existing large shale oil refineries, so far with American momentum producing other lighting fuels such as kerosene and photogen, were converted to oil refineries. Neither Łukasiewicz nor other Polish industrialists had such excellent conditions for the development of industry in the conquered country. So a few years later, in 1862, cheap, competitive American kerosene appeared

138 See H Höfer, *Das Erdöl und seine Verwandten*, Braunschweig 1912, Aup.III, p. 22–24 and 29–32. For example, a German scientist takes a firm stand on this issue: "*Since he (Łukasiewicz) was involved in oil distillation and built a suitable boiler, he purchased crude oil, and* ***in 1853 he introduced oil lighting in a general hospital in Lviv.*** *With unprecedented energy, he started his activity in a new branch of industry, initially running a refinery, which he built in Klenczany, but soon afterwards he established his own small factories in Gorlice, then in Jasło, later in Polanka, and then he started to extract crude oil in Bóbrka near Krosno. He is rightly considered to be the founder and father of the Galician oil industry, whose second, even happier beginnings bear fruit - regardless of later discoveries in America. When the first tests in 1854 were satisfactorily completed, the railway company Kaiser-Ferdinands-Nordbahn used Galician kerosene in winter 1858/59 to illuminate the interiors of its buildings; in 1865, after American kerosene entered the Viennese market in 1862, Galicia was pushed out of the market and forced to improve its products, the railway introduced oil for general use. In 1854 Professor B. Silliman Jr. also conducted more detailed research on crude oil, used fractional distillation to purify the oil for lamps using sulphuric acid and in this way he defined the principles of crude oil processing, which are still observed today, to* ***Colonel E.L. Drake, who, after overcoming certain difficulties and mockery of his fellow citizens on 27 August 1859 at the Willardfarm on Oil Creek near Titusville (Pennsylvania), succeeded*** *in drilling to a depth of 22 m and extracting significant amounts of oil (400 gallons - 124 q per day).* ***The given date is not only the birthday of the American oil industry*** *(in bold – author), but the breakthrough in all cultural countries counts from it.*" – Höfer sets a clear chronology of events, making American discoveries local, not global.

in Europe. However, it was still the result of Łukasiewicz's pioneering discoveries.

Archives, seemingly dry papers, however, sometimes hide an unexpected dose of human weakness. A shocking aspect of this story is two patents. One is the entry in the register of the Austrian Patent Office on the issue of a patent under the number 399 dated 22 November 1853, issued to the names of Jan Zeh and Ignacy Łukasiewicz from Lviv for their own paraffin candles, which confirms the cooperation between the two in a joint laboratory on products of crude oil distillation. The next item in this list is a patent registered under number 400 dated 1 December 1853[139], one week later, for the method of distillation of crude oil and its use for technical purposes. This is puzzling, but rather evident, proof of Zeh's disloyalty to his accomplice.

At the beginning of 1854 we find Łukasiewicz in a provincial pharmacy in Gorlice in a town in West Subcarpathia and, apparently, this was a very sensible move. The remaining reserves of crude oil and kerosene were purchased from the shareholders of the former Mikolasch company. For a few more years he produced kerosene in his laboratory, and until the end he stuck with being famous as the owner of the pharmacy where the first distillation of fractionated crude oil was performed.

The horizon of Jan Zeh continued no further. He also departed from Mikolasch's pharmacy at the end of 1853. His shop with kerosene and petroleum products on Krakowska Street prospered for some time. Zeh tried to promote his products at an exhibition in Munich in 1854. In 1856 he took out a patent for grease for machines and carts. However, his pioneering achievements ended with his participation in the first distillation of crude oil. He explained this not very convincingly in his only article on oil:

> "By applying for the privilege, although I wanted to ensure freedom of choice for the rectification and sale of cleaned kerosene, I never dreamed of a monopoly in this field of national industry."[140]

In what mood and for what specific, direct reasons Łukasiewicz left the capital city of Lviv is not known. Did he learn about the disloyalty of his partner? Why did he abandon prototypes of his lamps in Mikolasch's attics? Why did he not take his notebook out of the drawer? He never betrayed the reason.

---

139 See Austrian Patent Office, Vienna, p. 40, item 400: *"Zeh Johann, Master of Pharmacy in Lviv, Invention to purify natural mountain kerosene with chemicals in such a way that it can be directly used for technical purposes. Privilege document, 2 December 1853. Priv. Duration 2 years, secret"*.

140 Zeh, *Pierwsze objawy..*, p. 393.

He did not have the funds to set up his own pharmacy at that time. It is certain, however, that of the witnesses to and participants in the first successes only he did not lose sight of the bigger goal. While others remained at the stage of monetising pioneering discoveries in their cantors and pharmacies, Łukasiewicz soon drew new horizons for the whole world. With the invention of the lamp, the career of crude oil had only just begun, and Łukasiewicz was to set the next direction on this road.

II. The One Who Spurred the World

Pharmacy and dreams of oil. Entrepreneur and manager. An employer from another era. In the fight for the interests of the oil industry. For the common good – political activity. A good man – philanthropist and social activist. Chorkówka – the charm of the Polish court. Honours and distinctions. The memory of Łukasiewicz. Łukasiewicz – known unknown. Father Ignacy is a model for modern generations. Teamwork. Energy policy and Łukasiewicz's legacy.

Pharmacy and dreams of kerosene

The surroundings of the town of Gorlice at the foot of the Carpathian Mountains were known for oil leaks, which local landowners also tried to use, especially when the novelty of the oil lamp spread across the country and hopes for using oil for lighting purposes increased. One of the most active in this field was Prince Stanisław Jabłonowski, who in the so-called Empty Forest was engaged in obtaining asphalt from rock oil. He even proposed to Łukasiewicz the establishment of a company, but this one did not come to fruition[141].

Initially, Łukasiewicz's primary source of income in Gorlice became the pharmacy he leased, but at that time he also employed himself as a chemist, zoologist and botanist. He was an analyst and toxicologist. He studied the bodies of the dead and participated in the work of the Gorlice Forests Commission.

The work in the pharmacy not only gave Łukasiewicz a livelihood, but also brought him knowledge of people and their problems, the prevailing relations, as well as building his social position.

In 1855, in the fight against the cholera epidemic in Gorlice with great sacrifice, Łukasiewicz treated the sick regardless of their condition and religion, which earned him the extraordinary gratitude of the population of the town and its surroundings. The gratitude was so great that two years later, in 1857, when they heard of his move to Jasło, the Jewish *kehilla* in Gorlice not only insistently asked him to stay in the city, but also offered to cover part of the rent paid for the pharmacy[142].

---

141 Barbara Pratzer, *A pieniądze zamieniał na ideały* [And he turned money into ideals], (Warsaw: BMW, 2004), p. 20.

142 Brzozowski, *Ignacy* pp. 77–78; Kazimierz Kachlik, "Ignacy Łukasiewicz – od nafty świetlnej do rafinerii nafty" [Ignacy Łukasiewicz - From Kerosene Lighting to Oil Refining], *Nafta* no. 7–8 (1982), p. 82.

Figure 9: A crude oil distillery from the 19th century. Source: Museum of Pharmacy, Jagiellonian University Medical College

As a pharmacist, Łukasiewicz also had the opportunity to further improve the process of distillation of crude oil. He purchased the raw material for research from nearby Sekowa, and was accompanied by a younger colleague from the time of his studies, Walery Rogawski. The experiments, conducted initially in one of the pharmacy chambers, and later in the basement, were excellent preparation for later industrial activities. It was still, even for experts on the subject and pioneers of distillation, very dangerous work.

And here, in Gorlice, there were spectacular experiences. Years later, one of these explosive moments was described by Rogawski in his memoirs:

> One time, and it was on a Sunday, when people were going to church for the morning service, something in this boiler in these pipes did not close or broke, the liquid spilled out, and having been ignited from a nearby fire – exploded. The kettle and all the instruments were in flames, endangering the pharmacy and the entire building. We started to pour water over the boiler…., but this strange liquid not only did not go out, but under the influence of water it burned even more. We

Figure 10: Leaking crude oil. Source: Subcarpathian Museum in Krosno

finally came up with an idea and managed to suppress the fire with blankets, duvets and eiderdowns[143].

Łukasiewicz's experiments in processing oil into so-called new camphine, as he liked to call the new, flammable distillate, quickly became famous in the area, and in 1854 Tytus Trzecieski, who had been on peregrinations around the country for a long time in search of a chemist who would help him analyse and possibly use the leaks, which, as he suspected, were oil, arrived at his pharmacy.

He had also travelled to Lviv – the seat of universities and a thriving trade, where he was informed that only the young pharmacist could be such a person, but he had just moved to Gorlice.

Tytus Trzecieski was the owner of the village of Polanka, which had been ruined in the Peasant Uprising of 1846, near Krosno, where there were leaks proving the presence of oil. The meeting with Łukasiewicz resulted in the establishment of a company under which the partners were to extract and

143 Jan Józef Cząstka, "Ignacy Łukasiewicz twórca polskiego przemysłu naftowego" ["Ignacy Lukasiewicz the creator of the polish oil industry", in: *Muzeum – skansen przemysłu naftowego im. Ignacego Łukasiewicza w Bóbrce* [The Ignacy Łukasiewicz Oil and Gas Industry Museum in Bóbrka] (Cracow: Przedsiębiorstwo Doświadczalne Naftochem, 1974), pp. 2–3; Roeske, *Ignacy* ... pp. 53–54.

refine crude oil. Trzecieski's estate was adjacent to the villages of Bóbrka and Zręcin, owned by the landowner Karol Klobassa. These areas also seemed promising to the couple of oil enthusiasts. When Trzecieski approached Klobassa on their behalf for permission to use them, he, as Łukasiewicz himself testified, was to answer him:

> "I know about this oil, that it is there – you want to distil it with a pharmacist... I won't be mixed up in it."[144]

This slightly disrespectful attitude towards the plans of the innovative company is testimony to the state of awareness of what oil was at that time, and how valuable oil could be. Klobassa was an educated, worldly man, and yet in this respect he did not show any particular intuition. Even such an entrepreneur and, in fact, also a chemist – the pharmacist Mikolasch from Lviv – was discouraged from distilling oil and focused on the production of medicines. It was not unlike in the case of Jan Zeh's short-lived romance of only a few years with oil, who after a period of prosperity with his own trade in lamps and kerosene, stopped at the pharmacy stage. The essence of Łukasiewicz's work consisted in the first ever conscious search for oil deposits and the first ever drilling of oil wells.

Initially, Trzecieski and Łukasiewicz were only to pay rent to Klobassa. This situation was to change after a few years, when the shafts in his Bóbrka were already working. The sources of rock oil in Polanka, Trzecieski's property, did not meet the expectations of the two prospectors, and it was only in Bóbrka that the world's first crude oil mine was opened in 1854. Before the oil trafficking began, Bóbrka was a poor village with very few cultivated fields, and it seemed that all its wealth was a fir forest. The forest areas, between two small mountains, stretched from east to west for a few thousand steps[145].

At the time of the birth of oil mining, extraction methods were still very simple. Trenches and wells were dug and the oil accumulated on the water's

---

144 Jan Józef Cząstka, "Dzieje przemysłu naftowego w Krośnieńskiem" [History of the Oil Industry in Krosno Region], in: *Krosno. Studia z dziejów miasta i regionu. Tom II (1918-1970)* [Krosno. Studies on the History of the City and Region], ed. Józef Garbacik (Cracow: Państwowe Wydawnictwo Naukowe, 1973), pp. 8–9; Szczęsny Morawski, "Ignacy Łukasiewicz i ropa galicyjska (zeznanie własne)" [Ignacy Łukasiewicz and Galician oil (own testimony)] in: *Światek Boży i życie na nim* [God's World and Life in it], ed. Szczęsny Morawski (Rzeszow: Księgarnia Jana A. Pelara, 1871), p. 56.

145 Kazimierz Chłędowski, "Źródła nafty w Bóbrce w Jasielskiem" [Sources of Kerosene in Bóbrka in the Jasło Region], *Tygodnik Ilustrowany* no. 325 (1865), p. 252.

Figure 11: "Franek" manually dug well. Source: The Ignacy Łukasiewicz Oil and Gas Industry Museum in Bóbrka

surface was collected. The first shafts so obtained were called "*kopanki.*" Oil was measured in *garnieces*[146]. When significant depths of mines were reached, from a few to several dozen metres, oil was extracted with wooden or tin buckets on the line of a manual hoist operated by two workers. The raw material together with water was poured into vats or barrels. After the introduction of boreholes (wells), plunge pumps were used to extract, initially working on the basis of human strength, which over time was replaced by steam engines[147].

The first years of the later El Dorado in Bóbrka were, however, an arduous and not very profitable enterprise. In 1856 Łukasiewicz, still leasing the pharmacy, applied for an additional position as a cashier of the magistrate in Gorlice. He did not receive this job, probably because of his political past.

146 *Garniec* is the basic unit of capacity of the system of so-called Cracow measures dating back to the Old Polish times and in force in Galicia until 1857, when they were replaced by Austrian measures.; 1 *garniec* = 3.844 l.

147 "Wydobywanie ropy naftowej w kopalni w Bóbrce" [Extraction of Crude Oil in the Bóbrka Mine], *Wiek Nafty* no. 3 (2004), p. 21.

However, Łukasiewicz's hard work and constant research on the industrial method of oil distillation allowed him to take further steps.

In Ulaszowice near Jasło, on the estate of Franciszek Trzecieski, Tytus' brother, Łukasiewicz built a small distillery, which was to purify the raw material extracted in Bóbrka. It contained two and later three boilers, and the annual turnover of petroleum products was 1–2 thousand guilders. This meant the creation of the world's first oil plant, consisting of both a mine and an oil refinery. However, the business still functioned poorly, and Karol Klobassa for several years in a row neglected to collect rent from his partners for the exploitation of his land. According to Łukasiewicz's account, the landowner refused, with the words "I shall not take it! You yourself have had nothing from this, yes! Only outlays."

However, further exploration brought results and the new shaft brought more oil. Its efficiency was so high that for some time it ensured the use of the distillery in Ulaszowice. Despite this, the absorbing and difficult activity in the mine was full of problems, and the company still did not bring the expected profits[148].

At the same time, the first years after leaving Lviv were an important period in Łukasiewicz's private life. In 1855, thirty-three-year-old Ignacy fell in love with his niece, Józefa Honorata Maria Stacherska, 15 years younger, who was the daughter of his sister Emilia of the Łukasiewicz family.

The close relationship between the fiancées meant the need for many treatments and considerable expenses in order to obtain the civil and ecclesiastical permits to marry. Efforts in this matter lasted more than a year. The wedding took place only on 20 April 1857 in Gorlice; the witnesses were Walery Rogawski and the district doctor in Jasło, Alojzy Krziż. In February 1858, a daughter Marianna was born to the Łukasiewiczs. The child died of lung paralysis after less than two years. The despair and depression caused by the death of their only child cast a huge shadow over the life of Łukasiewicz and his wife. It nearly caused the abandonment of the parties in Subcarpathia and even the entire oil business, which soon was to gain momentum[149].

Despite the difficulties in the operation of the mine and distillery, Łukasiewicz continued to work hard on the success of his oil project. He consistently conducted research to improve the distillation of kerosene, and did a lot to promote the products obtained from it. At the end of May 1858 he took part in the National Agricultural and Industrial Exhibition in Jasło. It was the first general Galician economic exhibition held in the province.

148 Morawski, *"Ignacy Łukasiewicz"*, p. 57.
149 Brzozowski, *Ignacy..*, p. 107.

Figure 12: Ignacy Łukasiewicz. Source: The Ignacy Łukasiewicz Oil and Gas Industry Museum in Bóbrka

During the exhibition, Łukasiewicz presented his offer of petroleum products sold by the company.

These were not only crude oil and lighting camphine, but also greasy oil for lubrication of machinery and leather, lubricant for roofs, fences and all wooden equipment, protecting from rotting and vermin (today we would call it an impregnate), asphalt melted and mixed with sand, and powdered non-extinguished lime, which was a raw material for floors in hallways and corridors, but also used on foundations and as an agent for insulating

Figure 13: Diploma from the exhibition of 1858. Source: Museum of Pharmacy, Jagiellonian University Medical College

house walls from moisture, and so-called *gudrine*, which was a softer type of asphalt used as the raw material for manufacturing paraffin candles. Łukasiewicz's presentation was highly appreciated by the jury of the exhibition. The Cracow Economic and Agricultural Society granted him a public distinction. The diploma was based on the products presented by the pharmacist, which were already well known in the vicinity as "among the most perfect" of all the products offered at the exhibition[150].

Participation in the exhibition was something completely natural for Łukasiewicz. At that time he lived in Jasło, where he leased another pharmacy from 1857.

At the beginning of the following year there was an opportunity to buy out a pharmacy in Gorlice, but Łukasiewicz did not obtain the approval

150 Andrzej Laskowski, *Związki Ignacego Łukasiewicza z Jasłem w świetle materiałów archiwalnych, dawnej prasy i literatury* [Relations between Ignacy Łukasiewicz and Jasło in the Light of Archival Materials, Old Press and Literature] (Cracow: 2005), pp. 32–35; "O wystawie w Jaśle r. 1858." [On the Exhibition in Jasło in 1858], *Tygodnik Rolniczo-Przemysłowy* no. 45 (1858), pp. 356–357.

of the Austrian authorities to obtain the concession. Thus he parted with Gorlice, and Walery Rogawski became the owner of the pharmacy.

In Jasło, the pharmacist and novice entrepreneur quickly became, as in Gorlice, an important and respected figure. In 1857, he became a member of the committee for the construction of municipal barracks, where the aforementioned agricultural and industrial exhibition took place a year later. Although the army did not live in the barracks for a long time yet, and later quickly left them, the completion of their construction was considered a great achievement for Łukasiewicz in the city[151].

As in Gorlice, only a few years' stay in the city brought Łukasiewicz universal respect and recognition for his work for the common good and the local community[152].

In 1859 Łukasiewicz could afford to buy his own pharmacy for the first time. The National Government announced a competition for the opening of a pharmacy in Brzostek and Łukasiewicz put forward his candidacy. This time, unlike in Jasło, he received the "personal right to open a public pharmacy in Brzostek." The decision of the National Government approving Łukasiewicz's concession was issued on 15 February 1859. He took over the pharmacy at the end of April and set it up quickly, and the inventory he personally compiled and kept shows that he invested 2.6 thousand guilders in equipping it.

According to this extremely interesting list of estate and movable property, Łukasiewicz's pharmacy company consisted of the outbuildings, where medicines were made, a laboratory, a room next to the pharmacy, a laboratory station, attic and basement. Its devices were mostly wooden and

---

151 Brzozowski, *Ignacy..*, pp. 78–79, 92–93, 116; Laskowski, *Związki..*, p. 32.

152 This was confirmed by the fact that on 12 June 1861 he was awarded an honorary diploma of a citizen of the city. Mayor Jakub Głuchowski and members of the City Department wrote in the diploma awarded at that time: "To the Honorable Mr. Ignacy Łukasiewicz. The Municipality of Jaslo, considering that you through the time of living in our city have many times participated in various matters of ours, always with exemplary zeal and civic sacrifice were ready to respond to the trust placed in you. Considering, in particular, that during the construction of barracks for the army in 1857, elected a member of the building council not only from accepting this arduous duty, you did not evade and to the full satisfaction of the commune inspected all the accounts, keeping the most diligent accounts and through this gave our commune a considerable service. In order to celebrate your merits, we grant you the Diploma of Honorary Citizen of Jasło." See Wiesław Hap, "Życie i działalność Ignacy Łukasiewicz na tle początku świata i jasielskiego przemysłu naftowego" [The Life and Activity of Ignacy Łukasiewicz against the Backdrop of the Beginning of the World's and Jasło's Oil Industry] Rocznik Jasielski vol. 5 (2003), p. 14.

consisted of "frames in the basement and attic, together with equipment and 240 drawers with drawer signatures."[153]

The pharmacy equipment used to prepare medicines included: "1 drum from Lviv, 2 Viennese drums, 2 buckets, infusers of 6 and 12 ounces and 1 libra, 6 small strainers (for filtration), copper kettles bigger and smaller, a teaspoon of bone, a wooden machine for pills and an iron machine for pills, tin graduated cylinders for 3, 6, 8, 12 and 24 ounces. Iron mortar for pills, brass mortar for pills, mortar and pestle x 13, large mortar, porcelain mortar x 8..."[154]

In the cellar Łukasiewicz kept ointments, spirits, syrups, aromatic waters, oils, tinctures, and in the attic, mainly herbs. Among the drugs listed in the *Inventory...* there are many that did not yet appear in the pharmaceutical inventories of the time, as well as specifics, prepared in the laboratory, which was reliably equipped with a large number of modern, for those times and today no longer seen in pharmacies, instruments and chemical apparatuses:

> "Test tubes x 40, extinguishing device, Marsh apparatus, 7 funnels, 6 tubular retorts, 6 tubulated balloons and 6 untubulated balloons of capacity to 6 libras, 6 cooking flasks to 6 libres, Wulf bottle with neck at bottom, 4 3-necked Wulf bottles and 6 2-necked Wulf bottles, 12 boiling jars, 12 dripping bottles [...] Weter funnel, 2 *giftsager* suction pumps, 2 *sztuce* of 3 libras and 2 of 6 libras" and of bigger apparatuses "2 copper alembics with galvanised condensers, libica for cooling, apparatus for hydrogen sulphides."[155]

As it turned out, only four months after the opening of the pharmacy in Brzostek, Łukasiewicz sold it to Porfir Zieniewicz, with whom he had begun his studies in Cracow in 1850[156].

This was due to his next investment in oil extraction and processing. In the same year of 1859, Łukasiewicz entered into a partnership with the brothers Eugeniusz and Apolinary Zieliński, who offered him joint exploitation of the rich deposits in Klęczany near Nowy Sącz. According to the agreement, the Zieliński brothers built a large distillery in Klęczany and guaranteed the supply of oil for its needs. Łukasiewicz was to finance its equipment and ensure the implementation of the technology of refining the raw material.

Together with his involvement in the company, Łukasiewicz decided to go to Vienna with a plan to promote the utility of kerosene. He also purchased larger quantities of modern oil lamps, which he intended to provide for promotional purposes to purchasers of kerosene in Galicia. The short

153 Roeske, *Ignacy..*, p. 65.
154 Roeske, *Ignacy..*, p. 65.
155 Roeske, *Ignacy..*, pp. 65–66.
156 Roeske, *Ignacy..*, pp. 62–63.

Figure 14: Honorata Łukasiewicz. Source: The Ignacy Łukasiewicz Oil and Gas Industry Museum in Bóbrka.

stay in Vienna did not bring any significant commercial effects, and after returning to Klęczany, where the distillery was already working at full steam, Łukasiewicz came into conflict with his shareholders against the backdrop of the company's methods of operation. As an altruist and a man thinking in terms of more than just mere business, he did not want to agree with the Zielińskis' ruthless aspiration for quick enrichment. He looked at the development of the production and sales of kerosene much more widely, taking into account the achievement of general economic and social objectives. The fundamental differences of opinion led to the break-up of the company in the summer of 1859, Łukasiewicz did not even agree to accept the management of their own refinery in Klęczany in the Zieliński brothers' company[157].

In this way, in the name of his attachment to the ideals to which he remained faithful for the rest of his life, he was again faced with a difficult choice concerning his life's further path. The situation was all the more complicated because after the sale of the pharmacy in Brzostek, only the leased pharmacy in Jasło gave him a certain income, and the fire at the distillery in Ulaszowice (probably in 1859) put the future of the company with Tytus Trzecieski into question. All the more so as the activity of the mine in Bóbrka had already deteriorated as a result of the long illness and death of Trzecieski's wife. Anna Węgleńska Trzecieska passed away at the age of 36, leaving her husband with three still young children. Łukasiewicz talked about this difficult period:

> "Then his wife is dying! – a whole 5 months he was not at home – there was also no supervision; because I only looked once a month. More often I could not move away from the pharmacy and distillery. The wells did not produce only thick heavy paraffin, from which very little pure camphine was coming out. There were no profits. Additionally, the distillery was out of order..."

The distillery fire was not only a failure for the oil business, but also aroused the dilemmas of this charismatic man, as the reconstruction of the distillery was opposed by local peasants for fear of further fires[158]. Łukasiewicz, always forgiving to others and sensitive to people's pains, alone in fresh trauma after the death of his child, felt this catastrophe particularly painfully.

The unfortunate entrepreneurs were financially supported by Karol Klobassa, the owner of oil-bearing land, who felt sorry for them. As

---

157 Waldemar Bałda, "Klęczany – naftowa historia" [Klęczany – an oil history], *Wiek Nafty* no. 1 (2012), p. 13; Brzozowski, *Ignacy*.., pp. 79–81, 97–98, 107–108.

158 Włodzimierz Bonusiak, *Życie i działalność Ignacego Łukasiewicza* [The Life and Activity of Ignacy Łukasiewicz], (Rzeszow: Wydawnictwo Towarzystwa Naukowego w Rzeszowie, 1985), p. 96; Laskowski, *Związki*, p. 28; Morawski, "Ignacy Łukasiewicz", p. 57; Marian Wolski, "Teofil Tytus Tomasz Trzecieski", *Wiek Nafty* no. 3 (2009), p. 42.

Łukasiewicz remembered, at this critical moment Klobassa not only refused to accept the rent for the land, but also suggested to Trzecieski that both should invest a thousand guilders each and continue to look for oil. In return, Tytus had to tell him that he would forget about the kerosene if the next investment did not work.

Trzecieski agreed. It turned out that "they didn't extract two hundred! The kerosene spilled over 60 pots a day, 100 pots a day and up!"[159]

However, before this happened, Łukasiewicz, damaged by his failure in Bóbrka, the death of his daughter, the losses of the company with the Zielińskis and the burning of the distillery in Ulaszowice, was already close to giving up on kerosene. At the persuasion of his wife, he planned to move to the Kingdom of Poland, which was under Russian rule, and to start life anew. He bought horses and carts to take all his belongings. According to some sources, he had already started packing and taking things to the carts. Fortunately, Tytus Trzecieski, reliable in friendship and cooperation, acted energetically. With his actions and arguments, he prevented his friend from taking the desperate step. Not only did he offer him a flat in Polanka, but he also threatened that his departure would cause the collapse of the entire oil industry, or at least could stop its development for a long time, which certainly hit a sensitive point in his friend's soul. After all, both were positivists in blood and bone and the spirit of the age.

In April 1861, the shaft called "Wojciech" in Bóbrka splendidly and profusely flowed with oil, and this fact changed the life history not only of Łukasiewicz, but of the entire oil industry in Subcarpathia. In view of the prospect of serious income, the same year saw the establishment of a company extended by a third partner – the owner of the Bóbrka land, Karol Klobassa. It is admirable that this company, which was concluded for ten years, was never confirmed in writing and that the cooperation between shareholders was based on mutual respect and trust. "Despite the fact that the common interest counted hundreds of thousands of Rhine guilders turnover, there was never the slightest misunderstanding between them,"[160] a contemporary and close friend of Łukasiewicz testified. Such were the times, such were those people.

---

159 Morawski, *"Ignacy Łukasiewicz"*, pp. 57–58.

160 Władysław Anczyc, "Ignacy Łukasiewicz", *Kłosy* no. 876 (1882), p. 229; Władysław Bełza, "Kopalnia w Bóbrce" [The Mine in Bóbrka], in: *Iwonicz i okolice* [Iwonicz and Its Surroundings], ed. Władysław Bełza (Lviv: Gubrynowicz i Schmidt, 1885), p. 110; Zygmunt Bielski, "Ignacy Łukasiewicz, wynalazca nafty świetlnej" [Ignacy Łukasiewicz, Discoverer of Lighting Kerosene], *Przemysł Naftowy* no. 23 (1932), p. 591.

Entrepreneur and manager

The year 1861 was therefore a turning point in Łukasiewicz's activity, and the further course of his life and employment was determined by the fate of the entire oil industry in Subcarpathia, which, constituting 100 % of the oil industry of the Austro-Hungarian Empire, was an important, beating source of oil in this part of the world. At this stage, dynamic preparations for the exploitation and refining of the crude oil supplied by "Wojciech" started. At the same time, it turned out that subsequent searches brought equally promising results.

Łukasiewicz lived with his wife in Trzecieski's Polanka and took steps there to create a distillery that would meet the current needs. Due to the urgent need for rapid development of the discovered deposit, a building was hastily erected, in which 20 distillation apparatuses were installed[161].

It was another historical moment and a point on the map of oil history. Polanka became the new centre of the oil industry. The distillery, although at first still housed in a modest building, was soon a school for new students of the modern oil refining craftsmanship. In Polanka, Łukasiewicz "was in his element: he taught, explained and worked together with people; for we must know that in the beginning, despite a good wage, the rural people were reluctant to work in the plants because of the unpleasant smell of dirty oil, so as to encourage them, Łukasiewicz filled the kettles himself, carried the pots with oil, and soon overcame this repulsion."[162]

At that time, several hundred pots of oil were to be extracted daily in Bóbrka, which, compared to the just a few pots in the first years of the mine's operation, meant huge progress. In the mid-1860s there were 700–800 pots obtained from the 14 wells in operation at that time. As Kazimierz Chłędowski, who visited the mine in 1865, described the mine, kerosene appeared in these wells only at a depth of 12–23 metres. On the other hand, the water appearing in the wells could be easily turned to the side or pumped out. There was a strong smell of kerosene around, and someone who was not used to it could not breathe the air for an extended period. The workers toiling in the well were supplied with fresh air by fans. Crude oil from Bóbrka was sold in large part to Prussia at that time, and the rest was processed into pure kerosene by the nearby factory in Polanka[163].

---

161 Cząstka, "Dzieje przemysłu" [History of the Industry], p. 12; Andrzej Laskowski's research gives the date of the establishment of the distillery in Polanka as 1859, i.e. it could have been established immediately after the factory fire in Ulaszowice, see Laskowski, *Związki*, p. 30

162 Anczyc, *Ignacy Łukasiewicz*, p. 229; Cząstka, "*Dzieje przemysłu*", p. 12.

163 Chłędowski, "*Źródła nafty*", p. 252.

The work in the distillery was dangerous, because it was a place exposed to fires. As Łukasiewicz's former Lviv partner Jan Zeh recalled, "oil distilleries grew up like mushrooms after the rain and burned down every now and then."[164] In a situation where there were both fireplaces and oil purification equipment in one room, dramas often occurred.

And so it happened in Polanka, too. A thriving distillery burned down in 1865. However, this did not hamper the development of the company, as at that time the company's shareholders were already wealthy people. Their annual income was to reach 40 thousand Rhine guilders, and for such an amount two large villages could be purchased.

In the same year 1865, Łukasiewicz bought the village of Chorkówka with the adjacent farmstead of Leśniówka from Count Napoleon Bobrowski and decided to build a new refinery there. The refinery in Chorkówka was much bigger, more modern and, extremely importantly, safer to operate. The buildings of the new plant were made of brick, while the furnace with boilers was separated by a wall from the radiators with receptacles. Its unprecedented industrial appearance at that time dazzled visitors to Chorkówka[165].

"The high factory chimneys, heavy puffing of steam boilers, kerosene fumes and clouds of smoke floating above the roofs all together make an intoxicating impression on everyone who, straight from the quiet Bóbrka mine, fell into the very whirlpool of factory life here." The refinery, which processed 30,000 cetnars of crude oil annually (over 300 wagons), purified the raw material coming from Bóbrka and other mines in Jasło County[166].

With the establishment of the new refinery, the company's shareholders, who cared about the company's development, built a 7-kilometre-long beaten road to the mine in Bóbrka. This was extremely important at a time when railways had not yet been laid in the area, and the only means of transporting oil and distilled products was horse powered. However, thanks to the care of the partners, it could take place on well-maintained roads of excellent quality. The raw material and finished products were transported in special low and heavy-duty carts, in large wooden vats, which were rolled out of the mine and into the refinery on wide wooden ramps[167].

The mine, which had been personally managed by Ignacy Łukasiewicz since 1861, under his management became the most modern and the largest mine in the whole country and was yet more proof of the outstanding organising skills and diligence of the former pharmacist from Gorlice and

164 Laskowski, *Związki*, p. 29.
165 Brzozowski, *Ignacy..*, pp. 135–137.
166 Bełza, "Kopalnia w Bóbrce", p. 119; Bielski, *"Ignacy Łukasiewicz"*, p. 590; Brzozowski, *Ignacy..*, pp. 135–137; Laskowski, *Związki*, p. 26.
167 Bełza, "Kopalnia w Bóbrce", p. 119; Brzozowski, *Ignacy..*, p. 137.

Figure 15: Transport of steam boilers to Ignacy Łukasiewicz's distillery in Chorkówka (1865–1870). Source: Subcarpathian Museum in Krosno

Figure 16: Installation of steam boilers at the distillery in Chorkowka. Source: Subcarpathian Museum in Krosno

Jasło at that time. From the very beginning of his work in Bóbrka, he had surrounded himself with specialists whose role was to help develop mining with the use of newer and newer technical achievements. The greatest contributions to the construction of Bóbrka's power were made by Henryk Walter, a long-term drilling advisor to Łukasiewicz, Juliusz Noth, a German geologist and Albert Fauck, a German from Słupsk, a constructor and developer of drilling methods.

Henryk Walter, a graduate of the Austrian mining academy in Loeben, came to Bóbrka in May 1862 and had co-operated almost since the beginning with the director of the mine. It was he who ushered in manual impact drilling into the mine by means of an iron auger constructed together with a village blacksmith. A few years later he brought from Wiener Neustadt in Austria the improved Fabian drilling shears, which significantly accelerated the construction of wells, which had already permanently replaced the previously dug ditches or holes. At the end of 1862, the first manual impact borehole was drilled in Bóbrka, and the new drilling method was quickly adopted by local workers. Thanks to the impact drilling, it was possible to safely reach seams located at greater depths, which directly translated into greater extraction capabilities for the mines[168].

In the 1860s, Juliusz Noth came to Galicia and became interested in the local oil fields. While conducting geological research, he paid attention to the mine in Bóbrka, commonly regarded as the best organised and run. He went there to learn about the extraction methods used. He quickly became Ignacy Łukasiewicz's most important geological consultant at that time. On the basis of research, he determined the further direction of exploitation of the local deposits, which resulted in further oil-bearing boreholes drilled northwest of the already exploited areas. In 1868 Juliusz Noth published a paper devoted to the development of the mine in the first years of its existence, and later became a partner of Łukasiewicz in his investment in Ropianka near Dukla. Importantly, he was the inspiration for Albert Fauck's involvement in Bóbrka[169].

---

168 Władysław Wachal, "Stan techniki przeróbki ropy naftowej w okresie działalności Ignacego Łukasiewicza" [State of the art of crude oil processing technology in the period of Ignacy Łukasiewicz's activity], *Wiek Nafty* no 3 (2001), p. 5.

169 Jan Józef Cząstka, "Kopalnia ropy naftowej w Bóbrce (w 125-letnią rocznicę jej założenia)" [Bóbrka Crude Oil Mine (on the 125th anniversary of its foundation)], *Wiek Nafty* no. 1 (2004), pp. 23–24; Kachlik, "Ignacy Łukasiewicz", p. 85; Józef Zuzak, "Zarys dziejów najstarszej w świecie kopalni ropy naftowej w Bóbrce" [History of the world's oldest oil mine in Bóbrka], *Wiek Nafty* no. 2 (1994), p. 15.

Figure 17: Vehicle for measuring the fluid extracted from a crude oil well from the 19th century. Source: Anna Kozicka-Kołaczkowska.

Albert Fauck arrived in Bóbrka the latest, in 1870, although he had already been in Galicia for three years. He had American experience and was the first in 1867 to use the percussion drilling method on a rope line (the so-called Pennsylvanian method) to a depth of 250 metres, and was also the first to use a steam engine to drive a drilling rig in the Polish oil industry. In such a situation it was not surprising that Ignacy Łukasiewicz, who was still looking for technical innovations and new methods of work, invited Fauck to Bóbrka. Although it turned out that American drilling methods did not fully work on the Carpathian deposits and further tests with them were abandoned, Fauck improved the free-fall Fabian shears used in the boreholes, and then replaced manual percussion drilling with machine drilling based on steam energy, thanks to which Bóbrka once again became a leader among the surrounding mines. In 1874 there were four steam engines in Bóbrka.[170]

170 Cząstka, "Dzieje przemysłu" [History of the Industry], p. 13; Tadeusz Wais, "Kopalnia ropy naftowej w Bóbrce. 160 lat działalności" [Bóbrka Oil Mine. 160 years of operation], *Wiek Nafty* no. 4 (2014), p. 6.

Figure 18: Oil barrels, exposition of Bóbrka. Source: Anna Kozicka-Kołaczkowska

Łukasiewicz and his closest associates joined forces, both at work and in private. All four of them not only met regularly socially, but also after the end of direct cooperation they started joint ventures outside of Bóbrka.

With time, Adolf Jabłoński, together with whom Łukasiewicz used to learn pharmacy, became his right hand in the mine. This anti-Russian conspiracy participant, tsarist prisoner and then an insurgent in 1863 from the Kingdom of Poland, found himself in Bóbrka in 1870. As an insurgent and a former student of pharmacy, he was warmly welcomed by Łukasiewicz and initially employed as a technical official. In 1872 he and the son of Karol Klobassa, Wiktor, went to the United States to learn about American experience. Upon his return, he took over the technical management of the mine, while Łukasiewicz remained its chief director. Jabłoński introduced better drills and casing pipes in Bóbrka, and also improved the system of closing water in the shafts with the so-called Jabłoński bell[171].

Meanwhile, the exploitation of the mine in Bóbrka provided further space for Łukasiewicz's unprecedented invention. In 1868, two of the mine shafts

171 Jan Józef Cząstka, "Ludzie kopalni ropy naftowej w Bóbrce" [People at the Bóbrka crude oil mine], *Wiek Nafty* no. 4 (2004), pp. 12–13; Kachlik, "Ignacy Łukasiewicz", p. 85.

"Karol" and "Idzi" encountered a strong tide of mineral water containing large amounts of free carbon dioxide at a depth of about 90 metres, and the connection between the two wells was confirmed. At the request of Karol Klobassa, water samples were examined by Prof. Stopczyński from Cracow, who announced the results of his work in 1869.

The water from Bóbrka turned out to be unusual both in terms of composition and therapeutic properties. It contained sodium chloride, sodium iodide, sodium bromide, ferric carbonate, calcium carbonate, magnesium carbonate, sodium carbonate and silic acid, among others. The excellent therapeutic properties of the Bóbrka waters could only be compared with those from Iwonicz and Rabka, while they were much higher than the waters available in well-known German, Austrian and French spas. In terms of carbonic acid and sodium carbonate content, Bóbrka's waters exceeded all the then popular European spas, and in the case of iodine and bromine, it was replaced only by the health resort in Rabka[172].

In such a situation it was decided to set up a water treatment plant at the mine. It consisted of four buildings. In the most modern of them there were rooms for health baths. Nearby there were also facilities for guest rooms for patients. Using the properties of the local water, inflammatory conditions of the respiratory tract, lymph node diseases, but also various neuralgias, dermatological diseases and syphilis were treated in the establishment. Water was used on site and sold to hospitals in bottles. In 1870, 176 people were treated in Bóbrka, but the implementation of further plans to expand the health resort was preceded by a study of potential resources. These were made by means of intensive pumping of water from the shaft. Unfortunately, as a result, after five months the water started to turn brown, and its place was taken by crude oil, which ended Bóbrka's short career as a balneological centre[173].

It is difficult to obtain precise data on the size of the mine's production in the first period of Łukasiewicz's, Trzecieski's and Klobassa's company. In 1867, approximately 6.0 thousand cetnars of oil were obtained, in 1868 – 6.7 thousand cetnars and in 1869 – 7.4 thousand cetnars. Between 1861 and 1868 Bóbrka accounted for as much as a quarter of oil production in Galicia. In 1868, the mine already had 84 shafts from which oil was extracted, and more than a hundred employees worked there[174].

---

172 Barbara Olejarz, "Wody mineralne z Bóbrki. Zapomniana historia zakładu wodoleczniczego" [Mineral Waters from Bóbrka. The Forgotten History of the Hydrotherapy Facility], *Wiek Nafty* no. 3 (2014), pp. 14–15.

173 Jan Józef Cząstka, *Kopalnia ropy naftowej w Bóbrce, najstarsza na ziemiach polskich* [Bóbrka Crude Oil Mine, the Oldest on Polish Soil] (Cracow: Akademia Górniczo-Hutnicza, 1989), p. 72.

174 Bonusiak, *Życie i działalność*, p. 99.

Szczęsny Morawski, a well-known writer of those times, described his impressions from his visit to the mine as follows: "With each pump two blackened cyclopses draw day and night. Black oil flows into a barrel just next to the main gutter to which it flows, and having passed 500 fathoms it reaches the tank."[175]

As another writer visiting Bóbrka reported, "The mine is exemplary in every respect, everything is simple, the second follows so naturally from the first, so applied to local needs, that the idea arises that this mine and its entire facility must be a production of the national mind and that no foreigner has put his hand here. The administration was led here by Mr. Łukasiewicz, from whom many could learn; mining equipment, the method of digging, drilling, etc, was introduced by a talented miner, our compatriot Mr. Henryk Walter."

The mine was thus not only characterised by the largest production, but was also managed and arranged in the most exemplary fashion.[176]

The profitability of production in these first years was very high. As a close neighbour of the mine's shareholders recalled,

> "[when] wells were drilled, huge oil production turned out to be possible, and because the petroleum prices were still very high at that time, because the Caspian mines were not known, so huge sums of money flowed to the coffers of the three partners. Klobassa, a widower, took his second wife, a neighbour, Helenka Bobrowska (....), and lived like a sybarite in Zręcin; Trzecieski purchased a very expensive farm, and Łukasiewicz bought Chorkówka from the Bobrowski family, established a kerosene distillery there, the first one in Galicia, made his own fortune and did much good."

In the period 1861–1863 oil was sold at 6–8 Rhine guilders per cetnar, while kerosene was at 24–36 Rhine guilders, and in the early 1870s oil sales prices fell due to increased competition (both domestic and foreign – an inflow to Europe of American oil) to 5 Rhine guilders per cetnar, while kerosene was then sold for 10–14 Rhine guilders. On the other hand, the cost of oil cetnar production in Bóbrka at that time was only 2.10 Rhine guilders, which still proved the high profitability of mine work (in other plants in Galicia this cost was usually much higher). Therefore, it was not surprising

175 Bonusiak, *Życie i działalność*, p. 99; Szczęsny Morawski, "Ropa w Bóbrce" ["Crude Oil in Bobrka"], in: *Światek Boży i życie na nim*, ed. Szczęsny Morawski (Rzeszow: Księgarnia Jana A. Pelara, 1871), p. 64.

176 Jan Dębski, *Ignacy Łukasiewicz. Narodziny przemysłu naftowego* [Ignacy Łukasiewicz. The Birth of the Oil Industry] (Warsaw: Ludowa Spółdzielnia Wydawnicza, 1955), p. 109; A. Teleżyński, *Olej skalny, jego zastosowanie w przemyśle i życiu codziennym* [Rock Oil, Its Use in Industry and Everyday Life] (Lviv: Kornel Piller), 1870, p. 14.

that the income of the company's shareholders was growing fast and that Łukasiewicz's investment opportunities were so great in the mid-1860s, when he could afford both the purchase of land assets and the construction of a new petroleum refinery[177].

The economic importance of the company is proven by the fact that it took part in the Universal Exhibition in Paris in 1867, where it presented crude rock oil and its distillation products, i.e. purified kerosene, heavy oils and asphalt, in the "Raw Products" section.

Stefan Bartoszewicz stated at the beginning of the 20th century that the company established in 1861 brought its shareholders the astronomical amount of 1.5 million guilders (Rhine guilders) profit.

However, many of Łukasiewicz's expenses were hidden from the eyes of his neighbours, some of whom were sensitive to them, especially to those of his wife. In 1863, all shareholders supported the January insurgents fighting in the Kingdom of Poland against the Russians from their oil revenues. Ignacy Łukasiewicz donated significant sums of money to equip the units fighting against the Russians from Galicia to the Kingdom. In his posthumous memoirs, Władysław Anczyc wrote that in 1863,

> "Łukasiewicz also suffered a lot because of known reasons, and frequent searches invaded him and disturbed him; nevertheless, he went through this critical time unscathed and only materially suffered a great deal of losses, because his generosity and hospitality knew no limits."[178]

According to another author, the searches "carried out with all the official ruthlessness by twahe sergeant-major of the gendarmerie from Krosno, a man of Czech origin by the name of Szpaczek, who, in search for weapons intended for the insurgents, resorted even to taking out the floors in the rooms"[179] were particularly burdensome.

In 1868, Ignacy Łukasiewicz made an extraordinary decision against the backdrop of huge oil revenues. This leader and moving spirit of the

177 Anczyc, *"Ignacy Łukasiewicz"*, p. 229; Kazimierz Chłędowski, *Pamiętniki. Tom I. Galicja (1843-1880)* [Memoirs. Volume 1. Galicia] (Cracow: Wydawnictwo Literackie, 1957), p. 157; Edward Windakiewicz, *Olej i wosk ziemny w Galicji* [Oil and Earth Wax in Galicia] (Lviv: E. Winiarz, 1875), p. 6.

178 Anczyc, *"Ignacy Łukasiewicz"*, p. 229.

179 Anczyc, *"Ignacy Łukasiewicz"*, p. 229; Stefan Bartoszewicz, "Historia i stan ekonomiczny przemysłu naftowego w Galicyi", *Nafta* no. 19 (1906), pp. 294–295; "Nieco wiadomości o pionierze polskiego przemysłu naftowego Ignacym Łukasiewiczu zaczerpniętych z notatek śp. Józefa Piętniewicza, em. kier. szkoły powszechnej we Frysztaku" [Some information about the pioneer of the Polish oil industry, Ignacy Łukasiewicz, taken from the notes of the late Józef Piętniewicz, retd. primary school headmaster in Frysztak], Jasło, December 1931, p. 3.

company, without whose expertise, knowledge, consistent work and organisation the whole thing would have collapsed many times, decided to leave it. He explained his decision by the fact that he did not need and could not accept such great profits, all the more so because the partners had children and he did not. He declared that he would give up his shares in the company, committing himself to buying oil from friends and cleaning it. And, as Władysław Anczyc wrote in 1882 "a strange struggle for today's times between noble people began: one wanted to keep the contract, the other, although he still had the right to take a third part by virtue of the agreement, did not want to accept it."

It ended with Łukasiewicz remaining in the position of director of the mine in Bóbrka with remuneration in the amount of 6,000 Rhine guilders per year, which after a few years was reduced to 2,000 per year at his request[180]. In the mid-1870s, the mine employed about 150 workers, 23,000 cetnars of oil were extracted annually and its impact on the local population was considered "very beneficial" in a report from 1875.

The deepest shafts in all of Galicia were located in Bóbrka at that time, which once again confirmed Łukasiewicz's constant striving for the development of black gold mining technology. Out of 111 shafts in the mine at that time, one was as deep as 784 feet, i.e. 230 metres. The scale of this progress was evidenced by the fact that initially in Bóbrka the shafts were dug to the depth of 50–200 ft, then hand drilled to the depth of 400 ft, and after the introduction of steam engines it became possible to further deepen the well. In 1881, 40 shafts were in continuous operation, the deepest of which was already 319.5 metres, and the most efficient were shafts between 100 and 150 metres. Also at the end of Ignacy Łukasiewicz's life, the mine in Bóbrka was still "one of the most efficient and rationally run in the whole of Galicia."[181]

The state of safety in the mine was highly rated. According to Edward Windakiewicz, who described the Galician oil industry in the 1870s, in Bóbrka the management board took very good care to protect workers from unfortunate accidents. A characteristic feature of the mine was stronger ventilation than elsewhere, thanks to which there was enough fresh air in the shafts to prevent the risk of an accident. During the three years preceding Windakiewicz's visit to Bóbrka, only three fatal accidents occurred in the mine, which, with such a large number of employees and the dangerous

---

180 Anczyc, *"Ignacy Łukasiewicz"*, p. 229; Cząstka, "Dzieje przemysłu", p. 12.

181 Windakiewicz, *Olej i wosk ziemny*, pp. 6, 46; Rudolf Zuber, *Nafta i wosk ziemny w Galicyi* [Crude Oil and Earth Wax] (Warsaw: Drukarnia J. Bergera, 1883), p. 20.

work in the mine, was considered to be low. For comparison, during the same time there were 29 deaths per thousand employees per year in the Borysław mines, i.e. significantly more than in Bóbrka[182].

While managing the mine, Łukasiewicz was constantly learning and improving his qualifications in order to be able to manage the mine in the most effective way possible. He put great emphasis not only on further modernisation and improvement of mining technology, but also on proper preparation of people employed in the mine for work. Not only did he send young people abroad, but he also personally tried to teach his workers, especially drillers, how to work. Szczęsny Morawski, who described the mine in 1871, even quoted a lot about Łukasiewicz's approach to workers and the mine as a form of encouragement for those drilling shafts in the ground. Both then and nowadays, it is difficult to imagine a director who has a habit of approaching a borehole and cheering for a simple worker at the bottom "Pummel, dear child!.. Pummel!"[183]

Ignacy Łukasiewicz's approach to work was equally significant in 1877, when he found himself in Cracow's "Czas" ("Time"). His publicist regretted the low level of work in many oil mines and the incompetence of the people managing them, and quoted as an example of honest and successful work "one of the first oil industrialists in Galicia, Mr. Ignacy Łukasiewicz in Chorkówka." He recalled that he had devoted many years of his life to hard work before reaching a position that he could already enjoy at that time. He pointed out that Łukasiewicz initially played a dual role in the mine. He taught himself and his workers. He was often seen when he sat over the wells in Bóbrka and taught drillers and encouraged perseverance. He also brought in professionals to acquire skills from them and then pass them on[184].

Understanding the need to create a separate education for the developing oil mining industry, Ignacy Łukasiewicz, together with Karol Klobassa, attempted systemic action in this respect. In the mid-1870s both of them took the initiative to establish a lower mining school in Bóbrka, modelled on the one-year mining school at the salt pans in Wieliczka, which had been in existence for several years. These efforts were supported by the head of the District Mining Office in Lviv, Henryk Walter, who in 1878 intervened in this matter with letters addressed to the autonomous Galician government of the National Department and to the National School Council in Lviv.

---

182 Władysław Szajnocha, *Górnictwo naftowe w Galicji wobec ustawodawstwa górniczego* [Oil Mining in Galicia Versus Mining Legislation] (Cracow: Wł. Anczyc i Spółka, 1881), p. 59; Windakiewicz, *Olej i wosk ziemny*, p. 48.

183 Morawski, *"Ropa w Bóbrce"*, p. 62.

184 Rudolf Müldner, "O naszym przemyśle naftowym" [On Our Oil Industry], *Czas* no. 140 (1877), p. 1.

Unfortunately, the attempt to establish such a school was opposed by the commune of Bóbrka, which consistently blocked Łukasiewicz's initiative; it also found no understanding on the part of the school inspector in Jasło[185].

Despite this failure, the Bóbrka Mine still played an extremely important educational role. Many students of the art of drilling came and were even employed as night guards to absorb knowledge about what they had previously learned only theoretically. As Eng. Ferdynand Jastrzębski, a graduate of the University of Mining in Loeben, recalled, the plant run by Łukasiewicz at that time was widely considered to be

> "the most exemplary managed and organised mine, so much so that every miner with academic education only then had a chance to succeed in obtaining a job if he could demonstrate his apprenticeship in Bóbrka."

Importantly, the author of the memoirs was not disappointed with such a practice in any way. Years later he stated that "during my over six-month stay in Bóbrka, I found out that the mining equipment was reliably the best in comparison to the then state of other mines in the Jasło Basin, because it consisted in their continuous improvement, for which the costs were not spared to undertake an increasing number of new trials, based on experience."[186]

Łukasiewicz "with a heartfelt sincerity taught, explained, at the mine and distillery he kept for a few weeks for practice, happy that his thought was embodied and a great future was provided to his beloved country." Thanks to this, technical progress in oil mining, the symbol of which was the mine operated by Łukasiewicz, gradually spread to other plants, and the Bóbrka solutions became increasingly better known and widely used, which not only had a positive impact on the entire Galician mining industry, but was also another confirmation of the contribution Ignacy Łukasiewicz had made to its development. Thanks to his methods of work, "from the mine in Bóbrka came a host of miners, who spread all over the country abundant in oil essential information to further disseminate it."[187]

Leaving the company with Trzecieski and Klobassa in 1868 formally closed an important stage in Łukasiewicz's industrial activity, although it should be remembered that until the end of his life he worked in the mine in Bóbrka, which had belonged only to Karol Klobassa since 1870 (when Tytus Trzecieski resigned from participation in the company). What is important, however, is that Łukasiewicz could now devote more time both to the refinery in Chorkówka, which had caught his eye, as well as to investments

185 Cząstka, *Kopalnia ropy*, pp. 76–78; Laskowski, *Związki*, p. 41.
186 Jastrzębski, "Fragmenty z pamiętników", pp. 372–373.
187 Anczyc, "Ignacy Łukasiewicz", p. 229; Müldner, "O naszym przemyśle", p. 1.

in new oil projects, which constantly attracted him and aroused his interest. In 1868 in Ropianka, he not only founded a company with Juliusz Noth, but also a second, which included Tytus Trzecieski, Stanisław Glazor, Seweryn Smarzewski, Henryk Walter, Alojzy Alth and William Stocker. A few years later, when only three shareholders remained (Łukasiewicz, Glazor, Trzecieski), it became a large enterprise employing 75 workers and having 69 shafts with a production of 7.5 thousand cetnars per year. At the same time, Łukasiewicz was also the owner of 10 oil wells in Podgórzyn near Gorlice, and in 1874 he established further companies in Ropianka (with Apolinary Osiecki) and in Ropa near Gorlice (with Karol Klobassa, Feliks Stokowski and Mikołaj Fedorowicz), where a new distillery was also built[188].

The scale and scope of Łukasiewicz's oil investments are best evidenced by his ownership registered in 1881, i.e. a few months before his death. At that time, he was the owner or co-owner of mines in Ropianka (together with Apolinary Osiecki), Semereczna in Krosno district, Nowosielce Gniewoszowe and Uherce Mineralne in Sanok district, and Solina. The most extensive was the mine in Ropianka, where at that time 81 wells were in operation and 55 workers were employed, and the remaining projects were in the start-up stage, which only confirmed that until the end of his days Łukasiewicz was looking for new investment opportunities in the oil industry and invariably invested the money he had earned in subsequent projects[189].

Along with the development of crude oil production in Bóbrka and the increasing processing of raw material in the refinery in Chorkówka, but also with the pressure of competition, both the quality of the finished product and Łukasiewicz's marketing capabilities became more and more important. Łukasiewicz worked continuously on improving the distillation process and obtaining finished products. His successes, on the other hand, were indisputable, which was widely reported to the public by experts on the subject. As was written in 1877, "not only mining and drilling were raised by Mr. Łukasiewicz in our country, but also the refining of the raw material. The oil refining plant in Chorkówka is currently the first in the country."[190]

His achievements in this respect were so excellent that the refinery in Chorkówka was treated as a model for other plants built in Galicia, and its owner was recognised as the founder of the "Łukasiewicz School of Refining," whose solutions were widely practised until 1914. In 1873,

---

188 Włodzimierz Bonusiak, *Szejk z Galicji. Ignacy Łukasiewicz 1822-1882* [The Sheikh of Galicia. Ignacy Łukasiewicz] (Rzeszow: Libra, 2007), p. 125.

189 Stanisław Znamirowski, *Przegląd stanu kopalń nafty i wosku ziemnego w Galicji w drugim półroczu* 1881[Overview of the Oil and Earth Wax Mines in Galicia in the Second Half of 1881] (Tarnów: Józef Pisz, 1882), pp. 22, 28.

190 Müldner, "O naszym przemyśle", p. 1.

30,000 cetnars of crude oil were processed in the refinery, mostly from the Bóbrka mine, but also from other smaller extraction centres in the area. Refined crude oil in Chorkówka was mainly used to produce lighting oils, which included in total 65 % of the processed raw material, 15 % of production was heavy oils, 5 % *gudrine* and the remaining quantities were unavoidable losses in the technological process. As production increased, the refinery in Chorkówka was expanded and modernised, although its development was limited by the raw material supply capabilities of nearby mines. The poor communication conditions in Galicia and difficult terrain forced the creation of a larger number of small refineries located close to individual oil-bearing areas.

The quality of kerosene obtained in Chorkówka was much higher than the competing American and Romanian products. Romanian kerosene contained a lot of aromatic hydrocarbons and while burning it gave off smoke in the same way as badly purified American kerosene. Łukasiewicz's technology in this respect was at an incomparably higher level than the Romanian and American technology. As a result, despite the influx of kerosene from both countries into the Austrian market, the sale of products from Chorkówka remained unthreatened. Approximately 3,000 tonnes of crude oil were processed there annually. Initially, the distillery was operated by a dozen or so workers, later their number reached 20[191].

The excellent results in the innovative field of Łukasiewicz's oil companies posed a challenge for the world of industry, science, and especially competition and the business world. Chorkówka became the goal of journeys by the most outstanding specialists in the field of refining, interested in Łukasiewicz's technological solutions applied there, and it was visited, among others, by the outstanding Austrian refiner Gustav Wagemann.

Particularly interesting in this respect, however, was the appearance in Chorkówka of representatives of John Rockefeller's Standard Oil Company, who were building his oil empire in a ruthless manner. Their meeting with Łukasiewicz, during which Rockefeller's envoys wanted to persuade the host to give them the secrets of such effective refining of crude oil for an appropriate remuneration, became legendary.

Ignacy Łukasiewicz, whose approach to business activity was completely different from that commonly applied in the predatory 19th-century capitalist milieu, from the beginning behaved in this situation in his own way – he allowed the people from Rockefeller, already called the King of Oil, to learn the technological process used in Chorkówka for free, proving his

191 Brzozowski, *Ignacy Łukasiewicz*, pp. 138–140; Kachlik, "Ignacy Łukasiewicz", p. 85; Windakiewicz, *Olej i wosk ziemny*, pp. 51–52.

Figure 19: Tanks for oil, exhibition in the historic mine in Bóbrka. Source: Anna Kozicka-Kołaczkowska

great unselfishness and the fact that, as always, the good and development of the oil industry was at his heart[192].

Ignacy Łukasiewicz was not only an excellent chemist, organiser and tireless explorer of new sources of crude oil. In his business activity, he also played an excellent role as today what we would call the head of marketing of his projects. From the very beginning, he was fully aware of the need to promote petroleum products and conquer new markets for them. He was pleased with the development of the oil industry in the world, because thanks to this, the possibilities of distribution of its products were developing. Łukasiewicz presented them not only at local exhibitions, but he also tried to do it on a much wider scale. Many times he took care of more than just his own products. He took part in important trade fairs organised in the capital city of Vienna and Lviv, the capital of Galicia, thus becoming an ambassador and a symbol of the success of the Polish oil industry.

In 1873 he took part in the International Exhibition in Vienna. As one of the fourteen exhibitors associated with the oil industry, he presented

192 Bielski, "Ignacy Łukasiewicz", p. 591; Pratzer, *A pieniądze zamieniał*, p. 62.

the products of the Bóbrka mine and petroleum, gasoline and other petroleum products manufactured in Chorkówka. At his stand, there was also a collection of stones and earth deposits in which rock oil was found, as well as specimens of earth wax from the mines in the village of Ropianka, owned by Ignacy Łukasiewicz and Wiliam Stocker. He supplemented the exposition with "highly valuable, accurate geological maps from Bóbrka and Ropianka," which confirmed the enormous geological experience that Łukasiewicz had gained during his dozen years of work in the oil mining industry. Together with him, Albert Fauck took part in the event and presented his patented drilling machine. The exhibition ended with the success of entrepreneurs from Galicia. The international jury awarded Ignacy Łukasiewicz with a medal for kerosene and asphalt and a diploma for merits to the oil industry, while Albert Fauck received a medal "for the machine used in the oil mine in Bóbrka."[193]

In September 1877 Ignacy Łukasiewicz took part in the national agricultural and industrial exhibition in Lviv. It was a national event presenting Polish economic achievements. Łukasiewicz was a co-organiser (together with Henryk Walter and Julian Grabowski) of a separate oil pavilion, a symbol of the importance that oil mining had already achieved in the economy at that time. As was written in the official body of the exhibition, "the oil industry is one of the most important parts of the national mining industry, and is also the most well represented alongside saline mining."

The creators of the pavilion tried to publish a fairly complete picture of the oil industry. In the second issue of the Exhibition's body, published on 6 September 1877, it was written that

> "Messrs. I. Łukasiewicz and Klobassa, who sent numerous and beautiful specimens of crude oil from Bóbrka and products of its distillation from Chorkówka in Krosno; moreover, these gentlemen still put up an oil well in order to visually present to the visitors the manner of obtaining kerosene."

A significant change was made to the discussion of the remaining expositions in the oil pavilion "next to the excellent petroleum distillates of Mr. Łukasiewicz stand ..." which is probably the best proof of the quality and prestige that the products of Ignacy Łukasiewicz's work had, constituting a landmark of the exhibition. Thus, nobody could be surprised he was

193 Brzozowski, *Ignacy* .., pp. 148–149; Sławomir Łotysz, "Galicyjska nafta i wosk ziemny na Expo 1873. Relacja Agatona Gillera z Wiednia" [Overview of the state of the Galician oil and earth wax mines in Galicia for Expo 1873. Report by Agathon Giller from Vienna], *Wiek Nafty* no. 1 (2008), p. 4; Roeske, Ignacy Łukasiewicz, p. 68.

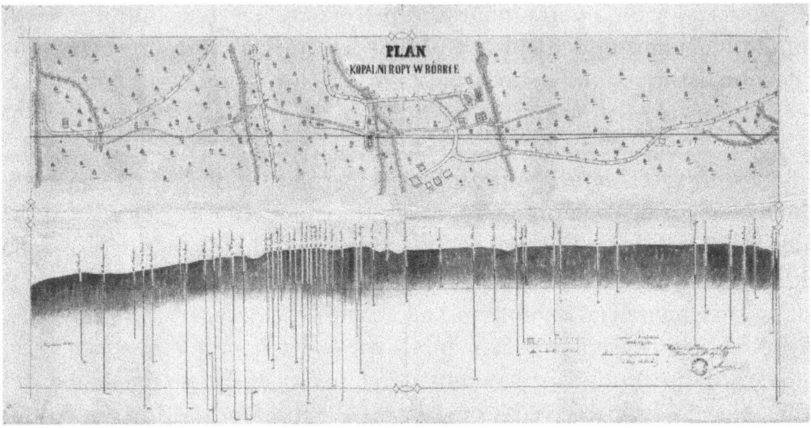

Figure 20: Plan of the mine at Bóbrka. Source: The Ignacy Łukasiewicz Oil and Gas Industry Museum in Bóbrka

awarded an honorary medal for petroleum products by the judges of the exhibition[194].

Ignacy Łukasiewicz's diligence was almost legendary. He got up very early, went to the factory and often took care of the bills, which he kept himself, until late at night. When friendly people asked him to spare himself a little and rest, because he did not have to work so hard, he answered that "a man in the world is like a soldier on guard, as long as he lives, he has to work, and what he will work, he will not take to the grave; it will be useful for other people."[195]

The universal esteem that Łukasiewicz enjoyed and the perfectly functioning companies run by him translated into the public perception of his successes. This was confirmed, among others, by the words of the journalist of "Czas" who wrote that "if there is a man who would not enjoy the good success of the Bóbrka mine and the refinery in Chorkówka, it is probably the same as managing a small capital in a short period of time without any worries or troubles to make a good fortune. Anyone who knows the whole course of the establishment of these plants must admit that every good citizen wishes all good fortune to such and similar procedures."

194 J. G., "Pawilon dla nafty i wosku ziemnego" [Pavilion of Kerosene and Earth Wax] *Wystawa Krajowa Rolnicza i Przemysłowa. Organ Komitetu Wystawy* no. 2 (1877), pp. 1–2.

195 Anczyc, "Ignacy Łukasiewicz", p. 229.

Incidentally, the author mentions only a few more oil industrialists from the Subcarpathia region who deserved to be praised. These were Józef Znamierowski, Wojciech Biechoński, Mikołaj Fedorowicz and Wiliam Stocker. The majority of refineries operating in the country were of a similar opinion. Apart from Chorkówka, the plants in Siary and Męcin were to be positive exceptions among them[196].

Ignacy Łukasiewicz understood the relationship between the development of industry and sales opportunities for petroleum products of his own manufacture. According to his first biographer Ludwik Tomanek, in 1854 he managed to sell his first 300 cetnars of kerosene to Vienna. From the refinery in Polanka, the majority of the goods were initially sent to a forwarder in Tarnów, who then sent kerosene by rail to the west. In order to ensure some overproduction of products, Łukasiewicz also entered into talks with the authorities of the Austrian North Railway regarding the supply of kerosene for lighting purposes. After arduous negotiations, a three-year agreement was signed, which became a prelude to the entry of kerosene from the Subcarpathia region to the Austrian railways on a large scale. By the end of the 1850s, it had pushed out the much more expensive lighting material previously used, i.e. *hydrocarbure*, which was obtained by distillation of bituminous shale imported as ship's ballast from Scotland to Hamburg[197].

In the initial period of exploitation of the Bóbrka field, Łukasiewicz, striving to maximise sales, distributed petroleum products both in wholesale and retail trade. He sold them to shops in Warsaw and Tarnów, and in Cracow. For a short time he even had his own shop. Later, with the increase in production, only wholesale trade was important for him in his business. In the 1870s, finished petroleum products from Chorkówka were sold primarily for railway purposes, and the main, although not the only directions of sales were Prešov, Tarnów and Przemyśl. Transport costs were different, the lowest in the case of Tarnów, and the highest in Lviv, where the products of the excellent refinery in Chorkówka also found their way[198].

---

196 Müldner, "O naszym przemyśle", no. 140, p. 1; no. 141, p. 2.

197 Bartoszewicz, *Historia i stan ekonomiczny* [History and Economic State], pp. 294–295; Ludwik Tomanek, *Ignacy Łukasiewicz. Twórca przemysłu naftowego w Polsce, wielki inicjator – wielki jałmużnik* (Miejsce Piastowe: Michalineum, 1928), pp. 31–32.

198 Bonusiak, *Życie i działalność*, p. 103; Laskowski, *Związki*, p. 43; Windakiewicz, *Olej i wosk ziemny*, pp. 51–52.

An employer from another era

Ignacy Łukasiewicz was not only a discoverer, innovator, entrepreneur and oil-rich industrialist. He was also a man who was ahead of his time not only in matters related to oil. This was particularly evident in the way in which workers employed in his subordinate companies were treated, including their families. An example of his new approach to people employed in his plants, seen nowhere else at that time, were the working conditions created in the mine in Bóbrka, run by Ignacy Łukasiewicz for many years.

When the mine was being established, it was mainly based on the work of Hungarian and German miners, who slowly put the locals to work and after some time left. The work of the local workers was initially inefficient, but gradually improved, and later its effectiveness was so good that they were able to do what they had needed for a month in a week.

The level of discipline in Bóbrka was high, and order and discipline were something natural, "which makes a very nice impression on the visitors to this plant." The work in the mine usually lasted 12 hours, 24 hours at the pumps, after which the workers had a correspondingly longer rest period. Łukasiewicz was an enemy of drunkenness. Drinking alcohol at work or falling into the claws of an addiction resulted in dismissal from the mine. The highest number of mine accidents occurred due to alcohol. Strict discipline in this respect allowed for a significant reduction of this phenomenon in the mine operated by Łukasiewicz[199].

Despite the importance of external workers in the first period of the mine's existence, it is worth noting that local workers were already engaged in exploration and extraction works at that time. Even the names of a few "pioneers" of Bóbrka's mining brothers have survived, who, thanks to their conscientiousness and acquired skills, gained the trust of Ignacy Łukasiewicz and his associates, over time becoming the middle-ranking technical staff of the plant. In 1854, Jan Horytiak was employed for the still very primitive exploration work, and soon afterwards he was appointed the supervisor of mining works in the mine. Another of these first local workers, Tomasz Jabłecki, contributed to the production of a primitive auger, which was used for crushing rocks at the bottom of the mines. Another young and talented worker, Kasper Łacki, passed through various levels of specialisation in Bóbrka. He started as a labourer working on digging, then he worked on drilling and finally as a turner and locksmith. And since Ignacy Łukasiewicz always appreciated talented and hard-working people, this

199 Windakiewicz, *Olej i wosk ziemny*, pp. 48–49.

simple worker not only gained his great trust, but also became a frequent guest at his home[200].

In 1866 Ignacy Łukasiewicz undertook another extraordinary initiative for those times. He set up a fraternal help fund in the mine, to which all employees were forced to belong. Each of them paid a contribution of 3 % of their salary, in return for which the employees were provided with "all medical care in case of illness and means of subsistence during its duration." In the event of the death of an employee, the fund paid the funeral costs, and the widow and children were provided with "supplies as long as deemed appropriate by a separate committee of three members chosen from among the workers." After working in the mine for twenty years, a worker could receive a pension of 2 Rhine guilders per month, and in case of disability caused by an accident at work he was entitled to a life annuity. The fund's capital gradually increased, and at the moment of Łukasiewicz's death it amounted to about 12 thousand Rhine guilders. From this capital it was also planned to pay for the vocational training of workers' sons, and even to fund the dowries of their daughters. Already in 1868 the fund paid benefits to 21 people, and it had a hundred and several tens of members. Three folk schools in Bóbrka, Chorkówka and Zręcin, established by the owners of the mines and refineries, also became an element securing the needs of workers and their families[201].

By creating the fraternal help fund and covering his employees by compulsory insurance against illness, but also against old age and accidents at work, Łukasiewicz was ahead of his time. At that time there were no such safety measures not only in the Austro-Hungarian Empire, but also in Europe, where only Otto von Bismarck initiated such a system on a large scale in the 1880s. Although the fraternal help funds were traditionally known in the mining industry and Łukasiewicz used them as a model, the scope of insurance he introduced was extremely wide and rare at that time in Europe. The provisions of the Austrian Mining Act of 1854 provided for the organisation of voluntary fraternal funds only, but in the oil industry these were not applied until 1884. The scope of cover was also narrower, as they covered only sickness and invalidity insurance[202].

---

200 Cząstka, *Kopalnia ropy*, p. 111.

201 Bełza, *"Kopalnia w Bóbrce"*, pp. 113–114; Andrzej Krajewski, *Krew cywilizacji. Biografia ropy naftowej* [The Blood of Civilization. A Biography of Crude Oil], (Cracow: Wydawnictwo WAM, 2018), pp. 78–79; Morawski, *"Ropa w Bóbrce"*, p. 65.

202 Karol Chylak, *Ubezpieczenia społeczne i zaopatrzenia emerytalne w II Rzeczypospolitej* [Social Insurance and Pension Provision in the Second Republic of Poland] (Warsaw: Wydawnictwo Polskie IHOO, 2017), pp. 45, 60, 62.

The best confirmation of the exceptional importance of the forms of protection offered to workers was the fact that in independent Poland, the pension insurance available to workers in Bóbrka came into force throughout the country only in 1934. According to contemporary opinion, "with this act, Łukasiewicz erected a monument more durable than a bronze statue in the hearts of his workers who, in recognition of extraordinary merits towards them and human misery in general, bestowed on him a filial attachment and called him Father Łukasiewicz, under which name the whole country knew him."[203]

The situation of workers in Bóbrka contrasted with what happened in other factories and mines, which is best evidenced by the description of working conditions created by Łukasiewicz in 1872 in "Gazeta Lwowska." Here it is:

> In Bóbrka the workers have an advance payment fund and a kind of savings bank set up by the manager and owner of the petroleum distillery in Chorkówka, Mr. Łukasiewicz. The relationship that Mr. Łukasiewicz has established between the numerous workers and craftsmen is one of the most beautiful in the country: every worker will find help in need and rescue in the event of illness, and a certain patriarchalism runs through the whole device and becomes a contrast to factories and plants where the manager, only by exploiting the last forces of the worker, does not take care to improve his moral and material existence at all. Mr. Łukasiewicz is not a utopian in his ordinations, but a strictly practical man, who has been able to bring order to his factories, featuring in the best foreign plants of this kind, and to gain the most sincere respect from his neighbours. As far as the development of every enterprise, even the oil company has been dependent on the nobility of the man at the head, the best proof of this is the many other mines in the Carpathian foothills, where the large amount of oil extracted serves the people only to pay the leaseholder and is an impulse for great demoralisation, while the vicinity of Bóbrka is rising materially, and the poor people use their earned salary well.[204]

The wages offered in Bóbrka also differed from what workers could count on in other mines. On the one hand, visitors to Bóbrka drew attention to an orderly system of wages closely related to the work performed, and on the other, they pointed to the admirably exceptional treatment of workers in Bóbrka, which contrasted strongly with the situation in other mines, both in Galicia, Russia and the United States. Władysław Szajnocha's account, dating back to the early 1880s, brings a gloomy picture of the relations

---

203 Bielski, "Ignacy Łukasiewicz", p. 592; Chylak, *Ubezpieczenia*, pp. 45, 60, 62.

204 Kazimierz Chłędowski, "Nafciarze" [Oilmen], *Dodatek miesięczny do Gazety Lwowskiej* no. 2 (1872), p. 75.

prevailing in other plants, where the fate of the worker was usually a matter of indifference to the owner. He was not obliged to support an employee who was involved in an accident, or his family; he did not guarantee access to a doctor and care, to put it bluntly: "a worker who is ill or injured does not exist for the owner of the mine." The memorial prepared before the oil congress in Lviv in 1877 tolled the alarm that safety relations in Galician oil and wax mines were lamentable. The best proof of this was the fact that in Borysław itself, on average, the authorities learned about one case of the death of a worker per week. In turn there were many such accidents about which the authorities never learned. In 1881 Szajnocha wrote: "Anyone who was in Borysław, this California of Galician oil-making, who saw the workers there, beings who barely looked deserving of the name man, who looked at the work there, but not at the occasion of special celebrations, but on a normal working day, he, if he went out unharmed, will have a foretaste of Dante's inferno."[205]

Meanwhile, in Bóbrka "on the face of each of the workers can be seen satisfaction, everyone here works crisply and eagerly, because everyone knows that only real work will be appreciated by the administrator." As Szczęsny Potocki wrote in 1868, "the people learned to work. He who used to be following the oxen to Multany and Wallachia, today prefers to work tarred. The workers are from Bóbrka, Wietrznica, Niżna Łąka and Wrocanka." A report from a few years later indicates another important benefit of working in oil mines for the poor people who had previously belonged to the landowners. It was a significant increase in access to education for rural children. In 1872 Konrad Coghen stated that "a poor mountain peasant, who had never dreamt of schools, today begins to send his children to lower secondary schools, and even to middle schools."[206]

Presented in a paper on the Galician oil industry published in 1875, the characteristics of the relations say a lot about Łukasiewicz's activity in Bóbrka and the working and living conditions of workers created thanks to

---

205 T. M., "W sprawie kongresu naftowego" [On the Crude Oil Congress], *Wystawa Krajowa Rolnicza i Przemysłowa. Organ Komitetu Wystawy* no. 2 (1877), pp. 2–3; Szajnocha, *Górnictwo naftowe*, pp. 40, 58; Windakiewicz, *Olej i wosk ziemny*, p. 52.

206 Morawski, "Ropa w Bóbrce", p. 65; Teleżyński, *Olej skalny*, p. 15; Aleksander Zyga, "Pisarstwo polskie o początkach przemysłu naftowego w Galicji" [Polish Writings on the Beginnings of the Oil Industry in Galicia], In: *Krosno. Studia z dziejów miasta i regionu. Vol. III* [Krosno. History of the city and region. Vol. III], ed. Stanisław Cynarski (Rzeszow: Krajowa Agencja Wydawnicza, 1995), p. 181.

him, as well as modernisation processes taking place in the area. Edward Windakiewicz wrote in it that

> the influence of this industry on the local population is very beneficial. All you need to do is to see the roads, the cultivation, the dwelling of small and larger owners, and finally the people themselves on the way to Bóbrka, in order to receive such an impression, as if you had moved into a better cultivated area of Germany or France. Good roads, trees planted out of the woods and meadows covered with water trenches everywhere, clean and extensive buildings and factories, peculiarly in Zręcin and Chorkówka, well fed and cleanly dressed landowners, all this is good, because everywhere you can see a certain wealth, which also when visiting the mines caught my eyes; I saw that during lunchtime workers cooked meat and bacon, which I have never seen before in Galicia. There is also a school in Chorkówka, and a Social Insurance Fund, so that a peasant, who usually sends two people to work, is quite liberated from the hands of usurers. Ignacy Łukasiewicz, the director of the mine and the owner of Chorkówka, has the greatest merit in all of this. He goes tirelessly and in every respect to people and is their true father[207].

In the times of the "wild" capitalism that was only being born in these areas, Łukasiewicz's attitude as an employer was not only pioneering, but also aroused the admiration of observers. It also confirmed what he showed on the occasion of the unsuccessful company with the Zieliński brothers in Klęczany. For him, investment meant not only the pursuit of maximum profits, but also the achievement of many other goals, which were not noticed by "conventional" industrialists. These included technical and organisational progress, securing appropriate working conditions, but also affordable prices of the product sold to consumers, supporting their interest in purchasing oil and oil products. Only then was it possible to further develop the oil industry, which was so important for Łukasiewicz, and the poor population could gain access to cheap lighting and thus live in more civilised conditions.

In the fight for the interests of the oil industry

Ignacy Łukasiewicz was deservedly recognised as the founder, organiser, but also the leader of the oil industry. Therefore, in a natural way, he became the leader of a group of entrepreneurs trying to take care of the interests of the whole industry. He saw very well that it was necessary to coordinate the actions of all oil companies in Galicia. Bearing in mind the differences and sometimes conflicts, he understood that there were common interests, and their realisation was possible only through the creation of an organisation whose task would be to defend these interests in every possible field

207 Windakiewicz, *Olej i wosk ziemny*, p. 52.

of activity. And it was just Łukasiewicz who had taken consistent action to organise such an association, despite the fact that for several years he met with the apathy and reluctance of some oil entrepreneurs who did not see the potential benefits of joint action[208].

The first meeting of oil industrialists was organised by Łukasiewicz on 14 December 1872 in Gorlice. Already at that time the concept of establishing an association of oil industry entrepreneurs appeared, and the result of the assembly was the adoption of "the need to establish an association between all entrepreneurs in the country with the aim of mutual support." Two years later, draft laws concerning the oil industry were presented to the national Parliament, but they were not approved. Meanwhile, the entrepreneurs were increasingly concerned about rising taxes and the emergence of government proposals to impose a consumption tax on kerosene. Combined with increasing foreign competition, this meant new threats to the Galician oil industry. In June 1876, oil industrialists met in Jasło, during which a memorandum was passed to the government in Vienna, indicating the negative effects of the proposed consumption tax and insufficient customs protection in the face of the influx of cheap American oil to the country. In the same year, during the regional agro-industrial exhibition in Przemyśl, the proposal to create an association of oil owners and oil entrepreneurs, whose aim would be to represent their common interests, came up again[209].

This initiative resulted in a congress of the oil industry. It met on the occasion of the Agricultural and Industrial Exhibition in Lviv on 27 September 1877. The Congress, chaired by Ignacy Łukasiewicz, was to address the most serious problems of the Galician oil industry. These included the lack of working capital and the irrationality of exploration activities, which often misappropriated these scarce cash resources. As Ignacy Łukasiewicz's close collaborator, Teofil Merunowicz wrote in a communiqué inviting people to the meeting, "the lack of qualified mine managers and supervisors in the mining profession is also causing a great number of losses to oil mining companies. Although there are already mines in our country that have trained themselves good directors, when they have luckily found working and reliable people, and these mines prosper very well, there are not many

208 Oskar Löwenherz, "Ignacy Łukasiewicz jako organizator" [Ignacy Łukasiewicz as an Organizer], *Przemysł Naftowy* no. 20 (1932), p. 483.

209 Wojciech Biechoński, "Pierwsze lata działalności Krajowego T-wa Naftowego. Referat wygłoszony na Kongresie Naftowym w Przemyślu w r. 1882" [The First Years of Activity of the National Oil Association. A paper delivered at the Oil Congress in Przemyśl in 1882], *Przemysł Naftowy*, June (1929), pp. 300–301; August Gorayski, "Siła stowarzyszenia" [The Strength of the Association], *Przegląd* Naftowy, June (1929), p. 298.

of them – they can be counted on the fingers. The largest number of Galician mines still consume a lot of expensive pennies through the experiments of the self-taught or, finally, through the dishonesty of charlatans."[210]

These words were extremely bitter from the point of view of Ignacy Łukasiewicz, who organised the work of his mine and distillery so perfectly, but they undoubtedly reflected his views on the situation in the domestic oil industry. After all, Łukasiewicz was the initiator and president of the congress, and together with Julian Grabowski he played a leading role in it. Practically all the most important issues related to the functioning of the Subcarpathian oil industry were discussed during the session. The agenda included issues related to the methods of oil exploitation, its processing, commercial matters, legal and tax issues, safety issues in mines, the need to establish an association for the development and care of the oil industry and the need to establish a mining school for the oil industry in Subcarpathia[211].

In his introductory speech, Ignacy Łukasiewicz recalled the basic postulates he formulated at the meeting in Gorlice in 1872. He pointed out that in the face of the growing threat of foreign competition, the primary objective should be to create an organisation uniting Galician producers and representing their common interests. The industrialists gathered in Lviv also set up a three-person committee with Łukasiewicz, which was to be responsible for the implementation of the congress resolutions. They concerned the efforts to obtain government funds for new deep drilling projects in the Carpathian Foothills from Limanowa to Pechenizhyn and for the creation of a chemical testing station for the oil industry. The second fundamental demand was to find a legislative exemption from taxation for the domestic oil industry until it was able to compete with American and Romanian producers. The Commission was also expected to address the issue of appropriate legislation to allow for the prompt completion of the land measurements necessary to regulate the ownership of oil-bearing land[212].

Although it was not possible to achieve the implementation of these important demands for the oil industry, an important and lasting achievement of the Congress was undoubtedly the establishment of the National Oil Society for the Care and Development of the Oil Industry and Mining in Galicia, based in Gorlice. The National Government approved the statute of the Society on 8 June 1879, and on 12 April 1880 its first constitutional meeting took place, during which Ignacy Łukasiewicz, of course, was elected president. The formation of the association representing the interests of the

210 T. M., "*W sprawie kongresu*", pp. 2–3.
211 Biechoński, "Pierwsze lata", p. 301; T. M., "W sprawie kongresu", p. 3.
212 Roeske, *Ignacy Łukasiewicz*, p. 69.

whole, one of the most important industries in the country, was at that time an unprecedented event and once again testified to Łukasiewicz's perspicacity and his ability to create new solutions, necessary in the changing reality. As President of the Society, he undertook energetic actions aimed not only at finally regulating the principles of functioning of the domestic oil industry, but also at ensuring the implementation of its most important postulates[213].

The first objective and, as it turned out, a great success was the development of a project to regulate the legal relations of the oil industry and the tax exemptions to which it was entitled. Although it was not until 1884 that the *Sejm* passed the Oil Act, Łukasiewicz's merits in this respect were indisputable. In addition to this fundamental issue, which provided the basis for the industry's activities in the coming decades, Ignacy Łukasiewicz, as head of the Society, also made efforts to change the railway tariffs applicable to the transport of American oil, which were unfavourable from the point of view of the interests of the domestic oil industry. This objective was achieved in the autumn of 1881, when the authorities discounted domestic kerosene transport tariffs on the Chernivtsi, Karl Ludwig and North railways and aligned them with the level of rates in force on the Szczecin-Vienna line for competing American kerosene. In the last period of his life, Łukasiewicz also strove to create a large joint stock company for oil companies, in which he saw the future of the domestic oil industry[214].

Łukasiewicz's activity as President of the Oil Society did not last long. In the middle of 1881, when he felt his health deteriorating, he resigned from this function in favour of his friend August Gorayski. He was then awarded the title of Honorary Lifetime Superintendent of the Society. In the last months of his life, Ignacy Łukasiewicz rarely personally participated in the work of the National Oil Society. He lived in Chorkówka, worked professionally all the time, was active in the local community, but there was no time and strength to participate in the activities of the association. However, as Stanisław Olszewski, the secretary of the Society, recalled, Łukasiewicz at that time supported the organisation with cash benefits, and the mark of his views was clearly imprinted on the Society's activities, even after his death. The successors of Łukasiewicz, in addition to their efforts to ensure the ongoing interests of the oil industry, also took care to disseminate their expertise, as evidenced by the publication of the first oil magazine *Górnik*

---

213 Löwenherz, "Ignacy Łukasiewicz", p. 483; Tomanek, *Ignacy Łukasiewicz*, pp. 43–44.

214 August Gorayski, "Ignacy Łukasiewicz", *Górnik* no. 2 (1882), p, 17; Löwenherz, "Ignacy Łukasiewicz", pp. 483–484.

on Polish soil. Its first issue was published in January 1882, which coincided with Łukasiewicz's death (the magazine was published until December 1886). The National Oil Society also led to the adoption of the statute on self-help mining funds, directly derived from the idea of the founder of the Galician oil industry; it also played a fundamental role in statistical research on this industry (in 1890 the Society, as Łukasiewicz himself once did, moved its headquarters from Gorlice to Jasło)[215].

A part of the activities aimed at protecting the interests of the Polish oil industry may be included as Ignacy Łukasiewicz's aspiration to broadly promote domestic, but also his own achievements in this respect. He believed that the fact that it was on Polish soil that the first kerosene lamp was lit and that it was a Pole who made the groundbreaking discovery for the emergence of the oil industry should be promoted. In particular, he tried to emphasise the fact that Bóbrka was the first oil mine in the world, despite the fact that for some time it brought more worries to the owners than benefits. Therefore, Łukasiewicz decided that the world and posterity should be left a permanent sign of this priority, which was quickly forgotten during the period of rapid development of the American oil mining industry.

With all his modesty, he decided to exhibit an obelisk in Bóbrka in 1872, still existing today, with the inscription "To preserve the memory of the established rock oil mine in Bóbrka in 1854 – Ignacy Łukasiewicz." In this way, he confirmed his priority in the world's history of the oil industry, and at the same time indicated which of the company's participants made the greatest substantive contribution to this work[216].

For the common good – political activity

Ignacy Łukasiewicz was a man who was extremely active for the common good. Both in Gorlice and Jasło, in a short period of time, he contributed to the local communities. His great heart and willingness to help others led him also to politics, which he understood as a service to the country and society. It seems that it was only a matter of time, as Łukasiewicz's social position

215 Stanisław Olszewski, "Ważniejsze momenty z działalności Krajowego Towarzystwa Naftowego w czasie od sierpnia 1881 do maja 1900" [The Most Important Moments in the Activity of the National Oil Association in the Period from August 1881 to May 1900], *Przemysł Naftowy*, June (1929), p. 303; Zdzisław Wójcikiewicz, "Górnik – pierwsze czasopismo naftowe" [*Górnik* – the first oil journal], In: *Krosno. Studia z dziejów miasta i regionu. Vol. III*, ed. Stanisław Cynarski (Rzeszow: Krajowa Agencja Wydawnicza, 1995), pp. 119–120.

216 Brzozowski, *Ignacy..*, pp. 147–148.

and his character traits almost naturally predestined him to get involved in this field of public life. In 1868, when he entered the district council, he became a member of the self-government authorities of the Krosno district, where Łukasiewicz's partners, i.e. Tytus Trzecieski and Karol Klobassa, had already played an important role. Significantly, although Łukasiewicz initially entered the council as a representative of the landed gentry, after a few years he represented rural communities, i.e. the peasant population[217].

For 14 years in the county council Łukasiewicz was a member of several commissions; however, road issues were definitely the most important for him, particularly important also from the point of view of the development of the oil industry. He was a permanent road clerk, and he carried out his tasks in this respect, as with everything he did, very conscientiously. Not only did he personally supervise the execution of corvees in the whole district and watched over every major investment, but also from his own funds he added annually about 1000 Rhine guilders for public works.

He was also the best possible lender of the county, which could invariably count on an interest-free loan from Łukasiewicz for the construction of roads and river regulations. The account of this loan constantly amounted to about tens of thousands of Rhine guilders. Łukasiewicz took an active part in the construction of the roads Krosno–Żmigród, Krosno–Zręcin and Krosno–Kombornia. In 1879, he also started a campaign of planting fruit trees on roads, and he planted the sections from Chorkówka to Zręcin and Bóbrka at his own expense. Thanks to Łukasiewicz's activities, Krosno county stood out from the surrounding area due to the existing network of good quality roads and the degree of river regulation. His involvement in activities aimed at improving the state of transport infrastructure in the district, including financial ones, brought about a saying, which was still known long after his death, that "in the Krosno district roads are paved with Łukasiewicz's guilders." Without his zeal and generosity "in the Krosno district nothing would have been done for public transport, as in many other districts nothing has been done so far, with the most honest society in any case."[218]

Łukasiewicz was aware of the fact that he was not able to take care of everything on his own and that only working together could bring about positive results in the area of road construction in the district. Therefore, on his initiative, in 1879 the Krosno County Council adopted a resolution according to which each head of the commune, who was distinguished by

217 Brzozowski, *Ignacy..*, pp. 153–154.
218 Bielski, "Ignacy Łukasiewicz", p. 592; Teofil Merunowicz, "Ignacy Łukasiewicz", *Gazeta Narodowa* no. 7 (1882), p. 1.

zealous and effective performance of his duties, proper maintenance of communal roads and planting them with trees, was to receive an award in the amount of 100 Rhine guilders. He was, of course, the first to take care of the road and that is why the road to Chorkówka was a model in this respect. This is because it was "planted with selected fruit trees, and each tree was surrounded by a strong woven basket in order not to be damaged."[219]

The excellent development of roads in the Krosno district, thanks to Łukasiewicz, was the opposite of the situation in the Borysław Basin, located in the east of Galicia and much richer in deposits. At the beginning of the 1880s, Władysław Szajnocha wrote that

> despite the road law, national and government subsidies, the condition of roads in many oil producing regions is truly barbaric. For example, the road from Schodnica, one of the most profitable mines in eastern Galicia to Borysław, i.e. to the nearest railway station, can probably be compared with mountain passages in Nicaragua or Abyssinia, or with the paths in the virgin forests of South America. Despite the fact that the construction of a strong permanent public road on this relatively short stretch, because only 1.5 miles, which would cost ten thousand or so, it currently needs a few pairs of oxen at certain times of the year, which would extract the loaded carts with oil barrels from the bottomless mud!

In Łukasiewicz's world such a situation was unthinkable, and this also confirms his insight and understanding of the whole economic needs of the region and the country in which he lived and worked.[220]

His activity in district authorities found its natural continuation on the national level. The years of Ignacy Łukasiewicz's public activity were a time when Polish lands under Austrian rule had autonomy granted to them in the 1860s. The process of building its institutions was completed in 1873, and included both its own institutions headed by the National *Sejm*, the National Faculty and the National School Council, as well as the introduction of the Polish language as an official language and the Polonisation of administration and education at all levels. The highest autonomous body of Galicia became the National *Sejm*, established by the so-called February Patent of 1861. In accordance with the division of competences within the Habsburg monarchy, it received all the issues related to the so-called national culture, i.e. economic, educational, municipal legislation and welfare issues in the broad sense[221].

219 Dębski, *Ignacy Łukasiewicz*, p. 131; Roeske, *Ignacy Łukasiewicz*, p. 75.
220 Szajnocha, *Górnictwo naftowe w Galicji*, p. 69.
221 Andrzej Chwalba, *Historia Polski 1795-1918* [History of Poland, 1795–1918] (Cracow: Wydawnictwo Literackie, 2000), pp. 491–496.

This institution appeared to be an ideal place for Ignacy Łukasiewicz, so active in public affairs, who on 24 October 1876 was elected in the district of Krosno-Dukla-Zmigród as a member of rural communities to the National *Sejm* [Parliament] in Lviv. His victory in the elections was indisputable. Out of the 186 voters entitled to vote in the district, 176 of whom voted, he received 118 votes, which only confirmed the scale of popularity that Łukasiewicz enjoyed in the region. The second in the election, Andrzej Jodłowski, received only 44 votes, i.e. almost three times fewer than the winner. Ignacy Łukasiewicz made his parliamentary pledge on 9 August 1877, and a day later he was elected a member of the national cultural commission, which dealt with issues so close to his heart, raising the civilisational advancement of Galicia. During the elections, he received the support of 110 Members out of 112 who took part in the vote[222].

At the beginning of the next session of the *Sejm*, inaugurated in September 1878, Ignacy Łukasiewicz also became a member of the national cultural commission. Like several other members of this committee, he received the support of all the voters. He was already heavily involved in the oil industry at that time and did not intend to limit his parliamentary activity to the work of the national culture committee. He became one of the initiators of the idea to establish a mining committee within the *Sejm*, which was elected on 27 September 1878. Again in this election he received the unanimous support of MPs, while other candidates received significantly fewer votes. It was another confirmation of the esteem which Łukasiewicz enjoyed in parliament after just one year in the House.[223]

As in the case of the district council, Łukasiewicz's parliamentary mandate lasted until his death. In the *Sejm*, he belonged to the democratic faction, which remained in opposition to the conservatives who ruled indivisibly. In spite of the existing political differences, as his close collaborator wrote, "he was careful to ensure that the team of his political friends did not misappropriate the national interests even for a moment for cheap liberal doctrines. The national interest was always the lex suprema, and he

222 Stenographic report on the dissertations of the Galician National Parliament (cite SS), 2nd session of the 1st session of the 4th period of the Galician Parliament of 9 August 1877, pp. 53–54, 59; SS, 3rd session of the 4th period of the Galician Parliament of 10 August 1877, p. 72.

223 SS, 2nd session of the 2nd session of the 4th period of the Galician *Sejm* of 13 September 1878, p. 23; SS, 7th session of the 2nd session of the 4th period of the Galician *Sejm* of 25 September 1878; SS, 8th session of the 2nd session of the 4th period of the Galician *Sejm* of 27 September 1878, p. 166.

condemned misappropriation of this interest for liberal formulas as much as any other subordination of national affairs to party programmes."[224]

He was an extremely conscientious Member in the performance of his duties. Despite the established tradition of the late arrival of MPs and the postponement of the beginning of meetings as a result of this, Ignacy Łukasiewicz was always one of the first in place. And, as Teofil Merunowicz recalled, "he never went to the meeting unprepared. Although he did not speak in the Chamber himself, each raising of his hand in the *Sejm* was an important act, which he conscientiously considered it his duty to consider. He also always came to the meeting carrying under his arm motions and papers relating to matters on the agenda. Before that, at home, he had read them from beginning to end, and he emphasised the more important paragraphs, advising everyone scrupulously on doubtful issues. We really know very few such Members." Łukasiewicz was also very principled in his public activity. Despite the fact that until the end of his days he remained an ardent Catholic and carried out numerous pious foundations, the most important thing for him in the *Sejm* was the good of the general public and not the interests of the Church. When a proposal came up to grant a significant subsidy of 10,000 Rhine guilders for the bursa of the Resurrectionist Order in Cracow, he strongly opposed this, saying that public funds should not be allocated for such purposes. Despite his reticence and chariness in speeches, Łukasiewicz was also able to show his claws in *Sejm* polemics. In October 1878 he firmly defended the good name of the Krosno Prefinancing Association, of which he was a member of the supervisory board. He even accused Count Krukowiecki of calumny, which was not in line with the then standards of parliamentary debate. As a result, the Speaker of the *Sejm* felt obliged to reprimand such a deputy and politician as Ignacy Łukasiewicz had always been[225].

Entering the highest autonomous body of Galicia, Ignacy Łukasiewicz became involved in numerous activities, the primary objective of which was the economic and cultural improvement of the whole country. This was facilitated by the work of the national cultural commission. The most important initiatives aimed at accelerating the development processes in Galicia, in which Łukasiewicz took part at that time, include a submission particularly close to him for accelerating work on road development, but also issues related to the regulation of peasant mortgages, statutory protection

224 Merunowicz, "Ignacy Łukasiewicz", p. 2.
225 SS, 15 session of the 2nd session of the 4th period of the Galician *Sejm* of 9 October 1878, p. 415; Merunowicz, "Ignacy Łukasiewicz", pp. 1–2; Tomanek, *Ignacy Łukasiewicz*, p. 48.

of peasant farms inherited by minor orphans or support for the preparation of good Polish gymnasium textbooks. He also devoted a lot of attention to construction issues, which he considered to be another important element of raising the "national culture." It is worth mentioning at this point that Łukasiewicz put forward in the forum of the *Sejm* the idea of merging Polish and Ukrainian student self-help associations of the University of Lviv, which at the time was extraordinary in Galicia. He also consistently supported the training of specialists for his beloved oil industry. In 1880, he supported a petition by Polish students of the Loeben mining academy for subsidies for poor students of the university[226].

Over time, Łukasiewicz increasingly focused on the oil issues closest to him. In 1878 he became chairman of the Mining Commission, and two years later, in July 1880, he headed the Petroleum Commission, at the same time being a member of the Commission for the estimation of land tax. In the following year, he was again the head of the petroleum committee. In committees, he worked with great commitment, but more rarely appeared at meetings of the *Sejm*, where matters coming from the committees he managed were often presented by other members of parliament. His parliamentary activity corresponded closely with activities aimed at regulating the situation of the oil industry. An important step towards this was the establishment of the Mining Council in 1878 as a result of Łukasiewicz's efforts. It was to be an advisory body on oil matters of the National Faculty acting as the government. The Council consisted of Łukasiewicz, August Gorayski and Henryk Walter[227].

Łukasiewicz's activity in the mining committee quickly brought measurable effects to the oil industry. As early as in 1878, the *Sejm* adopted a motion of the committee to allocate 10,000 Rhine guilders for deeper oil drilling and to transfer 3,000 Rhine guilders to the National Faculty for the purpose of making geological maps of Galicia. The *Sejm* also supported Łukasiewicz's and the National Petroleum Society's proposal to take the oil industry under the care of the autonomous authorities. This was followed at the beginning of 1880 by the establishment of the National Committee for the Raising of the Oil Industry, which included representatives of the

---

226 Minutes of the 4th sitting of the 3rd session of the 4th period of the Galician *Sejm* of 9 June 1880, *Spis petycji po dzień 16 czerwca 1880 do Sejmu krajowego wniesionych*, no. 128; SS, 27 sitting of the 3rd session of the 4th period of the Galician *Sejm* of July 22 1880, p. 931; Boniusiak, *Życie i działalność*, p. 117.

227 SS, 14th sitting of the 3rd session of the 4th period of the Galician *Sejm* of 5 July 1880, p. 377; SS, 9th sitting of the 4th session of the 4th period of the Galician *Sejm* of 30 September 1881., p. 154; Biechoński, "Pierwsze lata", p. 301.

Austrian Governorship, the National Faculty and 10 industry representatives, headed by Ignacy Łukasiewicz. The tasks of the committee focused on activities for the domestic oil industry and on raising outside capital necessary for its development. At the same time, the *Sejm* earmarked further funds for drilling, geological research and even for publications in the field of oil mining; a permanent job was also created for a mining engineer at the National Faculty. In the sphere of declarations of intent only in the same year, appeals were made to the state authorities to exempt the Galician oil industry from taxes for ten years, and to local authorities and residents of oilfields to build and repair roads. Another postulate covered faster regulation of peasant mortgages necessary for the legal acquisition of land by oil companies. How important for Łukasiewicz the improvement of the condition of roads in mining areas was, was confirmed by another proposal he submitted in 1880 on the matter, this time aiming at the extension of three road sections (Harklowa–Żmigród, Peczeniżyn–Słoboda Rungurska, Borysław–Schodnica)[228].

Despite the personal involvement of the Chairman of the Mining Committee, in October 1878 it was not possible to push through the issue of comprehensive regulation of the oil industry in the *Sejm*. The project primarily sought to restore crude oil's status as a reserved mineral, thus saving Galician deposits from plunder and unprofessional economy. Ignacy Łukasiewicz tried to convince the majority of the Chamber to agree that oil should be subject to the rules of mining regalia, which could guarantee investors the certainty of their mining rights. As always, he thought about the future and further development, and that is why he persuaded Members of the House, who were to a large extent linked to land ownership, with the words:

> Today's state of affairs will not destroy us, but we need capital to raise the industry. We do not have and do not want to give this capital, and this is because an entrepreneur, if he does not have right to this, will not come with his kreutzers to transfer them to the company; and this security can only become possible through the regalia. Because what are the regalia? They mean that kerosene, like any underground ore, are not tied to land ownership, and that at the tabular property, the mines are a separate body of ownership. I can take debts on this body, I can sell my mortgage, and nobody can deny me the right to own and dig out kerosene.

He added that there was a lack of capital in the country and only new legal solutions could direct this capital to oil companies. Unfortunately, despite the fact that this was one of the longest speeches made by Łukasiewicz during the plenary session of the *Sejm*, he failed to convince the majority

---

228 Brzozowski, *Ignacy..*, pp. 157–159.

of Members of Parliament and the motion was sent back to committee with just one vote of the majority of the opponents[229].

Łukasiewicz's speech from the same year, resulting from his characteristic care for workers, should be considered particularly important. He demanded that the court's seat be moved from Podbuża to Borysław, which at that time was a rapidly developing centre of oil production. He argued that "there are many thousands of foreign people in the Borysław mines, who today work there without any supervision and legal guardianship, and (…) that if these twelve thousand shafts could speak, not only fatal accidents, but also crimes could be detected there."[230]

As a result of Łukasiewicz's work as a deputy and head of the petroleum commission, the above-mentioned petroleum bill came into being in 1881. Its assumptions were closely related to Łukasiewicz's own convictions and aspirations. The bill was aimed at protecting the interests of the Polish oil industry and enabling its further development, already threatened by foreign competitors. Fearful of such a development, Łukasiewicz tried to at least legally guarantee a safe position for local entrepreneurs. The project was a compromise and rejected the possibility, previously supported by Łukasiewicz, of the government granting the right to drill on the basis of the mining regalia rules. On the other hand, it allowed landowners or private entrepreneurs to exploit the land, but with the owner's legal consent; it organised the rules for creating oilfields; it also reserved the right to operate mines for persons professionally prepared to do so, and to subject all oil mining to appropriate supervision by the national authorities. And despite the fact that the act was only finally adopted by the *Sejm* in 1884, i.e. two years after the death of the founder of the Polish oil industry, it should be regarded as the last, posthumous success of Ignacy Łukasiewicz in his tireless actions for the benefit of oil mining. All the more so because as early as 17 October 1881, i.e. a few weeks before Łukasiewicz's death, the Chamber accepted in full the principles of the draft submitted by the oil commission (the Petroleum Act adopted in 1884 was amended in 1908, and the solutions adopted at that time were in force in the areas of former Galicia throughout the interwar period)[231].

229 SS, 19th session of the second session of the fourth period of the Galician *Sejm* of 14 October 1878, pp. 597–601.

230 SS, 12th sitting of the third session of the fourth period of the Galician *Sejm* of 1 July 1880, p. 345.

231 SS, 21st sitting of the fourth session of the fourth period of the Galician *Sejm* of 17 October 1881, p. 467; Bonusiak, *Życie i działalność*, p. 113; Józef Sozański, "Historyczny przegląd ustawodawstwa w górnictwie naftowym" [Historical Overview of Legislation in Oil Mining], *Wiek Nafty* no. 3 (1999), pp. 8–10.

A good man – philanthropist and social activist

Ignacy Łukasiewicz was not only an inventor, entrepreneur, creator of modern oil mining, leader of the oil industry, landowner, prominent resident of the Krosno county and member of the National *Sejm*. Ignacy Łukasiewicz was also a man open to the problems of others, constantly focused on helping and sharing the good he managed to achieve in life. As Teofil Merunowicz wrote, "his charity and generosity became proverbial in the circles in which he moved." Łukasiewicz, however, was not a good-natured spendthrift, wasting money for empty applause. He did not spend money without thinking, and anyone who abused his charity could no longer count on it. He used to say, "It's a hard duty to be rich. And a difficult art. It's a hundred times easier to be poor wisely than to use a penny by stupidity."

> "He learned wisely until his last breath, to God's glory and for the benefit of others to use the property he had earned with hard work. Making others happy and leading them to good was the main condition of happiness for him."[232]

Kazimierz Chłędowski, who knew Łukasiewicz well, remembered that

> He was a strange man; modest, or even exaggerated in modesty, with his eyes lowered and a gentle smile, always in a long greyish chamarre, always at work. He seemed as if he endured this riches, as God's permission, as a burden put on him by Providence, of which he had to account in the final judgement. He also did a great deal of good work; there was no citizen in the neighbourhood that Łukasiewicz would not have granted a loan, there was no bankrupt that he would not have helped. He said that it was easy to do good, if one has with what, and he did good, and reasonably, so that one can say that with time he raised the whole area economically.[233]

Despite the fact that Łukasiewicz gained a fortune of millions and undoubtedly was, in the opinion of today's people, a "do-gooder," a term which often possessed and still has pejorative connotations, he was an exceptional do-gooder. Modest to the point of exaggeration, he always considered the wealth he had acquired as social property, and it was his greatest concern to use it as soon as possible for various social purposes or acts of mercy. When he was told that not all those he helped were worthy of it, he used to say: I prefer to give to 99 unworthy, rather than to skip one worthy!

In a slightly imprecise way, but a good reflection of Ignacy Łukasiewicz's approach to sharing his assets with the needy, his generosity can be confirmed by an argument derived from commercial rather than interpersonal relations, i.e. the "weight" argument. After Łukasiewicz's death, nearly 10

---

232 Merunowicz, "Ignacy Łukasiewicz", p. 2.
233 Chłędowski, *Pamiętniki…*, pp. 157–158.

pounds of confirmations of receipt of various donations and letters of gratitude for support given were found in his documents[234].

It is difficult in one sentence to list all the fields of Ignacy Łukasiewicz's activity in charity. Support for those who came to him in need was only a small part of his commitment to the common good. In his aspiration to raise the civilisation of the country, he did a lot as the owner of Chorkówka for the peasants living in it. He supported them financially, ensured the development of education, set up municipal funds, financed medical aid and took care of those who cooperated with his enterprises. He was widely known for his foundations for the Church and the consistent support he gave to the institutions associated with it. Finally, as a zealous patriot, he did not forget about the Polish cause and the needy veterans of the struggle for independence, who always found a wide open door and hospitality in Chorkówka.

It seems that from the point of view of Łukasiewicz's invariably visible aspiration for modernity, progress and changes in social awareness, his activity aimed at peasants living in the area was particularly important. The attitude towards them at that time was unique, reflecting the character traits of Łukasiewicz and his ideas, consistently implemented. It also clearly corresponded with his approach to the workers employed in his companies, although the directions of action were slightly different. The arrival of Łukasiewicz to the estate he purchased in 1865 was already an announcement of the arrival of new times for the Chorkówka peasants. On the first day after the move, the new owner opened a one-class rural school and hired a teacher at his own expense. Although he "only" fulfilled the obligations incumbent on all court owners during that period, many of them evaded this obligation, the best example of this were the previous owners of Chorkówka, who did not run schools (in the middle of the 19th century, in the Austrian Partition, more than 80 % of children and young people were deprived of schooling opportunities)[235].

With time, Łukasiewicz built a new building in the village, in which the school operated in the following years. He did not stop at these activities in Chorkówka. Together with Karol Klobassa, he also built new school buildings in Bóbrka, Zręcin and Żeglce. Łukasiewicz also supported the establishment of folk schools in other towns in the region. He also applied for co-financing of their establishment in various parts of Galicia at the forum of the National *Sejm*, where he applied for aid for the construction of schools for the communes of Leżajsk and Korczyna. This is also another characteristic feature of Łukasiewicz's public activity. He was able

234 Bielski, "Ignacy Łukasiewicz", pp. 592–593.
235 Anczyc, "Ignacy Łukasiewicz", p. 229; Chwalba, *Historia Polski*, p. 197.

to combine his concern for the whole country with effective actions for the immediate vicinity, that is, this "small homeland" for which he felt particularly responsible[236].

The above-mentioned attempt to establish a mining school in Bóbrka failed, but as a symbol of Łukasiewicz's determination and activity for the development of vocational education, in 1875 his wife Honorata founded a lace school for rural girls in Chorkówka. The school was financed entirely by the Łukasiewiczs, run by his wife Honorata, and calculated at 12 students. With the help of an instructor brought from Poznań and local nuns, it was able to become a nationally known centre for the production of high-quality products. The lace products presented at the National Exhibition in Lviv in 1877 gained widespread recognition among visitors, and the success of lace made in Chorkówka resulted in the emergence of similar initiatives in other towns in the region (schools and lace production were launched by the Countess Potocka nee Działyńska in Rymanów, Princess Jerzowa Czartoryska in Jarosław and Princess Jadwiga Sapieżyna in Przemyska). Although the school in Chorkówka ceased to exist after Łukasiewicz's death, the tradition of lace-making in the area survived long after the liquidation of the institution. As Władysław Bełza wrote in the mid-1880s, "Mrs Łukasiewicz's enterprise has already produced such excellent results that today, under her supervision, they produce lace that is far more durable and perfect than Czech lace, and much more practical than Valenciennes."[237]

The owner of Chorkówka did not only care about the education of peasant children. He also did a lot of good for the improvement of the life situation of the inhabitants of the commune as a whole. The situation in this respect was very difficult at that time. Only a dozen or so years after the enfranchisement of the peasants, things were no better than before 1848. Although they became owners of land, they lost the right to servitude, i.e. free use of pastures, forests or ponds belonging to the owner of the village. The farms were fragmented. High taxes, including those related to the costs of indexation, took up a large part of the income of the peasants, who did not therefore have the means to make basic purchases, not to mention investments aimed at increasing the level of farming. Moreover, in the absence of access to capital, usury was spreading en masse, which in the current situation posed a great threat to the existence of many farms. As a result, the model of a poor,

---

236 Brzozowski, *Ignacy..*, pp. 155, 174–175.

237 Bełza, "Kopalnia w Bóbrce", p. 32; "Wróblowice, z księgi wspomnień B. Sulity" [Wróblowice, From the Memoirs of B. Sulita], *Kłosy* no. 804 (1880), p. 340; Tomanek, *Ignacy Łukasiewicz*, p. 60.

backward, fragmented and indebted peasant farm with no developmental capacity and condemned only to difficult vegetation became established[238].

Ignacy Łukasiewicz, who was well acquainted with the living conditions of the peasant population, did not accept this state of affairs and, just as in the case of solutions offered to workers, he also tried to introduce modern methods of supporting the functioning of agricultural holdings in the rural environment. In the footsteps of teachers or priests setting up self-help funds in rural areas, he decided to set up such a fund also in Chorkówka, and having adequate financial resources, he could count on the success of this type of undertaking. The initiative consisted in organising an institution providing interest-free short-term credit, repaid in small weekly instalments, thus releasing peasants in need from the claws of such threatening usurious loans that would threaten their livelihoods[239].

The fund also performed savings functions, and its contributors could count on a larger loan, taken out for a longer period, with a low-interest rate, also repaid in small weekly instalments. Everyone who wanted to join the fund had to pay a guilder (Rhine guilder). The importance of Łukasiewicz for the fund in Chorkówka is best evidenced by the fact that in 1872 its capital consisted of 400 Rhine guilders donated by the owner of the village and 242 guilders coming from peasants' payments. The fund in Chorkówka was then best equipped with capital from among the existing institutions of this type existing in the Krosno district. With time, Łukasiewicz also financially supported the creation of funds in the neighbouring villages of the Krosno district, and the idea of self-help funds was gaining more and more popularity thanks to their success. Although it was not Ignacy Łukasiewicz who was its creator and pioneer, it was his capital, commitment and dedication to his activity that enabled him to create relatively strong credit institutions in the countryside long before they were promoted throughout Galicia in the 1890s by Franciszek Stefczyk, rightly considered today as the father of the rural cooperative movement[240].

Łukasiewicz not only supported the establishment of municipal funds, but also took care of their current functioning, keeping an eye on their accounts "with immeasurable difficulty." He was the one responsible for

---

238 Wojciech Morawski, *Dzieje gospodarcze Polski* [Economic History of Poland], (Warsaw: Diffin, 2010), pp. 118–119; Jan Rutkowski, *Historia gospodarcza Polski, vol. II, Czasy porozbiorowe do 1918 roku* [An Economic History of Poland, Vol. II, Post-Partition to 1918], (Poznań: Księgarnia Akademicka, 1950), pp. 80–82.

239 Brzozowski, *Ignacy..*, p. 176.

240 Bonusiak, *Szejk z Galicji*, p. 146; "Kasy oszczędnościowe i pożyczkowe" [Savings and loan funds], *Dodatek miesięczny do Gazety Lwowskiej* no. 6 (1872), p. 138.

the fact that the loan funds established in the area were based on the above-mentioned principle of weekly payments, even though their administration and bookkeeping were the most labour-intensive. He believed, however, that the educational value of these institutions was the highest and "he adhered to this system as the most morally improving of the people, because this way of repaying the loans forces the owner to redeem the debt with earnings, and not with the income from his small farm." He motivated indebted peasants to take up additional gainful employment, so that they could obtain at least a small amount of money for small investments in an agricultural holding. Sometimes Łukasiewicz individually lent larger sums to people in need, offering favourable interest rates, or even, in special cases, interest-free, in order to save the debtor from falling into usurious credit. Generally, he was eager to lend money to others, as evidenced by the account by Kazimierz Chłędowski, a neighbour and also a multiple debtor of Łukasiewicz, who stated in his memoirs that when Łukasiewicz died, he left behind 60,000 Rhine guilders of other people's bills of exchange, which he ordered to be torn up[241].

Other measures taken by Łukasiewicz also served to improve the level of agricultural economy in rural areas. He supported the investment by, for example, loans to peasants transporting oil from the mine to the refinery and finished products from the refinery to the railway station. They were to be used for the purchase and maintenance of horses and carts, which served both the interests of industry and were used in agricultural production. In accordance with Łukasiewicz's view on this matter, the loans were repaid in small weekly instalments, which allowed for small repayments that were not very onerous for the debtor, who, after all, was working on a lucrative basis, to redeem the loan. Repayments were deducted from the weekly earnings of the carters[242].

The improvement of the peasant economy was also to be supported by the land reclamation of peasant land, which Łukasiewicz supported, which he also promoted in his political activities, as well as activities to promote new or more modern forms of management. Not only did he encourage the establishment of fruit orchards, but he even paid peasants for planting trees belonging to "more noble strains" at home or along the road. In addition, he imported and distributed seeds of various plants useful to farms, he gave a lot to place a rational fish farm and apiary farming in his area, he paid the costs of raising cattle breeding and getting the people used to respecting side

---

241 Chłędowski, *Pamiętniki*, p. 158; Brzozowski, *Ignacy Łukasiewicz*, p. 176; Merunowicz, "Ignacy Łukasiewicz", p. 1.
242 Anczyc, *"Ignacy Łukasiewicz"*, p. 229.

earnings. He wanted thus to teach peasants that they could change their fate by improving farming practices, but also by undertaking additional activities that not only brought cash for current expenses, but also formed the basis for the necessary investments in the farm. In this way he tried to lead the surrounding farmers out of the so far vicious circle of feudal barriers and seemingly irremovable problems[243].

Another important area of Ignacy Łukasiewicz's social activity was connected with the above-mentioned issue of "improving the morals" of farmers. He was a great enemy of the drunkenness among the peasants, which was very widespread in the Galician countryside. The drunkenness was developing largely as a result of the propination rights of landowners, who had the exclusive right to produce and sell alcoholic beverages to peasants living on their property. Even the enfranchisement of peasants, carried out by the Austrian authorities in 1848, did not mean that all duties towards masters had to be abolished. One of them was the propination privilege, which in the Austrian Partition was finally abolished only in 1889. However, Łukasiewicz rejected the income from propination, believing that the landowner had no use for what the peasant would drink in the inn, and that only the Jews managing the inns benefited from the drinking. For him, the income was what the peasant could earn in the field, and not the fact that he drank the money he had earned with difficulty. That is also why he fought against the old mentality, the bad habits of farmers, and took great care to ensure that the peasant people, with whom he was in contact, would be temperate, hard-working and thrifty[244].

Apart from what could today be called systemic activities, Łukasiewicz was also a philanthropist in the literal sense of the word. The amount of his assistance to those in need escapes any attempt at measurement. It is also difficult to say how many beneficiaries of this aid there were. In 1882 Władysław Anczyc stated that "no one knows how many blessings Łukasiewicz gave to whom and to whom he helped for a piece of bread, because he did it as quietly as possible and sometimes attributed the good deed he had done to others." As Łukasiewicz's biographer Ludwik Tomanek wrote, "during each stay in Lviv, the door of the room where he lived did not close, so many supplicants asked him for support." Friends even tried to hold him back from generosity, warning that some of these "stakeholders" were merely exploiting him. However, these warnings did not do much good. He helped

243 Merunowicz, "Ignacy Łukasiewicz", p. 1.

244 Tadeusz Lalik, "Propinacja" [Propination], in: *Encyklopedia historii gospodarczej Polski do 1945 roku*, Vol. 2 (Warsaw: Wiedza Powszechna, 1981), pp. 138–139; Merunowicz, "Ignacy Łukasiewicz", p. 1.

on an *ad hoc* basis, but also in every other possible way. He gave woods from the forest for the construction of houses, he treated, as once he had the Gorlice residents, neighbouring peasants for free (e.g. during the cholera epidemic in 1873) or supported talented peasant sons capable of undertaking studies. In the latter work, he lived to see a peasant medical graduate, parish priest and oil engineer[245].

He was therefore not without reason called Father Ignacy Łukasiewicz. Observing his goodness for others, it was commonly believed that "God did not give him his own brood, for he chose him to be a father wherever his feeling, will and strength could reach, and his moral fatherhood spreads and influences, like a revitalising dew. God gives him earthly abundance, although he uses the least of it."

As a Lviv columnist wrote in 1878, "the rural population, orphans, brothers and sisters – young people, industrialists, exceptional poverty, the elderly and handicapped, houses of God and public institutions – can testify to God and people how much Father Ignacy of Chorkówka did and does for them."[246]

> Even after Łukasiewicz's death, it was recalled that,
> Everyone in the district, in the country, called him the Father – under the wings of the late Ignacy fled, you would not find a single person who left him without consolation and reasonable advice, in poverty generous material help, in disease medicine, in misery shelter and a rare hospitality. Gaining knowledge and property with difficulty and effort, he knew their value; and it is no wonder that he used his property where and how it was necessary to use it. Although childless, he had a large family for whom he was the best husband, brother, uncle, guardian. However, he never closed his generous hand and wherever he saw a beautiful and noble goal, he rushed with the help needed.[247]

Chorkówka – the charm of the Polish court

A somewhat poetic description of Łukasiewicz's physiognomy was given in a letter by one of his pupils, judge Woysym Antoniewicz of Cracow, who exalted him thus: "He was a great character, this insurgent from 1846, creator of the oil industry, benefactor of the entire neighbourhood, Member of Parliament, and papal chamberlain. High, slim with the face of the patriarch, prematurely grey, with a wonderful, sagacious forehead, big grey eyes,

245 Anczyc, "Ignacy Łukasiewicz", p. 229; Brzozowski, *Ignacy* ..., pp. 176–177; Tomanek, *Ignacy Łukasiewicz*, p. 61.

246 Tomanek, *Ignacy Łukasiewicz*, p. 59.

247 J. P., "Ś. P. Ignacy Łukasiewicz" ["Late Ignacy Lukasiewicz"], *Strażnica Polska* no. 21 (1882), p. 169; Tomanek, *Ignacy Łukasiewicz*, p. 59.

full of expression, kindness and reason, energy and peace, which all gave a strange harmony, aroused sympathy and trust and represented such a beautiful Polish type that it was difficult to meet a more beautiful one."[248]

All descriptions from which one can learn something about Ignacy Łukasiewicz are very similar to each other and have one thing in common. Each of the witnesses describing this man draws attention to his modesty, evident at every step and in every situation. It was visible both in his everyday life and when he received great honours. It was present in his attitude and behaviour, as well as in his clothes. It was particularly visible in sublime moments. When Łukasiewicz received the dignity of the papal chamberlain, he accepted it with great calmness and even with some distance. During lunch he listened to all the praise about himself and the toasts in silence and some embarrassment. Finally, one of his neighbours stood up and emphasised that all of Łukasiewicz's apostolic work was based on Franciscan modesty and an aversion to all praise, eminence and human fame. Then the hero himself stood up and said "You alone have understood me."[249]

This modesty was truly conspicuous to anyone who met Łukasiewicz, while "the main rule of life for Ignacy Łukasiewicz was conscientious and precise fulfilment of his duty." As his friend and close collaborator Teofil Merunowicz wrote, "He had many of these duties of various kinds, but all his duties – so voluntarily accepted, and somehow imposed on him by the will of others, small and great, he fulfilled them willingly, diligently, sacrificially, with an accuracy to the minute. No duty was too small, too little significant in his case, but whatever he regarded as his duty once was already sacred in his case – he had to perform it, even with the greatest sacrifice of cost and personal effort."[250]

Łukasiewicz did not pay attention to his outfit and appearance. They were so insignificant to him that he could not even be persuaded to buy new clothes. When his wife bought him a new chamarre[251], considered at the time to be Polish national costume, she had to resort to trickery in order to get him to wear it. The servant placed the new outfit for Łukasiewicz instead of the old one, and only after a few days did he notice that he was wearing new clothes. On a daily basis, however, he wore a white smock, the colour of which he usually managed only in the morning, and later it became more and more blackened and smudged. An anecdote says that one day, when his old friend Walery Rogawski came to Chorkówka and Łukasiewicz was sent

248 Tomanek, *Ignacy Łukasiewicz*, p. 65.
249 Tomanek, *Ignacy Łukasiewicz*, p. 72.
250 Merunowicz, "Ignacy Łukasiewicz", p. 1.
251 Men's outer layer, similar to a *kontusz*

for, he came home in just such a working smock, saturated with the characteristic smell of oil distillates. To the reproaches of his wife, "Feh, Ignasiu, how you stink," he answered, "If I didn't stink, you wouldn't smell nice."[252]

With his wife Honorata, he had a successful marriage. Despite the misfortune of the death of their only child, which met them at the beginning of the relationship, they were together and supported each other to the end. They took up the daughter of Łukasiewicz's first teacher in Zaduszniki, Walentyna Antoniewiczówna, who they treated as their own child. As a reporter wrote about "Father Łukasiewicz" in 1878, "His wife complemented his thoughts. She filled the home fire with motherly love, gathering together orphans and the brood from under the straw thatches. Honorata Łukasiewiczowa nee Stacherska is a worthy wife for Father Ignacy." Although it was certainly difficult for her to play this role in the face of her husband's great diligence, who, paying no attention to anything, tirelessly dealt with the affairs of oil companies, the *Sejm*, the county or the works he had set up to support the local population. To his wife's request not to work so much for the sake of his health, he answered, "I have to hurry and work, because how do I know if my life is enough for me to fulfil all my plans and intentions?" Honorata herself gave him strong support in his work for the common good. As Anna Potocka nee Działyńska recalls, she told the guests about "the content of her hard-working life had been filled to the brim: about the school she was building, about the various needs of the people she was looking for help for, about the lace industry, which she intended to introduce first to our district, and whose later specimens, worthy of spider's webs, delighted the audience at exhibitions."[253]

A certain shadow on Łukasiewicz's marriage, idealised in some writings, falls from Ignacy's attitude in the period of the perhaps not too difficult, but chronic ailment of his wife. When Honorata fell ill in 1868 with stomach ailments resulting probably from neurotic causes, Ignacy for ten years did everything he could to help her with the disease, paying for the best doctors and sending her to the famous healing waters of Karlsbad, Franzesbad, Bardejov and Krynica. On the other hand, however, from letters written by Honorata and her mother, it appears that her spouse did not exactly take the disease seriously. Moreover, the correspondence of her mother, i.e. Łukasiewicz's sister, does not give a very positive picture of her spouse, who,

252 Tomanek, *Ignacy Łukasiewicz*, pp. 61, 66–67.

253 "Honorata ze Stacherskich Łukasiewiczowa w Pamiętniku Anny z Działyńskich Potockiej" [Honorata Stacherska Łukasiewiczowa in the Diary of Anna Działyńska Potocka], *Wiek Nafty* no. 4 (2005), pp. 47–48; Tomanek, *Ignacy Łukasiewicz*, pp. 52, 59, 64.

although financially helpful in treatment, was quite indifferent to his wife's suffering. Emilia Stacherska was greatly sorrowed by her brother for his attitude, and her husband even suggested that Honorata leave Ignacy and live with her parents[254].

It seems that the busy Ignacy simply lacked the time, strength and patience for his personal affairs, and sometimes for his wife and her problems. Emilia in 1873 even wrote about her daughter that "her life is too sad, what comes from dignity, abundance, she is never healthy, there is no real pleasure of unity in home life. Ignacy is a good, noble, decent man, but a cool husband, who has no understanding of her sufferings, he thinks that it is enough to give money and allow her to travel to the springs, as if to save her health. In every treatment, first of all peace, freedom in thought (...) we know Honorata's disposition, she needs heart, tenderness." Meanwhile, Ignacy, like many other eminent people dedicated to work and the good of the general public, could not give her all, affected by all his duties and actions for the benefit of others. He treated domestic affairs in a rough and superficial way, he had no heart for them, and, as Honorata wrote, "Ignacy does not like to listen to or read these household scenes." Honorata's condition was associated with some cooling of relations between the spouses and these were overcome, and the improvement of the situation occurred with the deterioration of the health of Łukasiewicz himself, noticeable since at least 1874. As the wife wrote to her sister, "Ignacy is constantly unhealthy," she described his condition similarly three years later, but on the other hand she was very pleased with her husband's joy at her return from treatment: "There's such content, such joy that I have come back, that it is nice to see." It seems, therefore, that his own illness has brought about a certain re-evaluation of Ignacy's attitude towards his wife and her problems, as well as helped to repair a little bit of the previously damaged marital relations[255].

Writing about Ignacy Łukasiewicz, one cannot forget his profound religiousness. He was a very pious man. This was evidenced both by his entire faith-related life and by numerous pious foundations. The most important of them was the neo-Gothic church in Zręcin, founded together with Karol Klobassa, and built according to the design of architect Aleksander Gebauer, for which the material came from a brickyard operating on Łukasiewicz's estate. He co-funded the construction of the church, but renounced the name of patrons and the associated privileges, leaving them to Karol Klobassa. He was famous for supporting many of the church's charitable initiatives, as well as providing kerosene to numerous monasteries and churches, both Catholic

254 Brzozowski, *Ignacy..*, p. 169.
255 Brzozowski, *Ignacy...*, pp. 169–170.

Figure 21: Ignacy Łukasiewicz's mansion in Chorkówka. Source: The National Digital Archives.

and Uniate, for free. Although constantly busy, he never forgot about religious practices, and "not being able to visit the parish church in Zręcin every day, he arranged a chapel in the palace he had built in Chorkówka, the wide doors of which went out to the great hall, so that when they were opened, a church was created, which could embrace many pious people." As Władysław Anczyc wrote, Łukasiewicz led his life "to show the pseudo-progressives that deep religious feelings do not interfere with progress or philanthropy, and he embarrassed the wise men who, by means of phrases fighting against the misery and darkness of peasants and workers, accusing all wealthier people of superstition, stinginess and backwardness, but do not even contribute one penny to remedying the evil."[256]

Ignacy Łukasiewicz lived in many places in his life, but his real home was undoubtedly Chorkówka, where he not only built a modern oil refinery, but also found his place on earth. Already in 1865 Aleksander Gebauer built

256 Anczyc, "Ignacy Łukasiewicz", p. 230; Roman Mierzecki, "Stosowanie ropy naftowej w Galicji w XIX w." [The Use of Oil in Galicia in the 19th century], *Wiek Nafty* no. 3 (1999), p. 7; Jerzy Szczur, "Karol Klobassa Zrencki senior 1823–1886", *Wiek Nafty* no 4 (2008), p. 39.

a magnificent one-storey palace for the Łukasiewiczs, which became their home for years to come. It was a house widely open to family members, friends and neighbours, as well as veterans of insurrectionary fights, or young people coming to learn about the practical arcana of work in the oil industry in Łukasiewicz's plants. Chorkówka was full of guests, but also permanent residents, who found a roof over their heads with Ignacy and Honorata Łukasiewicz[257].

Honorata's sister Maria Stacherska, who was sixteen years younger, lived permanently in the manor house. The Łukasiewiczs gave her hand in marriage in 1871, against the will of her father, to a pharmacist from Krosno, Wojciech Pik, whom Ignacy supported in business, and also worked together with him in the district council. The similar fate of both Stacherska sisters was striking. Maria, like Honorata, married a pharmacist and social activist who was over twenty years older. Nine children lived to see this relationship, one of them being Ignacy's favourite, Franciszek, who often stayed in Chorkówka and later became a well-known poet. After the death of her husband, Honorata's mother, sister and mother-in-law of Ignacy, Emilia, stayed at the Łukasiewiczs. Shortly afterwards Łukasiewicz also took in Alojza Stacherska nee Folwarczna with her children, the wife of Jan Stacherski, son of Emilia Stacherska, who, living beyond his condition, lost his fortune[258].

As in other Polish manors at that time, veterans of the struggle for independence found shelter in Chorkówka, many of whom arrived in Galicia after the failure of subsequent national uprisings. The most famous resident of the Łukasiewiczs' house was a former prisoner of the Austrian fortress Kufstein, Kasper Cięglewicz. This poet, a November insurgent, former conspirator and author of revolutionary poems for the Ukrainian people chased by the Partition authorities, met Łukasiewicz in 1869 in Paris, where he received the offer to live in his residence. And so that is what happened. Cięglewicz stayed in Chorkówka until Łukasiewicz's death, i.e. for nearly 13 years, serving at that time as a teacher at the school there and being a symbol of the support that the creator of the Polish oil industry gave to the heroes of fights with the partitioners. Apart from Cięglewicz, who left for Lviv after Ignacy's death, Adolf and Bonawentura Jabłoński, Józef Kluczyński, Kacper Kącki and Jan Horytiak, among others, also lived in Chorkówka for good[259].

---

257 Tomanek, *Ignacy Łukasiewicz*, p. 62.
258 Brzozowski, *Ignacy..*, pp. 166–168.
259 Brzozowski, *Ignacy..*, p. 171; Julian Horoszkiewicz, "Kasper Cięglewicz. Żywot, prace i cierpienia" [Kasper Cięglewicz. Life, work and suffering], *Gazeta Narodowa* no. 232 (1886), p. 1, no 233 (1886), p. 1.

In the open manor house in Chorkówka, many people always sat down at the table, and during meals, important social and political issues were often discussed, especially those close to Łukasiewicz and many of his household members, which still involved Ignacy, as well as the veterans of the struggle for independence living in his house. The atmosphere of the Łukasiewicz house was perfectly depicted in her memoirs by Anna Potocka nee Działyńska, the wife of the owner of nearby Rymanów. She remembered her first visit to Chorkówka in this way:

> As I was passing through the dining room, I was surprised by the number of places at the table in the house of a childless couple and I thought that I had found a reception. But when we sat down to afternoon tea, I found out that these were only places for the household and permanent guests, who slowly began to descend. First one two people were introduced: a crippled veteran with grey hair and noble face, a victim of the Citadel's cruelties in Warsaw. Behind him came a few more veterans, and further on, in contrast, youth comely and dashing, who came from various universities and professional schools abroad for holidays to the Łukasiewiczs, who not only provided numerous peasant youths with means for education, but also garnered them to themselves truly like to their family home. It was also among this group, apart from honour and gratitude, that filial confidentiality prevailed. The conversation soon revived. Discussions began. Father Łukasiewicz (as he was not called otherwise in the area), let the young talk, from time to time only with a benignant smile, experience and advice, braking too exuberant gusts. An orphan taken in and beloved, like their own daughter, poured us tea in the flower of youth and beauty, (...) All this together was a feature of a real Polish manor house and charm indescribable![260]

This was confirmed by another regular guest visiting Łukasiewicz, one of many mining apprentices, who wrote that every Sunday in Chorkówka there were meetings of local citizens and neighbours, mostly people interested in the oil industry, including mine managers: Eng. Suszycki, Osiecki, Directors Jabłoński, Noth and many others. For lunch there were several dozen people sitting at the table pulled out from the whole dining room. Mrs. Łukasiewicz was always affectedly dressed at the head of the table. He at the other end of the table humble, cheerful, good-natured.

At home in Chorkówka, Łukasiewicz also often received clients seeking advice and support in matters related to the oil industry. As a witness of many such meetings wrote: "Here I found Father Łukasiewicz busy with the oil industry, surrounded by oil workers at that time, seeking his advice on mining issues, and mostly on financial issues, which he usually solved in such a way that he opened the Wertheimer safe at his desk and provided generous

---

260 "Honorata ze Stacherskich", p. 47.

Figure 22: Łukasiewicz's style. Source: Anna Kozicka-Kołaczkowska

subsidies either to open a new mine, or to pay costs in the drilling, or for economic or purely charitable purposes."[261]

An important part of the estate purchased by Łukasiewicz remained an agricultural holding, which covered 414 hectares, 40 % of which was arable land. Despite the fact that it was located on not very good soils, it was, like all Łukasiewicz's companies, well and rationally run. Although it was not mechanised, which was due to the fact that the owner's basic interests were connected with industrial activity, it undoubtedly stood out from other estates in the area. Particularly his cattle and sheep breeding, as well as orchards, which became "his noble mania." As Łukasiewicz's biographer wrote, his purebred cattle, his pedigree sheepfold, orchards, apiaries, fishponds and his modern tillage were a true model for villagers and neighbouring landowners.

There were relatively many horses on the farm, but this was understandable in the context of the refinery's transport needs. As has already been pointed out, Łukasiewicz did not make any profits from his propination rights, leasing only two existing inns, nor did he run, which was unique among the landowners, distilleries or a brewery at that time. In Chorkówka,

261 Brzozowski, *Ignacy* ..., p. 172.

apart from the palace built by Łukasiewicz, there was the former manor house and outbuilding and about twenty brick farm buildings. Within the limits of the property, there was also the refinery, as well as the above-mentioned brickyard[262].

The man in the stained apron

From Łukasiewicz's photographs in his years as an adult, a fully mature, calm man with a pleasant, regular, thinking face looks out[263]. His contemporaries, who after many years see with their own eyes the already great and popular "Father Łukasiewicz," a benefactor generously donating to the whole area, a member of parliament, a pioneer in industry and a wealthy businessman, are not able to judge his presence differently than through the prism of his achievements and unusual characteristics. In 1878, one of Łukasiewicz's guests describes him as follows:

> *"On a road planted with poplars, we can see, walking towards the factories (....),a frail husband with his shoulders bent, hair hoary not by the years, but by unbending work, but dressed in a grey capote, on which we can observe that its owner is looking into every corner of his factory. Face and hand burnt from the sun and by no means caressed."*[264]

The hunched, greying, worn out "frail husband" parading in stained working clothing did not look like an idol or a millionaire, but everything, as understood by the author of this description, speaks in his favour, including the not too impressive silhouette and careless clothing. In turn, in other eyes this "husband" looks more impressive and fitting to the nimbus that surrounds him. He is not so much "frail" as slim and, it turns out, that it is not low at all, nor inconspicuous:

> *"Tall, slim, with the face of the patriarch, prematurely grey, with a clever forehead, with large, grey eyes full of expression, kindness and reason, energy and peace, which all, arranged for true harmony, aroused sympathy and trust. (...).And when he dressed up in a black Polish costume from 1848 for some celebration, there was something to see. Usually, however, he used to wear a white smock, which he could change every day for a fresh one and lubricate every day with either ink or non-distilled kerosene."*[265]

Reports of Łukasiewicz's careless smock are repeated in the memories of his friends and acquaintances. Perhaps, contrary to the solemn assurances,

---

262 Tomanek, *Ignacy Łukasiewicz*, p. 54; Brzozowski, *Ignacy...*, pp. 163–165.
263 Cf p. 96.
264 Tomanek, Ignacy Łukasiewicz, p. 181.
265 Ibid.

also in the concept of friends, this costume was not the ideal outfit of a man of this format and the position that he owed himself. His wife, Honorata Łukasiewiczowa, suffered a great deal because of this greasy smock and the fact that he did not attach importance to demonstrating personal splendour. Herself a great elegant woman, who liked to dress richly and proudly, she could not tolerate the smell of crude oil, which his working clothes were saturated with. She had to resort to trickery in order to exchange his favourite outfits. The servant changed them on a hanger in his office without his knowledge. In general, the blissful harmony in this marriage, sung of in the many memoirs of contemporary witnesses and friends, was often a little bit sugary. In the published diaries of loyal friends it could not have been different. On the other hand, from the letters of the family and its trustees, from time to time some mystery of this couple leaks out, not fully known, whether united by true communication and feeling, or perhaps associated in a slightly artificial way. The great difference in age, education and personal passions between Honorata and Ignacy was even the basis for assuming that the force behind their marriage was mainly Ignacy's sister Emilia Stacherska. The reason was to prevent her brother's great wealth from slipping out of the family's hands. Emilia herself was forced to spend the autumn of her life in Chorkówka, with this great brother and son-in-law, because of the profligacy of her son, who squandered his family property.

Meanwhile, the brother, to whom she entrusted the fate of her daughter, gave her reasons for dissatisfaction. Absorbed with countless duties and activities, Łukasiewicz did not betray too romantic or sensitive a nature, and it was Emilia who blamed him for allegedly leading to Honorata's depression and disregard for the diseases her daughter suffered from.

> *"Her life is too sad, what comes to her from dignity, abundance, there is never health, there is no true unity in home life. Ignacy is a good, noble and kind man, but a cool husband, who has no understanding of her suffering; it seems to him that it is enough that he give money and allow her to go to the baths."* [266]

In a letter from 1873, Honorata's mother complained to her second daughter Maria. This state had lasted for several years. She had already written in a letter five years earlier:

> *"What a pain, what a sadness for her heart (...) such unusual indifference from Ignacy. I resent him greatly."*[267]

Łukasiewicz's wife suffered from stomach pains and diseases, probably on neurotic grounds. This lasted almost ten years. Łukasiewicz provided for her

---

266 Brozowski, Ignacy..., p. 169.
267 Ibid.

trips, for treatments to Krynica, Bardejov, Karlsbad and Franzesbad, but, apparently, he did not betray special concern about his wife's condition.

It seems that for a psychologist this matter would be quite obvious. A man who, with self-denial, laboriously broke through in life, with all his energy directed towards professional and social work, could not mope about the nervous quivering of an individual living under the cloche of blissful luxury. He never allowed himself to weaken nor slow down his pace. He grew up almost in the back of the pharmacy, at the constant disposal of others, in a difficult time, far from his loved ones, and again in prison, from where, having struggled with difficulty, struggling with poverty, at all costs to get an education. Indeed, he had little understanding for mental dilemmas. He was also not a man about town, so popular at the time in Galicia, full of ruined Polish aristocrats and noblemen without professional occupation and work, which due to the economic backwardness was not there, enthusiasts of gambling able to pledge or sell off the last of the family estates.

Łukasiewicz himself, already as a wealthy man with a great social position, got up every morning at five o'clock, personally managed the home accounts and those of his numerous companies and social projects, never mind taking care of matters in laboratories, mines and refineries, on the *Sejm*'s affairs, and many others. There is no doubt that in their home full of family, residents and guests – those dozens of table-tops at the same time during lunch – the time that the husband could devote to Mrs. Lukasiewicz was extremely limited.

> *"Mrs. Lukasiewicz is always pretentiously dressed at the head of the table. He, at the other end, is modest, cheerful, and good-natured. After the black coffee, the obligatory whist often lasted until late in the evening."*[268]

Jastrzębski describes a typical Sunday meeting of the local nobility and the management of oil mines. So unconventional was he a leader. Quiet, serious, no lionlike man of fashion, and yet in the young years a provocateur of dangerous revolts against the authorities, who was able to carry off even distinguished ladies. Probably, little Honorata followed his lead with big eyes. Finally, a demanding yet caring boss for his employees and an unquestionable spiritual leader of the entire area in the appearance of a modest good fellow, a busy millionaire in a stained apron with chemically eaten hands. Good-natured and unshakeable, sensitive to human needs but not sentimental. Sometimes a bit comical, when he said "Lord of Goodness" to his favourite white box.

268 Brozowski, Ignacy..., p. 169.

Honours and distinctions

Ignacy Łukasiewicz was lucky that already in his lifetime he was not only successful in business, but also widely appreciated by society. This was proved both by his position among oil entrepreneurs and by the trust his fellow citizens placed in him in the elections to the District Council and the National Parliament. This was also confirmed by the numerous acknowledgements he had received since the beginning of his public activity, more and more in the last years of Łukasiewicz's life, when his merits for both the oil industry and society were already widely known, not only in Galicia, but also far beyond its borders.

In 1873, among the many cases of Łukasiewicz being honoured, granting him the title of Pope's chamberlain with the right to wear insignia and the Order of St. Gregory was particularly important. This honorary title was awarded to the creator of the oil industry at the request of the Bishop of Przemyśl, Maciej Hirschler, who thus appreciated Łukasiewicz's charity work and numerous pious foundations. Numerous cases of granting "Father Łukasiewicz" the honorary status of a member of social and professional societies were also visible proof of the universal respect he was given at that time. In 1869 he became an honorary member of the Pharmacy Society in Lviv and Fraternal Help for students of the Agricultural Academy in Prószków, Silesia, in 1875 of the Tatra Society, in 1877 of the Galician Pedagogical Society, in 1880 of the Galician Medical Society and in 1881 of the Fraternal Help for students of the Lviv Polytechnic[269].

The celebrations of the 25th anniversary of the lighting of the first kerosene lamp, organised by Adolf Jabłoński in Chorkówka in 1878, were of a special character. During the ceremony, which gathered eighty guests connected with the oil industry, as well as representatives of the clergy and administration, a laudation in honour of Łukasiewicz was delivered, and then a memorial album with photographs of the most eminent citizens of Galicia, who had participated in the preparation of the ceremony, was handed over to him. The most solemn moment was when Edward Dzwonkowski, a Member of Parliament, presented the gold medal with Łukasiewicz's likeness and the inscription "To the creators of the oil industry on the 25th anniversary – producers, 1878."

The medal was struck in Munich, its order was mediated by the outstanding Polish writer Józef Ignacy Kraszewski and silver and bronze copies of the medal were given to the participants of the ceremony. The medals also reached Polish and foreign scientific institutions, including the Czartoryski

269 Bonusiak, *Życie i działalność*, p. 119.

Figure 23: Medal for the 25th anniversary – head + Medal for the 25th anniversary – tail. Source: Subcarpathian Museum in Krosno

Museum and the Academy of Arts and Sciences in Cracow, the Ossolineum in Lviv, the Society of Friends of Science in Poznań and the National Library in Paris. Kraszewski, already fascinated with the development of the oil industry, also received such a medal and wrote to Adolf Jabłoński, who organised the celebrations: "The name of the Venerable Citizen, to whom the country owes gratitude, was well known to me, and I am glad that my fellow countrymen gave this gratitude such a proper and lasting expression. I will keep the expensive souvenir for my family." On the occasion of his jubilee, Łukasiewicz was also awarded the Order of the Iron Crown by the Emperor in 1878[270].

In a laudation given in Chorkówka, Łukasiewicz's friend from Lviv, Konstanty Miączyński, addressed the hero of the ceremony with these words:

> At the moment when our mining industry was lying fallow and the people, not knowing the wealth of their native land, were only drawing from agriculture modest means of existence, you, a noble man, endowed with the spirit of bold initiative, took the thought of using the treasures that our land hides in its depths. Taking as your slogan: perseverance and work, with iron energy, you fought against all barriers, you removed countless difficulties, walking boldly and fearlessly towards the once-targeted goal. God blessed these endeavours, and the successful outcome culminated in your undertakings so beneficial to the country and its people. Having unveiled the inexhaustible treasures of the earth, you indicated rich sources of prosperity for thousands of fellow citizens. Faithful to your accepted principle and your undisturbed, tireless work, you have gained an honourable position and rich property, which in your hand is only a new means to fulfill lofty and nobler deeds. Having given the mining industry a more rational direction, with a skilful hand you continue the work you have started and, as always, you are not only looking for profits for yourself, but you are willing to share the fruit of your work for the good of the country and to help your fellow countrymen. And you have become the creator of existence and happiness of many families. This is evidenced by your past and present life, which is one uninterrupted chain of beautiful deeds and sacrifices for the country and mankind.

These words can be perfectly complemented by an opinion published in 1880, according to which "Łukasiewicz is a man of great merit, to which the country, mankind owes much. He was the first to discover kerosene in Galicia and was the first to use it for lighting, as evidenced by the medal struck on the 25th anniversary of his activity in mining. He was also the first to create a new oil industry in Galicia, which, over time, will become a source of all new wealth for the country."[271]

Although sometimes modest to exaggeration, Łukasiewicz welcomed this act of recognition for achievements in the oil industry, all the more so as it came from other industrialists, which was undoubtedly particularly pleasing to him. In the following years, he worked with even greater commitment for the benefit of this industry, which, according to popular opinion, was considered to be his work and for which he still felt great responsibility. The effects of this work undertaken both at the forum of the *Sejm* and the

---

270 Anczyc, "Ignacy Łukasiewicz", p. 230; Bełza, "Kopalnia w Bóbrce", p. 114; Roeske, *Ignacy Łukasiewicz*, p. 81; Tomanek, *Ignacy Łukasiewicz*, pp. 73–75, 116–117.

271 "Wielmożnemu Ignacemu Łukasiewiczowi" [To the Honourable Ignacy Łukasiewicz], *Wiek Nafty* no 3 (2003), pp. 2–3; "Wróblowice", p. 340.

Oil Society became increasingly visible, as has already been mentioned before. Unfortunately, Ignacy's state of health was not the best then, and although he still did not slow down and worked in all fields of his activity, the moment came when the body refused to obey. As Anna Potocka nee Działyńska wrote, "the shoulders of this quiet worker leaned away from the burden of superhuman work, his noble face was pale and yellowish, carrying traces of sleepless nights and worries, and all those around, anticipating the misfortune, tried in vain to stop him and persuade him to rest and cure himself. Until He once fell like a broken oak, not to stand up again!"[272]

But still in the autumn of 1881 Łukasiewicz participated in the session of the National *Sejm*. Afterwards, he returned home and, although he felt weakened, he continued to work. He wanted to go to Vienna as part of the Petroleum Society delegation to seek the interests of the oil industry, but his state of health did not allow him to do so. He was, however, still active in the district. Until almost the last days of his life, he made efforts to raise the condition of the weaving industry and to establish a Society of Industrialists in Korczyna near Krosno. As even "Gazeta Narodowa" wrote after his funeral, the establishment of a commercial weavers' company was supposed to be the indirect cause of Łukasiewicz's death – "it even seems that the cold he brought back from the act of establishing the company, which took place in a cramped room, became the direct cause of his last illness. He died like a soldier at his post...."[273]

Despite his weakening, however, Łukasiewicz still worked in a "normal" way for him. Even "on 31 December the late Ignacy still worked until 11 o'clock in the evening, regretting his general impotence." However, he was already suffering from severe pneumonia and the New Year 1882 "met him almost unconscious." However, even "In a fever, unconsciously, he still spoke about district roads, about oil mines, about the improvement of mining tools; it seemed to him that he was in the Deputy's Chamber – which turned out to be what his mind was concerned with," wrote Władysław Anczyc. Only once before his death did he regain consciousness for a longer time and then he accepted the last rites. He died in his own house in Chorkówka on Saturday, 7 January 1882, and "a heavy roll of thunder struck the hearts of all noble people, when suddenly the news of the death of the distinguished man spread." He died, "having orphaned his poor wife and the people for whom he was the father."[274]

272 "Honorata ze Stacherskich", p. 48.

273 *Gazeta Narodowa* no. 11 (1882), p. 3; Gorayski, "Ignacy Łukasiewicz", pp. 17–18.

274 Anczyc, "Ignacy Łukasiewicz", p. 230; *Gazeta Narodowa* no. 6 (1882), p. 3.

Figure 24: Ignacy Łukasiewicz. Source: Subcarpathian Museum in Krosno

The funeral at the parish cemetery in Zręcin, where Łukasiewicz and Klobassa had founded a new church, took place on January 11. It gathered many guests who came from far away and crowds of local and surrounding peasants, who for the last time wanted to thank and say goodbye to "Father Łukasiewicz." At the last farewell, there was no shortage of

> kerosene producers from the most distant regions of the country, friends of the deceased, admirers, and a very large group of those who received personal benefits from him, teachers from the Jasło-Krosno school district in their entirety, several dozen clergy from both Catholic rites, members of parliament, citizens of the Sanok and Jasło lands, as well as burghers and rural people from a district of several miles around. Most of them were Municipal Councils, including the village mayors at the head. Various folk costumes were swarming around. Next to the serious figures of burghers in navy blue capotes and high peaked hats, the crowds of slightly dressed Ruthenian highlanders were progressing, who went by 4–5 miles to the funeral of the beloved gentleman from Chorkówka.[275]
> The correspondent of the "Polish Watchtower" wrote with emphasis:
> "as beautiful as was the life of the late Ignacy, so blessed was his death. One had to see the complex – in the court chapel on a catafalque surrounded by light and greenery, remnants of His earthly life. What an angelic peace rested on the face of this man, what an aureole of heavenly happiness surrounded, with grey hair covering the 60 year old man's skin; – with what holy reverie was received the sight of the so much beloved and respected Father, only he could imagine and understand who was present there. To die, and to die like the late Ignacy – that's a big difference. For if during your life you looked with pleasure at the excess of gentleness and love of His full face, then after death, sanctity emphasised it even more and poured into your soul the conviction that He did not die, but sleeps – the sleep of the blessed."

After the initial ceremonies, the coffin with Łukasiewicz's body, after being led out of the house, was moved in a mourning parade to the church in Zręcin, half a mile away. It was carried "all the way on the shoulders of local citizens, clerks and burghers from Krosno, miners from the distillery in Chorkówka and the oil mine in Bóbrka, firemen and peasants – numerous clergy in front, behind the noble family, friends, acquaintances, children from the school founded by the late Ignacy and maintained at his own cost, famous for its lace-making products. At last, peasants of both sexes, all in solemn chic, all on foot."[276]

Up to four thousand people were to take part in the funeral ceremonies, the order was supervised by the ceremonially uniformed hosts of volunteer fire brigades from Krosno and Rymanów, whom Łukasiewicz had significantly helped in equipping. The coffin was covered with laurel wreaths from the representation of the Krosno district, of which he was the most active

275 *Gazeta Narodowa* no. 11 (1882), p. 3.
276 J. P., *"Ś. P. Ignacy Łukasiewicz"*, p. 169.

member from the beginning of its existence; from the towns of Krosno and Jasło, of which he was an honorary citizen; from the Galician Oil Society, of which he was the founder and life curator; from parliamentary colleagues from the MP's progressive club; from the Pedagogical Society, of which he was an honorary member; from the landed citizenry with the inscription: As one of the noblest sons of his homeland[277].

During the ceremony, on behalf of his father who was abroad, Wiktor Klobassa spoke, who,

> with deep sorrow and heartfelt love, spoke of the man he called his second father." The Inspector of Jasło District School Council spoke, and raised Łukasiewicz's merits for folk education, and also August Gorayski, a friend of the deceased, who recalled Łukasiewicz's great merits as an farming activist and citizen. The coffin with the body of the deceased was placed in the tomb of the Klobassa family. In a way it could symbolise Łukasiewicz's whole life, who "never forgot about anyone, but forgot about himself and perhaps not expecting death yet (because his almost unexpected cold inflammation caused by his bed colds and despite the zealous medical care in the few days of his life the thread was cut prematurely) had no time, no place of rest to prepare for himself and his family."[278]

A poem written on the day of his funeral by one of his neighbours, Janina Łozińska from Potok near Krosno, published, among others, in "Gazeta Narodowa" on 14 January 1882, became a reminder of Łukasiewicz's last journey. It is worth quoting it as a testimony of the moment and feelings accompanying "Father Łukasiewicz's" farewell on that January day:

> "In the quiet cemetery today they have borne you! / A flower more beautiful than all the flowers / On your grave, grateful down pour/ Tears of the gentry and the peasant./ In quiet cemetery today they have borne you! / When spring melts the ice, / On the grave your flowers will bloom, / This symbol of the spirit of serenity./ And the wild rose will bloom luxuriantly here, / Loose their thorns; / As a sign of your work, because you worked vigilantly, / Arduously and faithfully./ The thorny path you yourself walked in life, / To others you sent flowers, / Into the heart and soul of each you looked – / Not into the garments./ So, when they bore you to the quiet cemetery, / This flower, above all beautiful flowers: / Over your grave as one poured down / Tears of the gentry and the peasant"[279].

The memory of Łukasiewicz

The death of Ignacy Łukasiewicz was a shock to all who knew him or were aware of his achievements. It also became an opportunity for attempts to

277 *Gazeta Narodowa* no. 11 (1882), p. 3.
278 Anczyc, "*Ignacy Łukasiewicz*", p. 230; J. P., "*Ś. P. Ignacy Łukasiewicz*", p. 170.
279 Janina Łozińska, "Pamięci Ignacego Łukasiewicza" [Memories of Ignacy Łukasiewicz], *Gazeta Narodowa* no. 11 (1882), p. 3.

permanently commemorate this outstanding figure. In the second issue of *Górnik*, published by the Oil Society, its President and Łukasiewicz's successor in this position, August Gorayski, in the introduction to his memoir of the deceased, pointed out that

> "in the late Ignacy Łukasiewicz the country has lost an extraordinary man, while the entire oil industry its creator and zealous guardian; – at the beginning of our writings we wish to express our deep sorrow about an unforgettable man, who sacrificed all his strength and means for the common good in every field of national work, we wish to recall the exceptional virtues and merits which, hiding in the shadow of modesty, should be put forward (...) If his memory is forever preserved by the invention of kerosene lighting and the opening of an excellent new economic area, he deserved it in equal measure with his virtues, deserved his love of his country and his neighbour, his diligence, his generosity, his fruits, which his qualities have now finally achieved."

Władysław Anczyc wrote in April 1882: "Hostile fate has struck a deep gash in the lean company of people who sacrifice their lives for the good of the country. Not soon will a second Łukasiewicz appear, not soon will another noble soul shine, such a righteous, selfless and noble character, such a man, proving his love of country by every deed."[280]

An attempt to permanently commemorate Ignacy Łukasiewicz was made by his parliamentary companions and friends who, in April 1882, established the founding committee of the Ignacy Łukasiewicz Memorial Foundation. It was headed by Franciszek Smolka, Feliks Zabłocki, Apolinary Stokowski and Teofil Merunowicz, who in their ideas directly referred to the values that Łukasiewicz was faithful to in his life. As the committee justified:

> "When the main and most pleasant field of activity of the late Ignacy Łukasiewicz was to work on the welfare of rural people, mostly through the rational use of the organisation of the county and commune on an autonomous basis, the signatories have decided to apply to this disposition his way of worship of his posthumous memory."

Therefore, the objective of the foundation was to annually award and reward mayors of rural communes distinguished by their merits for exceeding the fulfilment of ordinary duty. Like Łukasiewicz, they were to serve as a role model for others. The Committee invited the public to make voluntary donations to the Foundation, but it turned out that it was not possible to mobilise sufficient resources to implement the plans. Similar was the fate of the initiative of the Oil Society, which intended to create a foundation that would award a medal with the bust of Łukasiewicz to creators of inventions

280 Anczyc, "Ignacy Łukasiewicz", p. 230; A. Gorayski, "Ignacy Łukasiewicz", pp. 13–14.

or improvements contributing to the further development of the oil industry. Again, this type of initiative did not succeed in making a success of this kind of initiative[281].

This was not the best testimony to the permanence of the memory of Ignacy Łukasiewicz. In a completely incomprehensible way, this outstanding citizen was quickly forgotten, by which, half a century later, those who tried to rebuild his place in the pantheon of the greatest Poles could not be surprised. Why did this happen? The reason for this state of affairs could have been the fate of the estate left by Łukasiewicz in Chorkówka and his industrial plants. His death meant a dramatic change in the position of all those associated with him. According to his will, half of the estate was given to his wife Honorata, one-quarter to her mother and Ignacy's sister, Emilia Stacherska and the remaining part probably to Alojza Stacherska and Maria Pikowa nee Stacherska. In 1882, Honorata funded her husband's tomb in the cemetery and a bust in the church in Zręcin, after which she left for Cracow, where she lived until her death in 1897. For many years she was even believed to have been buried there, but Honorata's funeral took place in the church in Zręcin founded by her husband, and her mortal remains were buried at the local cemetery next to Ignacy[282].

Significantly, his heirs sold their assets in Chorkówka and the local refinery as early as 18 February 1882, i.e. only a few weeks after the death of Ignacy Łukasiewicz. The buyer was the well-known oil businessman Seweryn Stawiarski, who immediately transferred half of the shares to another manufacturer, the owner of a small refinery in Klecy near Kołaczyce, Władysław Fibich. The refinery in Chorkówka, the most important child of Ignacy Łukasiewicz, remained in operation for nearly a quarter of a century. On 31 December 1904, however, it burned down and was not reopened, and the owner opened a new plant a year later in Krosno. The fate of the manor house in Chorkówka, so vibrant with life and full of guests and household members, was no better. More and more valuable property, including Łukasiewicz's notes, were successively removed, and after moving the refinery to Krosno, the owners practically ceased to be interested in it. As a result, the palace became more and more dilapidated, and this was completed by World War II and the post-war period. To this day, nothing has remained of the Łukasiewicz manse except for the traces of the basement[283].

281 Tomanek, *Ignacy Łukasiewicz*, pp. 89–92.

282 Brzozowski, *Ignacy..*, pp. 185–186; "Honorata ze Stacherskich", p. 50.

283 Brzozowski, *Ignacy..*, pp. 186–187; Zdzisław Łopatkiewicz, "Seweryn Stawiarski. Materiały do biografii" [Seweryn Stawiarski. Biographical materials], *Wiek Nafty* no. 4 (2006), p. 49.

The fate of the mine in Bóbrka was different, but it must be remembered that its owner was Karol Klobassa, and Ignacy Łukasiewicz, who managed the mine, left an excellent successor in the person of Adolf Jabłoński. Under his direction, the mine continued to prosper, even when, after Karol Klobassa's death, it was taken over by his sons Wiktor and Stanisław. When Jabłoński died in 1887, Zenon Suszycki, who had previously headed a mining school established in 1885 in Ropianka, became its director. He modernised the mining processes by introducing the Canadian drilling method in Bóbrka, thus initiating the next stage in the development of the mine. In 1893, Stanisław Klobassa sold a 45 % stake in the mine to William Henry MacGarvey, a Canadian, for the gigantic amount of about half a million Rhine guilders. The investor also received the administration of the shares of Helena Klobassowa (Karol's widow) and Wiktor Klobassa, which meant that at that time the world's oldest thriving oil mine passed into the hands of foreign capital. In 1895 Bóbrka, like other MacGarvey mines in Galicia, became the property of the Galician Carpathian Oil Society, whose main shareholders were MacGarvey and its partner John Simeon Bergheim[284].

When the material remains of Ignacy Łukasiewicz's life and activity were so quickly wasted, it was hardly surprising that the devastation that took place in subsequent years in human memory. Adolf Jabłoński, publishing in 1884 his *Mining Oil*, dedicated this book

> "to the memory of Ignacy Łukasiewicz, the creator of the oil industry," and in the introduction he recalled that "Galicia is the homeland of the oil industry; here, in 1853, the light of oil, whose creator was Ignacy Łukasiewicz, flashed for the first time."

In the same year, portraits of Łukasiewicz and his two partners Tytus Trzecieski and Karol Klobassa were hung in the meeting room of the City Council in Krosno, and ten years later Stanisław Klobassa presented portraits of these three in the oil pavilion of the General National Exhibition in Lviv. And there, as it might seem, the memory of Łukasiewicz died[285].

As Oskar Löwenherz rightly wrote in 1932 in "Nafta":

> "We have forgotten about the silent Pole from Krosno, who was afraid of all the fame and honours, we have forgotten about his tombstone in Zręcin, in front of which only the inhabitants of the surrounding villages would nod their heads in memory of his selfless benefits."

---

284 Cząstka, "Dzieje przemysłu naftowego", pp. 17–19; *Krośnieńskie kopalnictwo naftowe. Wspomnienia i sylwetki* [Krosno Oil Mining. Memories and Profiles] (Krosno: Oficyna Wydawnicza Apla, 2005), pp. 15–16.

285 Bonusiak, *Szejk z Galicji*, p. 154; Adolf Jabłoński, *Kopalnictwo nafty* [Oil Mining] (Cracow: Wł. Anczyc i Spółka, 1885).

Figure 25: The oil field in Bóbrka 1911. Source: The Ignacy Łukasiewicz Oil and Gas Industry Museum in Bóbrka

And accusingly pointed out that, "Łukasiewicz was a Pole, and this explains to a large extent to us the fact that he was forgotten and not recognised. Our society usually underestimates its great people, not so much out of laxity and lack of interest, but out of disbelief that Polish thought and art could give truly epoch-making values to the world."

The spell of bitterness overflowed when in 1927 the Viennese daily *Neues Wiener Journal* published an article "On the 75th anniversary of the lighting of the first kerosene lamp," in which it was stated that Schreiner, the poor Jewish merchant from Borysław, was the first in history to distil kerosene and light a kerosene lamp. This text met with a strong reaction in Poland and actually helped to restore the memory of the work of Ignacy Łukasiewicz. In December of that year, articles on him appeared in the *Illustrated Daily Courier* and "Warszawianka." The first periodical, extremely popular in pre-war Poland, played an important role in the "rediscovery" of the memory of Ignacy Łukasiewicz[286].

286 Oskar Löwenherz, "Ignacy Łukasiewicz odkrywca nafty i twórca przemysłu naftowego" [Ignacy Łukasiewicz, Discoverer of Kerosene and Creator of the Oil Industry], *Nafta* no.8–9 (1932), p. 206; Tomanek, *Ignacy Łukasiewicz*, pp. 15–16.

The author of the text, Ludwik Tomanek, later the author of Łukasiewicz's first biography, recalling the development of the method of oil distillation and the invention of the oil lamp, sadly stated that even in former Galicia "the memory of the man who made both these discoveries has almost disappeared." Moreover, as a result of the article on the subject published in the sensational Viennese daily newspaper, "the world's public was completely misinformed about who was the creator of the first kerosene lamp and the first oil distiller." For the first time in many years, Ignacy Łukasiewicz was remembered on such a large scale in a relatively large article for a daily. As a result, on the initiative of the local starosta Emil Rappe, a Committee for the Commemoration of Ignacy Łukasiewicz, whose portrait was still hanging in the meeting room of the county department, was established in Krosno. On 30 September 1928, i.e. on the 75th anniversary of the lighting of the first kerosene lamp, the Committee organised a celebration in memory of Ignacy Łukasiewicz, combined with the installation of a foundation stone at his monument. The solemn Holy Mass was attended by many distinguished guests, including Count Wojciech Gołuchowski, the Voivode of Lviv, who represented the government. Additional meetings and banquets devoted to Ignacy Łukasiewicz and his contribution to the Polish oil industry and the Krosno region were also held in the city. In the same year, thanks to the Committee's efforts, the aforementioned biography of Łukasiewicz by Ludwik Tomanek was published. Its main purpose was to disseminate knowledge and popularise the figure of Łukasiewicz, and to serve this end, copies of the monograph were sent to school libraries[287].

Four years later, on 23 October 1932, on the ninetieth anniversary of the birth of the great inventor, industrialist and philanthropist, a monument to Ignacy Łukasiewicz was unveiled in Krosno. The plinth, made by an eminent artist, the President of the Union of Artists in Cracow, Jan Raszka, presented Łukasiewicz on a pedestal made of Volhynian granite. The city was beautifully illuminated and decorated, and on the eve of the holiday "a wonderful parade with lanterns," in which soldiers and young people participated, passed through its streets. The ceremony brought together many distinguished participants: "Just as Krosno is Krosno, this town so old does not remember such a numerous congress of guests, so impressive a participation of crowds of celebrations." It was honoured by, among others, the

---

287 L. T., "75 lat temu... pierwszą lampę naftową dla świata zapalił Polak" [75 years ago... the first kerosene lamp in the world was lit by a Pole], *Ilustrowany Kurier Codzienny* no. 342 (1927), pp. 2–4; Löwenherz, "Ignacy Łukasiewicz odkrywca", pp. 206–207; Barbara Olejarz, "80-lecie odsłonięcia pomnika Ignacego Łukasiewicza" [The 80th Anniversary of the Unveiling of the Monument to Ignacy Łukasiewicz], *Wiek Nafty* no. 4 (2012), p. 5.

ministers of industry and commerce and communication, high-ranking military, parliamentarians, professors, presidents of economic organisations, numerous representatives of pharmacy circles from Poland and abroad, the oil workers who were attending their congress in Krosno and living representatives of Ignacy Łukasiewicz's family. The ceremony was broadcast by Polish Radio, and its publicity in the media was a confirmation of the ongoing efforts to popularise the knowledge about the work and life of Ignacy Łukasiewicz. One of the speakers, the rector of the Mining Academy in Cracow, Zygmunt Bielski, appealed directly "that Polish society should keep in permanent memory the great merits of Ignacy Łukasiewicz for the development of Polish industry."[288]

The process of giving Łukasiewicz his rightful place in the pantheon of Polish heroes was continued in the post-war period. During the Congress of Polish Science held in Warsaw in 1951, his portrait was hung among the six most eminent people of science, next to the likenesses of Nicolaus Copernicus, Stanisław Staszic and Maria Curie-Skłodowska, and throughout the post-World War II period there were numerous forms of commemoration of Ignacy Łukasiewicz. Other monuments, busts and obelisks were erected both in southern Poland (in Gorlice, Glinik Mariampolski, Bóbrka, Chorkówka, Rzeszow), as well as in other centres related to the oil industry in Poland. In 2007, a monument to Łukasiewicz was unveiled in Płock, and a year later, a similar initiative on the 40th anniversary of crude oil production on the so-called Polish Lowland materialised in Zielona Góra. In 2013, an Ignacy Łukasiewicz bench was unveiled in Poznań, thus commemorating the 160th anniversary of the lighting of the kerosene lamp[289].

The most important and the most beautiful monument to Ignacy Łukasiewicz was the Museum of Oil Industry in Bóbrka, opened in 1961 – currently The Ignacy Łukasiewicz Oil and Gas Industry Museum in Bóbrka. It is there, where the world's first oil mine was established, that the only such museum was established, where oil is still extracted and the "Janina" shaft, which is still in operation manually. There is also the "Franek" shaft, which

288 Löwenherz, "Ignacy Łukasiewicz odkrywca", p. 207; "Polska uczciła pamięć pioniera nafty pomnikiem" [Poland honoured the memory of the kerosene pioneer with a monument], *Ilustrowany Kurier Codzienny* no. 296 (1932), p. 3; "Poświęcenie pomnika Ignacego Łukasiewicza w Krośnie" [The consecration of the monument to Ignacy Łukasiewicz in Krosno], *Gazeta Lwowska* no. 245 (1932), p. 3.

289 Bonusiak, *Ignacy Łukasiewicz*, pp. 124–130; Anna Szczuchniak, "Był żywym ogniem podwójnego światła – z dziejów pomników Ignacego Łukasiewicza" [He Was a Living Flame of Double Light – From the History of the Monuments to Ignacy Łukasiewicz], *Wiek Nafty* no. 4 (2013), pp. 24–25.

started operating in 1860, a wooden smithy from 1856, a reconstructed manual drill rig from 1862 and an administrative building, the so-called Łukasiewicz House from 1865[290].

Ignacy Łukasiewicz, who, thanks to his efforts to improve the situation of peasants and workers with public activity, was positively perceived by the communist authorities of post-war Poland, also received a number of other memorials. Several dozen schools of different types were named after him, including the Rzeszow Polytechnic, in 1982 in connection with the Year of Łukasiewicz announced on the hundredth anniversary of his death and the 160th anniversary of his birth, a special commemorative medal was minted. And a year later, a fifty-zloty coin with a likeness of the world pioneer of the oil industry appeared in circulation. The most important elements of the Łukasiewicz Year celebrations were ceremonies organised in Krosno and Bóbrka on September 10 and 11, 1982. The celebrations were complemented by the later unveiling of statues of Ignacy Łukasiewicz in Cracow and Gorlice[291].

Free Poland also did not forget about one of the most outstanding Poles. In 2003, on the 150th anniversary of the lighting of the first oil lamp, the Senate of the Republic of Poland, by a resolution "on the commemoration of the outstanding achievements of Ignacy Łukasiewicz, pioneer of the global oil and mining industry," proclaimed another Year of Łukasiewicz. Once again numerous symposia and conferences were held throughout the country, and the National Bank of Poland issued a series of coins dedicated to Ignacy Łukasiewicz[292].

---

290 Jacek Munia, "40 lat pracy Muzeum Przemysłu Naftowego i Gazowniczego im. Ignacego Łukasiewicza w Bóbrce w zachowaniu kultury technicznej i tradycji polskiego przemysłu naftowego i gazowniczego" [40 years of work of The Ignacy Łukasiewicz Oil and Gas Industry Museum in Bóbrka in Maintaining the Technical Culture and Traditions of the Polish Oil and Gas Industry], *Wiek Nafty* no. 2 (2001), pp. 16–19; O *Muzeum*: 27 Mar. 2018 https://bobrka.pl/o-muzeum/.

291 Adam Barzyk, "Wielkie jubileusze Ignacego Łukasiewicza" [Great Anniversaries of Ignacy Łukasiewicz], *Wiek Nafty* no. 1 (2002), pp. 3–4; Bonusiak, *Szejk z Galicji*, pp. 161–162.

292 Barzyk, "Wielkie jubileusze Ignacego Łukasiewicza", pp. 3–4; Bonusiak, *Szejk z Galicji*, pp. 161–163; *Uchwała Senatu Rzeczypospolitej Polskiej z dnia 19 lutego 2003 roku w sprawie uczczenia wybitnych osiągnięć Ignacego Łukasiewicza, pioniera światowego przemysłu naftowego i górniczego* [Resolution of the Senate of the Republic of Poland of 19 February 2003 on the commemoration of the outstanding achievements of Ignacy Łukasiewicz, pioneer of the global oil and mining industry]: 27 Mar. 2018 http://ww2.senat.pl/k5/dok/uch/034/321uch.htm.

Figure 26: Wooden smithy of Łukasiewicz in Bóbrka. Source: Anna Kozicka-Kołaczkowska

Łukasiewicz's memory has been preserved, but he has never become such a hero of collective admiration as Nicolaus Copernicus or Maria Curie-Skłodowska. Despite the many forms of commemoration of the work and life of Ignacy Łukasiewicz which are not mentioned here, the knowledge about him still remains relatively modest, and the aforementioned complaints from the 1930s, quoted above, that Poles do not value their outstanding compatriots, remain partially valid. It seems that the reason for this state of affairs lies in the fact that Łukasiewicz was and still is a regional and local hero. He lived during the Partitions, so he was appreciated in Galicia, but in other Polish territories, which at that time belonged to Germany and Russia, he remained an unknown or almost unknown figure. Just as the history of the Polish oil industry is rather a part of the history of Galicia, once the poorest district of Polish lands, so Łukasiewicz remains a local hero in the collective consciousness of Poles. He is still unknown beyond the borders of Poland, and that is why it is worth showing this wonderful, almost indomitable figure to the whole world.

Łukasiewicz – known unknown

Ignacy Łukasiewicz began his career as an industrialist modestly, leaving Lviv for Gorlice and taking over the lease of a pharmacy there. However, he did not abandon his dreams of oil and its distillation, as evidenced by his tireless experiments in this respect. Despite his initial failures, he was a great success, and his determination, great diligence, talent, skills, organisational capacity, drive for innovation and the ability to learn, so popular today, led him throughout his life not only to the pinnacle of wealth and business success, but also the fulfilment of his dream of kerosene. After all, his aim was to popularise kerosene and develop the oil industry in Galicia, and this goal was achieved during his lifetime.

When he was at the peak, however, he still remained the same as the man before. He was a great patriot who, however, wanted to serve his homeland not through conspiracies, but through persistent work on the civilisational uplift of the country. He was a politician who, in the public service, constantly cared for the development of his county, even with his own money, and later for the whole country, putting particular emphasis on factors that are nowadays considered necessary conditions for modernisation, i.e. the development of education and infrastructure. He was the leader and great authority of the oil industry, the founder of the association of Galician oil workers, tirelessly fighting for the interests of the industry, supporting ambitious entrepreneurs. Finally, he was "Father Łukasiewicz," a man to whom one could always come for advice or help, a social activist ready to help others and along with his growing wealth, able to do more and more for them – founding schools, places of worship, establishing loan funds, introducing social security for workers, treating sick peasants, helping the poor, taking the needy out of poverty and giving shelter to veterans of the struggle for independence.

At the same time, Ignacy Łukasiewicz was an extremely modest man, who earned money not so much for himself as for doing good. He did this not for applause, but out of his inner conviction that he should do so. It is no coincidence, therefore, that he did not leave behind as great a fortune as one might think after the momentum of his ventures and the scale of success he experienced in business. He managed to simply share such a large part of the money he had earned with others during his lifetime. Giving was for him an inseparable part of life, just like work or deep religiosity.

Although he had no children, he was the father of a large family, he was its patriarch in the biblical sense of the word. But his family life was to some extent the result of everything he did in his life. His work and all his additional duties certainly distracted him from his family, and great generosity for his loved ones could not always compensate for this, especially for Honorata, who had to share her husband not only with the mine in Bóbrka and the refinery in Chorkówka, but also with almost the whole area, for which he was "Father Łukasiewicz" and with the whole country, about the good he never forgot and for which he worked until his last days.

Discoverer and inventor, titan of work, innovator and teacher of youth, entrepreneur and millionaire, philanthropist and social activist, patriot and politician, but above all a humble, good man devoted to others. Such was Ignacy Łukasiewicz, and this is how he should be known to the world.

Father Ignacy is a model for modern generations

Ignacy Łukasiewicz built the foundations of the modern oil sector, which has contributed to the development of civilisation. His figure has become a permanent fixture in the economic history of the world energy sector. He combined the skills and courage of striving to discover new things and the researcher's passion with entrepreneurship, diligence, patriotism and the attitude of a philanthropist. He proved that being an innovator does not exclude social, simply human, attitudes.

The contemporary generation, which impatiently tries to pursue set goals or has problems with defining them, should take as an example Ignacy Łukasiewicz's diligence and tenacious experimentation skills. Even in the face of setbacks and life's difficulties, he did not give way in the pursuit of his goals. In an era of increasing competitiveness between individual economies, which are looking for their competitive advantages, the values represented by him are particularly valuable. These are timeless features that can be a source of inspiration for everyone.

When Łukasiewicz founded the world's first crude oil mine, which he managed, he showed an innovative approach, which consisted in continuous modernisation of the mine, as well as the refinery, which ensured the continuous development of his initiatives. Nowadays, it is noticed that such actions consisting in permanent improvement, searching for new solutions, optimisation of production and constant improvement of effectiveness are building business models for the best and largest companies in the world.

Łukasiewicz tried to infect other people with his entrepreneurship. He knew that oil had a potential that should serve the general public. He perceived the development of the oil industry as an opportunity to create new companies, which he encouraged others to do. He was not afraid of competition, but sought to create new entities that could cooperate with each other. The National Oil Society he founded in 1877, which first operated under the name *Towarzystwo dla Opieki i Rozwoju Przemysłu i Górnictwa Naftowego* (Society for the Care and Development of the Oil Industry and Mining in Galicia), supported the development of mining on Polish lands and was active in independent Poland until the outbreak of World War II. It led to the adoption of the Petroleum Law. We should learn from Łukasiewicz's attitude of moving away from a selfish, profit-driven approach to business.

Teamwork

Łukasiewicz was able to reconcile his own innovation and individualism in cooperation with others, especially while working in the mine in Bóbrka, when he cooperated with Karol Klobassa and Tytus Trzecieski. This confirms that outstanding talent develops in a collaborative environment, which we now commonly call teamwork. His attitude, full of a creative and unconventional approach to reality, combined with perseverance and diligence, is not only a role model to follow, but also a signpost for success. Nowadays, employers are looking for employees endowed with a set of features that were characteristic of Ignacy Łukasiewicz. He showed how to be courageous in thinking and cross the barriers that are created mainly in our minds. He saw what nobody else saw. He took risks where others withdrew. He travelled along paths that others had not walked before, and as history has shown, they became a path for others.

Responsibility in business

Ignacy Łukasiewicz had an innovative approach to employees. In his time he was able to create a unique business model in which man played an important role. He generously remunerated his employees who worked in oil extraction, introduced allowances for working in difficult conditions and also rewarded effective work confirmed by good results with bonuses. He used a motivational system that increased the efficiency of production and contributed to a better atmosphere among employees who often risked their lives during their daily duties. In this, he distinguished himself from many other entrepreneurs who did not use such incentives. In 1866, he established the first workers' fund in the Austrian Partition, which was an aid and insurance system for employees. Its assumption was to introduce the obligation to pay 3 cents of each earned Rhine guilder by each employee. The funds collected were used to cover the costs of treatment of employees in the event of illness, payment of the so-called sickness benefit for each day of incapacity for work, payment of the funeral costs of the deceased employee, as well as payment of disability pension and benefits for widows and orphans. He also introduced assistance in the event of a house fire. These rules were an innovation at a time when others did not apply them. They built public trust and a sense of stability. Łukasiewicz tried to reduce the risk of incidents leading to unfortunate consequences, which is why he introduced an absolute ban on drinking alcohol for mine employees. He supported safety at work on his own initiative, despite the lack of appropriate regulations in this area. He created good practices, which over time became common standards. So in this respect, too, he went beyond the epoch in which he lived.

After his death, he left his achievements, which should be multiplied by future generations. Especially in the era of robotisation and computerisation, knowledge and human capital are the source of competitive advantage. The world needs Łukasiewiczs.

Energy policy and Łukasiewicz's legacy

The energy policy of the 21st century is characterised by the need to constantly read the upcoming changes and take advantage of the opportunities resulting from many variables. Its fundamental objective is to achieve energy security and competitive advantages. The example of Łukasiewicz is a good point of reference for building energy strategies. It indicates that innovation in thinking, courage and consistency in action must be the guiding principles. Create your own standards and solutions and then disseminate them. At the same time, innovation should be multidimensional and must not be limited to work in a laboratory, but must also include work organisation, company management, business models, communication, cooperation and teamwork, as well as legislative solutions. The effectiveness of energy policy requires determination, consistency and even persistence. If we forget about these attributes and qualities, then The Ignacy Łukasiewicz Oil and Gas Industry Museum in Bóbrka will constantly remind us of them – a facility which is unique on a global scale where you can get to know the origins and with your own eyes see the cradle of the Polish oil industry in the world's oldest and still operating oil mine.

III Underwater Oil Pioneer – Witold Zglenicki

The Heir, Witold Zglenicki. 2,400,000 dollars in gold. In the Mazovian manor house and Subcarpathian Bóbrka. In Warsaw and St. Petersburg. In the footsteps of Stanisław Staszic. Flights and falls. The Assayist of Riga. In Baku, where heaven and hell exist. The Assayist Engineer. Learning and working, not throwing a hat. The inventor. Desiderata – the key to the unnoticeable Aladdin's cave. And yet they will make the sea sleep. 165 oil-producing fields, marine fields, platforms and breakwaters. Fields according to Zglenicki. The first maritime field, or when there are enough important matters on land. Technology, science and art. No happy ending. A legacy "for everlasting times". Shafts in the Gulf of Ilyich Lenin. And the oil is still pouring.

"Son – the writing will pass, but you will remember, grandson (…)
That's how he will read what you're reading today,
But he will remember me… because I will not be here!"[293]

Cyprian Kamil Norwid (1821–1883)

The Heir, Witold Zglenicki

The great Polish poet Cyprian Kamil Norwid, misunderstood and undervalued during his lifetime, expressed in his prophetic, poetic intuition the fate of the works of geniuses, rather common in history, which also became the participation of Witold Zglenicki – the great heir to the mission and work of Ignacy Łukasiewicz, who saw oil as a source of modern industry and society's prosperity. And even if Zglenicki did not even die prematurely, he would not have lived to achieve the right goal. His great idea of building a drilling platform at sea, which is today a symbol of the most modern technical thought of the mining industry, was realised only by his grandchildren's generation.[294]

293 Cyprian Norwid, *Pisma wybrane* [Selected Writings], Państwowy Instytut Wydawniczy, Vol. 1, Warsaw 1968, pp. 192–193.

294 Zija Ali Buniatzade, *Witold Zglenicki i problem morskiego wydobycia ropy naftowej w basenie Morza Kaspijskiego* [Witold Zglenicki and the Problem of Oil Extraction in the Caspian Sea], in: Historia rosyjsko – polskich kontaktów w dziedzinie geologii i geografii [History of Russian – Polish contacts in the field of geology and geography], (Warsaw, 1972), pp. 32–34.

The decision to build a drilling platform at sea had to mature for decades. First, in 1949, at a distance of 100 km from Baku in the Caspian Sea, the world's first platform, Neft Daslari, was built. In 1954, the Americans established their first platform, Mr. Charlie, in the Gulf of Mexico.

Meanwhile, a Pole – Witold Zglenicki – a graduate of the Warsaw School of Economics, engineer, miner, geologist, inventor, cartographer, discoverer of natural resources and charismatic continuator of the industry initiated by his compatriot Ignacy Łukasiewicz – had been striving for the construction of such a platform without success already fifty years earlier in Baku. As Łukasiewicz initiated the digging of crude oil disregarded by the industry, Zglenicki discovered it under the seabed and fought for its extraction.

His desiderata on this matter, officially submitted to the authorities for the first time on 29 July 1896[295], which had been submitted several times, included the first ever offshore oil rig design with a description of the oil transport technology. In it, the Pole documented his request for the allocation of maritime fields. It is a monument of pioneering rank and a worldwide artefact. A certificate of a breakthrough moment of civilisation and the genius of a Polish engineer, who alone caused a revolution in the perception of the world. He opened people's eyes to the infinite hoard of the riches of the Earth, which man, living on his planet for many thousands of years, had been looking for only in the place of the everyday trotting of the mainland.

Zglenicki's contemporaries, oilmen and businessmen with the most sensitive nose for good business mocked his offshore oil extraction initiatives to the end. They did not even believe, having in front of their eyes his expert opinions, geological research and hand-drawn maps of oil and gas deposits, so innovative was the idea. Although they begged him to make his maps of the continent available, they considered his projects of methods of extracting oil from the seabed to be unrealistic phantasmagoria. The authorities of the Russian state, which at that time ruled the oil fields in Azerbaijan, even when they finally agreed with the claim that oil has its deposits also under the seabed and believed after years of convincing in the sense of starting the exploration of underwater deposits, preferred to backfill the sea and build artificial land rather than implement the plans for Zglenicki's drilling platform at sea and the methods of oil extraction and transportation.

---

295 Andrzej Chodubski, *Witold Zglenicki "polski Nobel" 1850–1904* [Witold Zglenicki, the "Polish Nobel" 1850-1904], (Płock, 1884), p. 66.

2,400,000 dollars in gold

The beginning of the 20th century was not a sudden breakthrough, but a long farewell to the past. Ferments and political convulsions only in the second decade of the new century were to lead to a world war which destroyed the old map of Europe. In 1904, until 1918, Warsaw was still the backbone of the Russian Empire with its capital in St. Petersburg.

And the year 1904 was exceptional in the history of Polish culture.[296] Writers always hungry for grants, publishers and editors who could not offer their national monumental works, encyclopaedias, studies, dictionaries, or song collections, charismatic enthusiasts of scientific discoveries, students and doctoral students dreaming of scholarships and laboratories, academics, steadfast Poles determined to fight for the survival of the Polish language and nation throughout the entire 19th century – the age of the Partitions – the age of captivity – herolds of education, positive work and strengthening the spirit in the foreign state system, received an extraordinary inheritance. [297]

This record was donated to the Józef Mianowski Fund – a Polish social foundation established by the students of the Warsaw School of Economics for twenty-odd years, supporting noble, academic and literary goals from private donations, the only institution of this kind approved by the Russians in the Polish zone. At first this will seemed exotic and unrealistic to the members of the Fund Committee. It assured their participation in oil production revenues from plots of land in Azerbaijan.

The scepticism of the heirs turned into astonishment when it turned out that astronomical sums started to flow into the Mianowski Fund. By 1918, the revenue of the Fund from the oilfields in Baku amounted to 2,400,000 dollars in gold, which accounted for 63 % of all the amounts that the Fund had at its disposal since its establishment in 1881 until 1918[298]. In comparison, the sum of the Nobel Prizes of 1901 amounted to 150,000 Swedish kronor worth about 40,000 dollars, except that over time this amount lost its value. The value of extraction from Azerbaijani oil-producing fields – quite

---

296 Andrzej Chodubski, *Testament inżyniera górnika – geologa Witolda Zglenickiego (1850-1904)* [The Miner's Testament – Geologist Witold Zglenicki], "Teki Archiwalne" 1992, T. 22, pp. 39–63.

297 Adamczewski, *Zapis testamentowy Witolda Zglenickiego dla Kasy im. Józefa Mianowskiego, Studia z Dziejów Rosji i Europy Środkowo Wschodniej* [The Testamentary Record of Witold Zglenicki for the Józef Mianowski Foundation, Studies in the History of Russia and Central and Eastern Europe], PAN, XLVIII.

298 *Kasa Mianowskiego 1881–2011* [The Mianowski Fund 1881–2011], ed. Zasztoft Leszek, (Warsaw, 2011), p. 109.

the contrary. It grew until the Revolution of 1917, when the new state of the Russian Bolsheviks ceased to respect their legal obligations.

In 1907, when the matter of accepting the inheritance and dealing with formalities was still in progress, the total payments of the fund to beneficiaries amounted to 31,644 roubles. In 1912, with great prodigality, 84,546 roubles were spent on subsidies. A year later 160,671 roubles and in 1917, it was 177,939 roubles.[299] Despite this unprecedented increase in funding for scientists and their laboratories, creators and publishers, the Fund still held huge surpluses which could not be utilised. The environment of Polish culture in the Russian Partition, as never before, turned out to be too small and had no possibility to assimilate such huge resources. It was not possible to keep up with the cash inflow from oil, as the annual revenues were over ten times higher than the total expenditure of the Fund at the beginning of the century. This miracle was a phenomenon in the history of Poland, which has not been beaten to this day. The donor of these sums was Witold Zglenicki. Graduate of the Warsaw School of Economics, engineer, miner, brilliant geologist, inventor, cartographer, discoverer of natural resources and pioneer of crude oil extraction from the seabed, unwavering originator and propagator of drilling platforms against unbelievers, pests and scoffers.

It cannot be ruled out that without Zglenicki's donation of many works that Polish culture has at its disposal as national treasures of culture and science, only a few manuscripts would survive to this day in the nooks and crannies of the warehouses. This circumstance may not delight the world very much, but surely it should be known to the world that every drilling platform on the high seas, which today is synonymous with the power and richness of societies, state-of-the-art technology and human horizons has its origin and pedigree in the uncommon mind and the unbreakable personality of Zglenicki.

In the Mazovian manor house and Subcarpathian Bóbrka

Witold Leon Zglenicki was born on January 6, 1850 at the estate of Wargawa Stara, to the noble family of the Prus II coat of arms.[300] At that time, the Polish noble tradition with its customs and costumes still were still alive in

299 Andrzej Chodubski, *Witold Zglenicki, Polski Nobel 1850–1904*, Towarzystwo Naukowe Płockie, (Płock, 1984), p. 104.

300 Jakub Chojnacki, *Komunikat z dnia 30 listopada 1978 roku o ustaleniu daty i miejsca urodzenia Witolda Zglenickiego* – "Polskiego Nobla" [Communique from November 30, 1978 about the date and place of birth of Witold Zglenicki - "the Polish Nobel"], "Notatki Płockie" 1978, Vol. 4, No. 97, p. 36.

Mazovian manor houses. This colourful, robed world of noble culture was only wiped out by the fashion of the approaching new century.

Witold Zglenicki was the youngest of three brothers. Their initial education was taken care of by their father, who placed particular emphasis on making the children love reading. At the age of nine, Witold, like his older brother Bolesław, was sent to study at the gymnasium in Płock.

In the times of Zglenicki, in the middle of the 19th century, this school as an institution in the Russian Partition had the status of the Gubernial Gymnasium[301]. It was today's Marshal Stanisław Małachowski High School[302], the former Jesuit college of St. Michael, one of the oldest schools in Poland. Established in the 12th century, it has enjoyed a great reputation and a list of excellent graduates, headed by Paweł Włodkowic.

Płock, located near the family of Zglenice, was a place that Witold's father, Konstanty Zglenicki was particularly fond of. It also enchanted the young Zglenickis. Witold remembered the historical town, his friends and seven years spent in the Płock Gymnasium warmly. Without much difficulty, both in the sciences and the humanities, he became one of the prime students.

Meanwhile, since 1854, in a distant corner of Galicia, i.e. the part of Polish lands seized by Austria, at the foot of the Carpathian Mountains, pharmacist Ignacy Łukasiewicz and his partners created the foundations of the world oil industry, and his oil lamp began to change the lives of people all over the world. The oil industry in the Subcarpathia region was developing, drilling methods were being improved and increasingly efficient technical equipment was being constructed. From this Polish province, from Gorlice, Jasło and Krosno, crude oil started its instant conquest of the world. The fascination with the profile and discoveries of Ignacy Łukasiewicz and the enchantment with the phenomenon of crude oil was also to shape the life of Witold Zglenicki.

However, the reality outside the school walls was already maturing to the next, armed culmination of Polish rebellion. The uprising began at the end of January 1863 and its consequences did not leave Płock untouched. Few biographical sketches, anticipating the positivist, post-Uprising trauma to national armed uprisings, seem to underline in particular the fact that Witold Zglenicki did not take part in the January Uprising.

"Years later he remembered them with a hint of fear. He remembered the searches, arrests of guilty and innocent people, various kinds of abuses, beatings of passers-by with rifle butts, burning bonfires in the city, executions

301 Governorship

302 *Słownik geograficzny Królestwa Polskiego*, Vol. 8, pp. 285–308.

and political repressions,"[303] writes Andrzej Chodubski, author of Zglenicki's biography.

In January 1863, the junior secondary school student Witold Zglenicki was barely thirteen years old. He was not able to train with the insurgent, forest units. Besides, the "trigger" of the Uprising organised for months in the noble manors was the regulation on the conscription in the Tsar's army, the so-called *branka*, used by the Russians as a kind of repression and a way to eliminate the element of the nation most predestined to resistance. For young Poles, the *branka* meant many years of service most often beyond the Urals, in the Caucasus, under commanders both hateful and criminal towards their compatriots and homeland. It was tantamount to the ruin of life plans, or to a death sentence. The young Witold was not threatened by the *branka*. In 1866 he completed his junior secondary school education and went to Warsaw to study.[304]

In Warsaw and St. Petersburg

Witold Zglenicki entered the Main School[305] in 1866 and completed four years of studies at the Faculty of Mathematics and Physics. His older brother Bronisław had also chosen this course of study earlier. The studies were undertaken by Zglenicki without major problems and even material worries, which was not the rule under the Partitions among the students of the Polish intelligentsia. Zglenicki lived with his brother at a relative's lodging, not being forced to give tutoring, which we know so well from the descriptions of the lean student years of many well-known figures of that age. It is even known that the Zglenicki brothers' favourite but not financially catastrophic entertainment included playing cards. Their parents were already farming in the beautiful, significant estate of Dębe nad Narwią. The Zglenicki brothers spent every holiday in Dębe, hosting friends. After studying in St. Petersburg, the brother Bolesław undertook the management of the Dębe estate.

It is a fact that the Zglenickis did not stand out in the Main School by their attraction to the pen, which provided so many colleagues of the institute with fame and, in a short time, celebrity. The main reason was perhaps that they were not forced to earn like others, and their interests went, according to the spirit of the times, in a pragmatic direction. Witold Zglenicki wanted to

303 Chodubski, *Witold Zglenicki...*, p. 14.

304 Tadeusz Zglenicki, *Polski Nobel* [The Polish Nobel], in: Księga Pamiątkowa Zjazdu Małachowiaków 1180 – 1957/58, (Płock, 1959), pp.184–186.

305 Stanisław Fita, *Pokolenie Szkoły Głównej* [The Generation of the Main School], (Warsaw, 1980), p. 13.

become a mining engineer early on. Studying in St. Petersburg at the Mining Institute was a natural choice for him.

His brother Bolesław supported these plans and, as the heir to the estate of Dębe, undertook to finance them. It was a sensible step to become a mining engineer and to become involved in the mining industry. Especially in the system of a state in which for the conquered nation there were not many ways to a livelihood for the subjugated people. At that time, Poles could only conduct their scientific careers abroad, so they could be found in faculties and laboratories in France, Switzerland, Germany and Great Britain.[306] Working in the official apparatus of the invader was considered to be more or less collaboration, usually connected with the renunciation of national identity. The position of official was in fact difficult for Poles to access in the Russian state. In a country of liquidated universities, schools, cultural institutions with a margin of strictly censored literature, and in the lands of artificially inhibited economic development, many roads were closed to Poles.

The great Polish writer Eliza Orzeszkowa, unable to maintain the Miłkowszyzna family estate after the death of her husband, a Siberian prisoner, quotes some facts of life on this subject:

> Two professions introduced themselves to my imagination: teacher and telegraphist. The second seemed less troublesome to me, above all less like slavery. I knew from the newspapers that women were starting to be accepted into the telegraph offices. I knew four languages, I enjoyed life in Warsaw, full of bookshops, editorial offices, newspapers and my friends from my stipend. So when I had the energy, I went to Warsaw, went to some dignitary and asked him if I could get a job at the Warsaw telegraph office, presenting my language skills to him. I received the response (…) that only Russian women were accepted for telegraphic offices, and for Polish women they were inaccessible. I can't tell you the impression that this fact made on me (…)
> How so! In Warsaw we, Polish women, have no right to work![307]

In 1869 in St. Petersburg, the developing metropolis of the capital, more than 11 thousand Poles lived. There was a university established in 1819. In the 1870s, about 900 Poles studied in St. Petersburg, quite well organised, running their own self-help fund and student canteen, which was also used by Witold Zglenicki. After all, Poles remained under the constant supervision of police and university authorities and all their initiatives needed to be officially reported. They were subject to strict controls at the Kaunas border

306 I. Blum, *Polacy w Rosji carskiej i w Związku Radzieckim* [Poles in Russia and the Soviet Union], "Wojskowy Przeglad Historyczny" 1966, Vol. 3, No. 39, pp. 187–218.

307 Orzeszkowa, *O sobie* [About Myself], (Warsaw, 1974), p. 107.

crossing point. Among the most forbidden things were patriotic books by Polish Romantic authors.

His studies at the Mining Institute confirmed Zglenicki's belief in the appropriateness of the choice of mining and geology. The Institute was a well-equipped university with rich mineralogical collections and a library. Students even complained about the over-ambitious program, but Zglenicki was a graduate of the Warsaw Main School. He had an excellent university education, and he had no problem with studies. Poles constituted 15 % of the students of his university. Zglenicki kept with the Poles from both the Institute and the University. He lived in the dormitory of the Mining Institute.

Witold Zglenicki chanced to be at the University of St. Petersburg at the time when Dmitry Mendeleyev, whose lectures and publications were a great treat for students of science, was active. He attended Mendeleyev's lectures and managed to get to his chemical laboratory, where the level of knowledge and skills attracted the scientist's attention. Mendeleyev, the creator of the periodic table of elements, saw him as a future outstanding chemist. However, Witold did not take advantage of the fantastic opportunity that the Russian offered him. Born in exile, Mendeleyev[308], who came from the aristocratic spheres, was a remarkable subject of the Russian tsar and, after all, a scientist of great fame, but Witold Zglenicki dreamt of working under the direction of Ignacy Łukasiewicz in the Polish oil industry. He was not the only one. Many Polish students of the St. Petersburg Mining Institute had such plans.

Dmitry Mendeleyev himself also appreciated the importance of the oil industry. He even went on a reconnaissance to Baku. But it was to Łukasiewicz's refinery that American engineers, sent by John Davidson Rockefeller, were already coming to learn. It was Ignacy Łukasiewicz who the world, fascinated by the prospects for oil expolitation, admired. The opportunity to cooperate with Łukasiewicz seemed to Zglenicki a real one, all the more so because Polish graduates of the Mining Institute in St. Petersburg who were a few years older had already worked in their native lands in the mining profession. Wincenty Choroszewski, the first graduate of the Institute of Polish nationality, became the head of mining plants there. Aleksander Wyrzykowski was the head of the Warsaw Mining District.[309]

308 Nikolai Alexandrovich Figurovski, *Dmitry Ivanovich Mendeleyev*, (Moscow, 1961), pp. 7–19.

309 Jerzy Jaros, *Polacy w Leningradzkim Instytucie Górniczym* [Poles in the Leningrad Mining Institute], Kwartalnik Historii Nauki i techniki, 1972, Vol. 3, e. 505.

Many well-educated Russians were not sent to the Russian Partition because they were severely lacking in the Empire itself. In the territory of the Tsarist Empire, Polish deportees often carried civilisation and culture. Polish doctors, naturalists, geographers, and engineers were often the only educated cultural elite in its endless spaces.

Studies at the Mining Institute in St. Petersburg – considered the most expensive and the most difficult – were elite and rarely anyone managed to complete them on time. Despite the obstacles placed by the partitioners, Poles were given the opportunity to work in the industrial centres of the former Commonwealth, which were developing in an increasingly lively way.

In the footsteps of Stanisław Staszic. Flights and falls

Witold Zglenicki completed his studies in 1875 in first place and, in accordance with the regulations, was sent for an apprenticeship in the Mining Plants of the Eastern District of the Kingdom of Poland. Since 1870 the mining and metallurgical industry in the Kingdom of Poland had been subordinated directly to the Ministry of Finance in St. Petersburg. Now without the facade of self-governance, which just after Russia took over, the Old Polish Industrial Basin had somehow functioned in a residual form.

At the beginning of July 1875, Witold Zglenicki came to Suchedniów in the Kielce Governorate, in the former Eastern District of the Old Polish Basin with a two-thousand-year-old tradition of iron ore mining and metallurgy[310], in the vicinity of which the geology and wealth of deposits was described at the beginning of the 19th century by Father Stanisław Staszic.[311] The Mining Board of this old settlement welcomed him warmly, but the local mining and metallurgical plants were in stagnation due to retaliatory repressions for the participation of the population in the January Uprising. Work in the Mining Management Board – supervision and administration of the plants, correspondence, logistics, planning, technical improvements – became the perfect work for Zglenicki and he was quickly promoted. In the middle of the following year, Zglenicki was appointed the head of the Smelting Plants in Mroczków on the Kamienna River.

Mroczków was an important point in the Staszic Plan from the beginning of the century, which was to comprehensively develop the Zagłębie region. Staszic was an enthusiast of the extension of the Mroczków blast furnace for

---

310 Jan Zieliński, *Staropolskie Zagłębie Przemysłowe* [The Old Polish Industrial Area], (Wroclaw – Warsaw – Cracow, 1965), pp. 79–87.

311 Stanisław Staszic, *O ziemiorodztwie Karpatów i innych gór i równin Polski* [On the Carpathian Mountains and other mountains and plains of Poland], (Warsaw, 1955), pp. 20–70.

iron smelting, which at the end of the 18th century was built by the owner of the settlement, Prince Michał Radziwiłł. Unfortunately, in 1828 Mroczków became the property of the Russian government and the blast furnace was already technologically obsolete. Renovated, it broke down, and although it was restarted, at the moment of Zglenicki's arrival in Mroczków, it was a museum piece and of minor economic importance.

Nevertheless, Zglenicki was full of enthusiasm, determined to make this furnace the leading mining and metallurgical plant in the region. Inspired by the works of Stanisław Staszic, he was convinced that this could be done. He decided not only to modernise the blast furnace, but also to thoroughly research the natural resources of the region. He planned to assess the size of deposits and their suitability for use by industry in order to obtain resources to increase production. In this respect, he also assumed the improvement of the region's communication with an emphasis on clearing the Kamienna River.

He energetically set about the new order. The Mroczkowski blast furnace received an additional cylindrical bellows and a machine for heating air to 350 degrees. Initially, the head of the Suchedniów Mine was in favour of Zglenicki's plan, but it proved impossible to acquire new machines for the wobbling Mroczków plant. So Zglenicki arranged for supplies of old equipment, which the larger ones disposed of, but soon the old equipment ceased to come. For unspecific reasons, because the production at Mroczków was going well. Maybe that's why the overzealous engineer started to be ignored by the management over the course of time, even considered as an importunate fantasist, although in business relations he became known as a perfect employee and loyal to the authorities. He was even promoted several times in the rank hierarchy of the civil administration, but these formal approvals did not support his efforts to better equip his plants.

Despite his absorption with iron ore mining and metallurgy, the thought of crude oil did not leave Zglenicki. From the nearby mine in Bóbrka, although already in the Austrian zone, behind the Russian border of the Partitions, came more and more fascinating news. The oil industry in Subcarpathia had already managed to shock the world.[312]. In the world, moved by Łukasiewicz's achievements, a great race to oil deposits, a new gold rush, was already taking place.

Thus, Zglenicki took up oil exploration in the area available to him. After four years of his own work, following Stanisław Staszic's instructions, he

---

312 Andrzej Chodubski, *O zasługach inżyniera geologa Witolda Zglenickiego "polskiego Nobla"* [On the merits of geological engineer Witold Zglenicki, the "Polish Nobel"] in: Maciej Boryń, Andrzej Chodubski, Bartosz Duraj *Historia polskiego przemysłu naftowego* [History of the Polish Oil Industry].

published in Russian in the St. Petersburg Mineralogical Society a study entitled "*Sources of Oil in the Kingdom of Poland.*"[313] It was in oil that he saw an opportunity for existence in the present situation. A personal and national opportunity.

At the same time, the metallurgy of the Kingdom of Poland faced new challenges. Since the 1860s, production had been increasing, but the demand for iron was also on the increase. The age of steam engines and iron railways was in progress. The iron and steel plants in the Kingdom could not keep up with demand, but Russian law provided for duty-free imports of pig iron and iron products for machinery manufacturing plants. The Kingdom was eventually flooded with imports from Russia, Great Britain and Silesia. The fact that the mining and metallurgical industry of iron ore was still on its last legs here was, unfortunately, only due to the lack of sufficient transport infrastructure.

The collapse of domestic industry was ultimately caused by the parallel step of the partitioner – the privatisation of government plants. Since 1878, the Russian government had made it possible for private individuals to buy out the plants, which, in view of the consequences of the policy of disinheriting, confiscating and looting Polish property for several decades of this century, resulted in a massive import of foreign capital and the transfer of all the larger plants of the Old Polish Basin into its hands. For example, banker Samuel Anton Fraenkel bought the steelworks in Starachowice, Ostrowiec, Machory and Chmielów in the 1870s. In addition, the 1880s brought innovations in the construction of blast furnaces, with which old furnaces could not stand the competition.[314] Without technical investments, Zglenicki's plant in Mroczków also had no chance. The management in Suchedniów put it up for tender, but nobody even applied for the auction.[315]

Zglenicki, still full of faith and energy, intended to take part in this competition, but in his own way. In 1878 he bought smitheries in Błota and Pstążnica for 3,030 roubles. For comparison, the cost of studying at the Mining Institute in St. Petersburg was 100 roubles per year, which was affordable for hardly anyone. However, the owners of local, competitive plants did not intend to allow the enterprising engineer to spread his wings. Before Zglenicki managed to set up his own plants, they slandered him with the accusation that he used the government plant for private purposes,

313 Cf. Witold Zglenickij, *Sources of crude oil in the Kingdom of Poland, Records of the Mineralogical Society in St. Petersburg*, 1880, series 2, Vol. 15, pp. 25–30.

314 Natalia Gąsiorowska – Grabowska, *Z dziejów przemysłu w królestwie Polskim 1815–1918* [From the History of Industry in the Kingdom of Poland 1815–1918], (Warsaw 1985), pp. 184–191, 272–277.

315 *Słownik geograficzny Królestwa Polskiego…*, Vol. 10, 532, Vol. 15, pt. 3, p. 629.

as well as the violation of the Mining Code, which had no longer been in force for years, but this time the authorities dug it out of oblivion. This was the end of Zglenicki's engineering career. The St. Petersburg Mining Board accepted the request of the management in Suchdniów and suspended Zglenicki from his duties for a year, and then, at the request of the head of the Suchedniów plants Józef Lisiecki, Zglenicki was dismissed from his work on 14 February 1884 on charges of appropriation of property.

These allegations were hollow and made to measure. Zglenicki, probably, stood in the way of local, thriving, foreign competition, as well as Russians and their compliant subjects focused on facilitating foreign capital, and not on supporting Polish private industrialists in the Russian Partition.

For a long time Zglenicki could not come to terms with the accusations and exclusion from the profession of miner. He fought to regain his good name for six years. He settled in Kielce and worked in private workshops. Dreams of working under Łukasiewicz's supervision in the Subcarpathia region, where oil extraction and processing flourished at that time, finally lay in ruins. Ignacy Łukasiewicz died in 1882.

Witold's brother Bolesław gave him constant support in his troubles. Witold visited the familial Dębe on the bank of the Narwia every holiday. After the death of his parents, the role of heiress of the estate was assumed by Olimpia Załuska, married to his brother Bolesław, thanks to whom the Polish tradition of their noble nest lasted happily.

The Assayist of Riga

At the age of 40, in 1890, after years of efforts to clear his name and restore his right to practise in the profession, the Mining Institute appointed Zglenicki to work.

However, the Pole was not allowed to stay on Polish lands or work in the mining industry. He was sent to work in Riga[316], in the Assay Office. At that time Riga was the third largest city in Russia after St. Petersburg and Moscow[317] – the world's leading trader in wood, an important shipbuilding town and a former Hanseatic port. Witold Zglenicki had the opportunity to demonstrate his expertise and geological knowledge here. The trade in valuable goods went very well for him. There was no shortage of rich people and foreign, mostly German, capital in the city. The first pride of Riga at

---

316 Zija Ali Buniatzade, *Witold Zglenicki a problem wydobycia ropy naftowej z dna morskiego* [Witold Zglenicki and the problem of crude oil extraction from the seabed], "Notatki Płockie", 1978, Vol. 4, No. 97, p. 31.

317 Cf. Z. Łukawski, *Ludność polska w Rosji 1863–1914* [The Polish Population in Russia 1863–1914], (Wroclaw – Warsaw – Cracow – Gdansk, 1978).

that time consisted of 2 Latvians, several members of different nations, and 64 Germans.

The work entrusted to Zglenicki was an obvious proof of the untruthfulness of the acts he had previously been accused of and the full awareness of the authorities that the accusation of dishonesty was a lie and a slander. The position of head of the Assay Office required absolute trust in professional marking, reliability and professional valuation of goods and valuable items. Assayist Zglenicki soon gained respect, especially among German clients. The Mining Authority in the Russian capital followed his work and, two years later, submitted to him an honourable proposal to become the chief engineer of the Donetsk Mining District, which was already an important industrial centre at that time.

The reasons why Zglenicki refused this nomination from the Russian authorities are not known. In Tsarist Russia, however, the rejection of the proof of the grace of power was perceived as a brazen affront, the antics of an arrogant Polish nobleman, and could have had very dangerous consequences. One hundred years later, in 1984, this was expressed very diplomatically by his biographer:

> "So the idea of sending him to work in one of the northern provinces of Russia was born. The wide possibilities of working in Siberia were taken into account."[318]

However, Zglenicki was not able to employ the wide range of possibilities of working in the Siberian sub Arctic Circle. With the energetic intervention of Prince Benckendorff, one of the most influential and richest industrialists in Russia, a well-trained diplomat, a tsarist friend of the home and future ambassador of Russia in Copenhagen and London, the exile to Siberia was replaced by Zglenicki's being directed to the Caucasus, to Baku.

In Baku, where heaven and hell exist

The Caucasus was a place of deportation of rebellious, cocky civilians and soldiers considered by the Russian authorities to be degenerate units in the Russian Empire. There are known accounts, including the most widespread opinion of the writer Maxim Gorky, of the terrible living conditions of indigenous people living in Azerbaijan at the southern end of the Absheron Peninsula bordering the Caspian Sea.[319] The stinking earth penetrated by oil, slimy mud and polluted, salty water with a repulsive smell terrified the writer. Chekhov was supposed to have said that he would not live there and for a million roubles.

---

318 Chodubski, *O zasługach…*, p. 51.

319 Maxim Gorky, *Collected Works*, (Moscow, 1962), Vol. 2, p. 23.

Azerbaijan was not an unknown land for Polish exiles and travellers.[320] Many of them were fascinated by the world of the East. There are many accounts written since the 17th century by authors enchanted with the nature, the mysterious practices of religion and Oriental cultures created among the phenomena associated with volcanic oil and gas leaks, explosions of fire from the Earth's germinating depths and rocks. In the 19th century Azerbaijan was a melting pot of nations – Turks, Persians, Armenians, Tatars, Englishmen and Russians.

It is curious how much, at the time when Count Benckendorff was going to the Tsarist court in the case of Zglenicki, was Zglenicki's own suggestion in this very direction of soft exile. Azerbaijan, with its 60 km and 30 km long and 30 km wide oil production centre in Baku, located on the Absheron Peninsula and entering the Caspian Sea, became at once a dream land for Zglenicki as a discoverer, geologist, engineer and scientist.[321] The great paradox in the story of this Polish penal exile to the "land of eternal fire", which in the future was to distance professional academics, cartographers, mining experts and geologists many times, was the fact that for the rest of his life he worked here as a Assayist.

Witold Zglenicki received a position in the Assay Office in Baku at the end of 1891. This was a time when the southern part of the Absheron peninsula was experiencing a period of rapid development of the oil industry, which had started here about twenty years previously, at a time when the Russian authorities opened their doors to foreign private capital, when the Bóbrka mine and Łukasiewicz's first distilleries were celebrating their twentieth anniversary.

In the 1890s Baku, which at the beginning of the 1870s started from the level of prefeudal culture, had already been led by world tycoons. Three brothers Alfred, Ludwik, and Robert Nobel had been doing business there since 1873, when they founded a paraffin factory. Over time, the thriving Nobel Brothers Naphta Company, based in St. Petersburg, was founded. Drilling shafts in Baku grew like mushrooms after the rain.

When in 1873, i.e. 19 years after Łukasiewicz had excavated the first shaft in Bóbrka, there were only 9 shafts in Baku, in 1900 – already 1,710. Technologies had been improved, pipelines had been built, and tanker transport methods had been developed. Instead of barrels, river and sea tankers appeared. The purchase of oil-producing plots continued, and new types of industry were created, not only related to oil. At the time of Zglenicki's

320 Cf. M. Gralewski, *Kaukaz. Wspomnienia z dwunastoletniej niewoli* [Caucasus. Memoirs from a twelve-year captivity], (Lviv, 1877).

321 P.G. Bałajew, *Crude oil in the Land of Fire*, (Baku, 1969), p. 118.

Figure 27: Witold Zglenicki. Source: The archive of the Płock Scientific Society

arrival, Baku already accounted for 50 % of the world's output and 95 % of Russia's output. It was one of the three world mining centres next to the Galician and United States basins. For many people it became a city of fast careers and roads to fortune.[322]

Alfred Nobel[323] – a brilliant self-taught inventor and world tycoon from a Swedish family settled in Russia since 1837 – was the number one person among Baku oil workers. In Baku, the permanent representative of Nobel companies was Ludwik. Until the second decade of the 20th century, until the revolutions broke with private property in Russia, the Nobels were the most powerful oilmen in Azerbaijan.

During Alfred's stay in Baku, Witold Zglenicki and Alfred Nobel became friends. Their property status was incomparable, but they shared common interests – an idealistic cult of science, technical skills and mutual respect. Unmoved in his principles, the pedantic Zglenicki valued Alfred Nobel not

322 *History of Azerbaijan*, (Baku, 1960), Vol. 2, pp. 204–206.
323 E. Bergengren, *Alfred Nobel*, (London, 1962), pp. 76–84.

only for his talent, but also, contrary to the pronouncements about exploitation by the Nobels, for his modesty and care for his employees, care for good working conditions, housing and construction of schools and hospitals. Personally, Alfred Nobel spent his life travelling, never realising his dream of settling permanently in Sweden. He died in Italy, five years after Zglenicki's arrival in Baku, leaving the world with his unique legacy – a fund to reward the most outstanding people in science, culture and peaceful politics. His nephews tried to overthrow this record, but the Nobel Foundation has been fulfilling the will of its founder since 1901. The example of the charismatic Alfred Nobel made an impression on Zglenicki.

The Assayist Engineer

Witold Zglenicki did not waste his chance in the Azerbaijani Eldorado. He enjoyed the opinion of being an extremely dutiful, reliable and service-minded head of the Governor's Assay Office. In particular, he gained extraordinary trust from the Muslim population. He valued the nations of the Orient for their artistic sense and, in his opinion, the sense of aesthetics they shared with Poles, which he had the opportunity to learn about every day at work.[324] He was housed in a rather prestigious apartment in the centre of Baku, in one of the new buildings which, with the development of the oil basin, began to change the face of the city with their more European style. He earned well, lived ascetically, below the level of his income. His luxury was his geological passion and he spent much time, strength and money on it.

Already at the beginning of his stay, he came into contact with the Ukrainian family of the official Winogradov, who admitted to his Polish roots. In 1896 the son of Zglenicki, Anatol, was born from a romance with Maria – daughter of Winogradov, but despite pressure from Maria's family, Zglenicki did not decide to marry, never settled with Maria, or even betrayed to his family at home the fact that she and his son existed. Without losing hope of returning to his homeland, he wished his son to grow up in Warsaw, in the Polish tradition.

Learning and working, not throwing a hat

Zglenicki, with the pedantry with which he ran the Assay Office in Baku, on the striking of the clock indicating the end of work, left the office and moved towards his passion. It was his response to the challenges he was facing – as a miner's engineer and geologist – from the place where he found himself.

---

324 Cf. Stanisław Brzezinski, *Jaką drogą dostała się do polski sztuka perska?* [What was the Persian Art's Route to Poland?], (Warsaw, 1935).

Geological research and exploration work in Azerbaijan was already carried out on a significant scale. Since 1868, the Caucasian Geological Office had systematically printed the results of this work. In the *"Materials of Geology of the Caucasus"* the entire Areopagus of Russian scientists published. The big fish in the oil industry impatiently demanded information about the fields, also employing crowds of private geologists.

However, it was Zglenicki, who did his work as a hobby and on his own account on the basis of the analysis of the available works, the Caucasus mountain-forming patterns in terms of their history, geological structure and, above all, thanks to his own research and measurements in the field, who was the author of the first geological map of the Absheron Peninsula in the scale 1:420,000. Maps drawn since the 1840s by Polish exiles from previous generations using the triangulation method – Józef Chodźka[325] and Ignacy Stebnicki – proved to be particularly helpful in this work.[326]

Zglenicki meticulously penetrated the area of about 440,000 m2 of Azerbaijan. He analysed the geomorphology of the area from the Black Sea to the Caspian Sea, and in a north-south line from the Kuma and Mancha rivers to the Urmia Lake. The eastern ends of the Caucasus and the most oil-bearing area of the Absheron Peninsula were examined particularly thoroughly in terms of oil and gas content by Zglenicki. He personally traversed this peninsula hundreds of times in all directions.

At the end of the 19th century, the Absheron Peninsula was already home to three main oil production areas based on natural leaks from the interior of the earth: the Sanbuczyn region with its centre in Sanbucz, the Surachanski region around Surachan and the Bibi-Heybat region with the capital in Baku. In his search for and marking of the deposits of natural resources, Zglenicki did not limit himself to marking leaks seen with the naked eye, but used a complementary, analytical scientific method. He drew hundreds of maps, carried out multiple measurements, analysed, prepared and studied geological samples.

As an expert geologist, he was interested in all the natural resources of Azerbaijan. He also added minerals to the maps, but, in contrast to his work on oil and gas deposits, he kept these results to himself. He kept these maps

325 E. Berezowski, *Józef Chodźko – geodeta i geograf XIX wieku* (1800–1881) Józef Chodźko – surveyor and geographer of the 19th century (1800–1881), "Przegląd Geodezyjny" 1970, Vol. 10, pp. 430–431., Cf. A. Furier, *Józef Chodźko 1800–1881, Polski badacz Kaukazu* [Józef Chodźko 1800–1881, Polish researcher of the Caucasus], (Warsaw, 2001).

326 Cf. Andrzej Chodubski, *Aktywność kulturalna Polaków w Azerbejdżanie w XIX i na początku XX wieku* [Cultural activity of Poles in Azerbaijan in the 19th and early 20th century], (Gdansk, 1986).

and rich mineralogical collections at home, not betraying them to anyone, as if these findings were his most precious secret.[327] And the Caucasus is rich in iron ore, alumina, coal, copper, rock salt, manganese, barite, arsenic, as well as silver and gold.

The Polish engineer was entertained by diviners brought in large numbers by hungry oilmen from all over the world, most often from the USA, Pennsylvania. He claimed that while searching for oil and gas with a magic diviner's rod, they could equally successfully throw a hat and dig where it falls. He was a Warsaw positivist and relied on science.

The inventor

Count Benckendorff certainly never regretted his efforts at the tsarist court to save Zglenicki from Siberian deportation. In Baku, apart from his professional work, Zglenicki spent most of his time on private field research and mapping Azerbaijan's natural resources, but he also gained the reputation of an irreplaceable expert in mining techniques – drilling, transport and processing of crude oil. He never ceased to be a miner by passion, education and years of work experience in the industry of the Kingdom of Poland.[328] In Azerbaijan, he became a permanent consultant at the Benckendorff refineries and, over time, his partner. He also consistently advised several other friendly oilmen, including the Rothschild and Nobel brothers.

He was involved in making improvements to drilling technology and made a number of significant improvements in this area. At that time, the old methods of extracting oil from hand-excavated wells stopped being profitable. Technical innovations were very much sought after and immediately taken up, as the depths of the shafts were already considerable, reaching several thousand metres. The situation urgently required the invention of a way to keep the shaft walls as perpendicular as possible, both for reasons of efficiency and safety.

To control the curvature of the drilled shafts, the famously advertised as a miracle of the latest technology, the so-called Fernshtrem apparatus, was used at that time. In 1893, two years after his arrival in Baku, at a meeting of

327 Jakub Chojnacki, *O utrwalanie pamięci wybitnego Polaka inżyniera górnika – geologa Witolda Zglenickiego* [On preserving the memory of the outstanding Polish mining engineer – geologist Witold Zglenicki], "Notatki Płockie" 1978, Vol. 4, No. 98, pp. 23–30.

328 Z. Wójcik, *O niektórych polskich geologach i gleboznawcach kształcacych się bądź pracujacych na wyższych uczelniach Rosji* [On some Polish geologists and soil scientists studying or working at Russian universities], "Studia i Materiały z Dziejów Nauki Polskiej" 1976, series C. n. 21, pp. 64–67.

the local branch of the Russian Technical Society, the Pole proved, however, that the effectiveness of the Fernshtrem apparatus was a fake and presented his own design for such a device[329].

Baku oilmen soon found out that Zglenicki's apparatus was excellent, and that it not only detected curves and accelerated the pace of work, but, very importantly, it provided the opportunity to minimise fire catastrophes and surprise oil and gas explosions.

Desiderata – the key to the unnoticeable Aladdin's cave

The axis of Zglenicki's geological passion in Baku became the study of coastal areas and the area of the Caspian Sea itself. It took a long time to stop treating these explorations as the unreal delusions of a fantasist, but the Polish engineer was imperturbable. His independent research indicated that the most valuable and abundant deposits of oil and gas were located in the coastal areas. So he decided that it was worth exploring both the islands and the seabed. The observation of the water table confirmed the geological knowledge.[330]

It is hard to resist the impression that in this last decade of the 19th century Zglenicki was subconsciously already racing against time. Led by passion, on the wings of sensational results, he often worked around the clock. On his own, at huge cost, he engaged several research teams to help, which were active in many places at the same time. He carried out a thorough search in the Absheron Archipelago Islands and in the Gulf of Bibi-Heybat. He made a dizzying number of sketches of all islets and rocks protruding from the sea. He marked and described unusual phenomena on the surface of waters, underwater volcanoes, gas outlets. He spent nights on a boat on observations and measurements.

As a result of these searches, on 29 July 1896 an unusual request for the allocation of two sea fields in the Bay of Bibi-Heybat was received by the Office of State Resources of the Baku Governorate.[331] The same letter was sent to the Mining Department in Baku on 3 October 1896. The author of the letters – engineer Witold Zglenicki – expressed his eagerness to

329 W. Konstantinowicz Zglenickij, *Information at the meeting of the members of the Imperial Branch of the Russian Technical Society in Baku about the device for measuring curvature.* Papers of the Imperial Branch of the Russian Technical Society in Baku, 1893, September-October, p. 73.

330 Andrzej Chodubski, *Udział Polaków w poznaniu Morza Kaspijskiego* [The Participation of Poles in Exploring the Caspian Sea], "Nautologia" 1988, nr 4, pp. 3–10.

331 Chodubski, *Udział Polaków...*, p. 66.

thoroughly examine the seabed resources and to build oil wells directly at sea. In an attachment, the engineer presented a technical design for the offshore drilling rig and a description of the method of transmission of oil. The project foresaw the construction of a watertight platform at a height of 12 feet above the water level at the site of the oil eruption. The oil was to be transported from there to specially adapted oil barges or large iron tankers.

Zglenicki's first desiderata from 29 July 1896 is material evidence, a document of pioneering rank and the world artefact of a breakthrough moment for civilisation. And yet, at the turn of the 20th century, discoveries and projects even in Baku, one of the most advanced oil fields in the world at that time, were ignored, judged to be extravagant and unreal. The first two petitions and the drafts submitted were not approved by the authorities.

Zglenicki, confident of his accuracy, trekked a long way through bureaucratic obstacles, dogmatic, conservative opinions and shallow assessments of his concepts. He submitted his application to the Ministry of Earth and State Resources in St. Petersburg.[332] He aroused some interest, but here too officials would send back his writings and charts. The opinions of the expert committee of the Mining Department were multiplied. The judgments were negative, although the papers and arguments of the Polish engineer gradually began to circulate, creating an intriguing aura around the original proposals. They shocked, but also aroused interest. The lists of geological resources, both onshore and offshore, documented by Zglenicki were extremely detailed, reliable projects. No expert could deny the professional methodology and outstanding level, which led to the experts being even more confused. In black and white they led to the conclusion that the most valuable oil deposits were to be found in the coastal areas and under the Caspian Sea bottom.

And yet they will make the sea sleep

After two years of official resistance and unsuccessful efforts with the authorities, in February 1898, Zglenicki once again asked the Caucasian Mining Board for permission to erect a drilling rig at sea and again unsuccessfully. However, the news about the stubborn efforts of the Pole had already become famous among oilmen, all the more so as the authorities started to consider at that time enabling not only the lease but also the purchase of oil fields by private capital.

In July 1900 Zglenicki presented a detailed paper at the meeting of a special committee of the Mining Board. He discussed his plans and methodology for

332 Chodubski, *Udział Polaków…*, p. 67.

extracting oil from the seabed using drilling rigs on piles. The Commission tended to recognise the results of Zglenicki's geological explorations, but once again concluded that the concept of offshore oil extraction was unrealistic. Experts from the committee pushed for the idea of backfilling the Bay of Bibi-Heybat and exploiting the oil from the artificial land thus obtained. This option was supported by the St. Petersburg Ministry. The decision to backfill more than 330 hectares of the water table in the Bay of Bibi-Heybat would be taken one year later, but only five years after that would an international competition for the project be announced. In 1909 the primitive backfilling of the Bay would begin.[333] In 1910, when Zglenicki had long passed on, the expert Polish engineer Paweł Potocki would take the matter into his own hands.

Zglenicki in the meantime had proven gratis that this was the wrong, risky way. The cost of backfilling the Bay planned by the government was in millions. According to Zglenicki, it would be much higher. In addition, the plan did not provide for additional natural phenomena occurring in marine conditions, such as precipitation of water level, tides. The arguments did not change the position of decision-makers. The bay would be backfilled.

With time, the Russian authorities would allow Zglenicki to be glorified in the literature. In an illustration for a short story entitled "*The Dreamer from the Assay Office*"[334] by Lew Połonski there is a distant figure of a lonely passer-by. But Zglenicki, although he was an idealist and visionary, worked all his life with the sober, iron consistency of a realist and positivist practiser, and it was his free intellect as a scientist and researcher that allowed him to reach beyond the horizons of stereotypes, to uncover before people, and give them a world of inexhaustible wealth that they had not noticed before.

165 oil-producing fields, marine fields, platforms and breakwaters

The specific atmosphere of the Promised Land of Baku, the spontaneous development of the region, did not give peace to the community of private oilmen. The sharks of the local industrialists' community were increasingly voicing their fears that, due to the publicity of Zglenicki's revelations and the fact that in May 1900 the regulations on the possibility of buying oil-bearing fields on the Absheron Peninsula and its surroundings came into force, they might lose their priority and the chance to participate in new initiatives.

333 P.M. Lisiczkin, *Distinguished activists of national science and technology related to oil*, (Moscow, 1967), p. 227.

334 Cf. L.A. Połonskij, *Absheron Threshold*, (Moscow, 1966), pp. 31–40.

Immediately afterwards, in the same year, the entrepreneurs associated in the Council of the Congress of Baku Petroleum Engineers asked Zglenicki to delineate the 100 most promising oil fields on the Absheron Peninsula. Within two weeks, Zglenicki had made available a detailed analysis of the oil-bearing capacity of the whole Peninsula area and published it in the journal "*Nyeftianoye Dielo*"[335]. It was a sensational phenomenon, because the Pole not only made the results of his work of many years, long and at great, private expense available, but also made a professional division and evaluation of resources and did it for free, for the public good. Private geologists imported to Baku from all over the world for their expertise in the fields offered for sale were receiving great gratuities. And the competition of private investors became increasingly fierce, all the more so because the Russian state, which even at the end of the 19th century forbade private research on oil deposits, had not conducted such research so far, despite the spontaneous growth in income from the oil industry.

Zglenicki's study was the world's first scientific work of this type and has been used continuously throughout the 20th century. Zglenicki included 165 plots of land allocated by him on the Absheron Peninsula, taking into account the oil-bearing capacity of their area and in connection with gas resources and other geological phenomena. He additionally separated 30 fields in the surrounding counties, 15 fields in the Puta Lagoon and 20 fields in the Caspian Sea in the richest areas of Bibi-Heybat.[336]

He described and distinguished more than thirty geographical regions as part of the study, taking as a criterion the division of deposits that occurred in them. He defined specific areas, towns and villages, and the types of resource that occurred. He also included mud volcanoes in his research. He did not agree with the general opinion that they were destroying oil deposits. On the contrary, he claimed that they indicated them. He presented rich material, supported by abundant documentation, of the geological structure of areas and deposits.[337] He advised that exploration work should be carried out first on the Peninsula, as a particularly resource-filled area. However, in an area outside the Peninsula with smaller mineral deposits, he also indicated precisely the places that were favourable for exploitation in the long

---

335 Witold Zglenickij, *About towns on the Apszeroński Peninsula and outside its borders, where plots for oil exploration should be created, on the basis of temporary rules approved by His Highness on May 14, 1900,* Crude Oil Business, Vol. 15 of 20.08.1900, p. 837.

336 P.M. Lisiczkin, *Distinguished activists of national science and technology related to crude oi*, (Moscow, 1967), p. 227.

337 Chodubski, *Udział Polaków…*, pp. 69–72.

term. He proposed a concept of allocating fields in a way that was easy to merge within the investment of a single owner.

Above all, this work is groundbreaking on a historical scale due to the fact that it included a thorough analysis of the oil-bearing capacity of numerous islands and the Caspian Sea seabed. Tirelessly, he publicly argued for it to be explored by means of platforms fixed on piles. First, at least experimentally, on fifteen fields. He proposed to build a breakwater in order to protect the platforms against sea winds. He pointed out from which volcano, and from which slope, stones should be obtained to build the breakwater. He calculated that the construction of the entire investment would take one year. He was invariably critical of the concept of artificial land.

Fields according to Zglenicki

Success came with the beginning of the new century. In 1901, the special committee of the Mining Board, which had met for the nth time, accepted Zglenicki's report as an official plan of oil fields without corrections. [338]The wolfish appetites of the Baku oil sharks, impressed by the new perspectives, had reached their zenith.

When the Committee of the Mining Board published the list of oil fields allowed to be exploited by private investors, this decision became, in a rather predictable manner, first of all, the prey of the existing oil Croesuses. New fields were acquired by the Nobel brothers, the Rothschilds, and other local industry leaders. Six fields were also won by a Polish company. The Rylski millionaire family was known in Baku for its charity work and Polish employees. The Rylskis were one of the few families with whom Zglenicki, appreciating diligence and honesty towards their employees, maintained a social acquaintance in Baku.

The activity of the Polish engineer gained him authority and publicity. Grateful for his geological work in Persia, he was also valued by the Persian sheikh of that time. The ruler asked the tsar of Russia for permission to award Zglenicki the highest Persian Order of the Lion and the Sun.[339] Zglenicki was a Russian citizen and still an official of the Russian Assay Office. The tsar gave his consent.

---

338 Andrzej Chodubski, *Udział Polaków w poznaniu Morza Kaspijskiego,* "Nautologia" 1988, Vol. 4, pp. 3–10.

339 Liebfeld, *Działalność inżyniera Witolda Zglenickiego w Baku w latach 1890-1904 i jego zapis testamentowy dla Kasy im. Mianowskiego* [The activity of engineer Witold Zglenicki in Baku in the years 1890–1904 and his testamentary record for the Mianowski Fund], "Kwartalnik Historii Nauki i Techniki" 1977, Vol. 2, p. 314.

Zglenicki received two fields near Surachan from the Russian authorities in response to his petition and as a token of appreciation for his merits. One for his exclusive use in Karachchuri and after a year the other, to a company with Count Benckendorff in Chile. At the same time, he received carte blanche for conducting geological surveys in every area and for every kind of investor – private and public.

The first maritime field, or when there are enough important matters on land

The following year 1902 brought Zglenicki his most beautiful reward – the awarding of the most desirable for years sea field in the Bay of Bibi-Heybat, No. 29, together with the permission to build a breakwater with an area of about 220 hectares.

Soon other land allowances appeared for him: four fields in the Baku Governorate and two with deposits of copper ore and Glauber's salt outside the border of the Governorate.

However, in order to implement the projects, it took a lot of money to build the plants. Zglenicki started to obtain it. He was particularly keen to launch extraction from the bottom of the Bay of Bibi-Heybat, but it was this project that was the least appealing to the imagination of potential lenders. His Rothschild friends disappointed him. Many potential lenders offered Zglenicki large sums for buying back fields on land, even without checking and measuring their resources, but Zglenicki refused. The Rothschilds proposed serious sums only for their cooperation in searching for other fields, but he was no longer interested in this. He wanted to start extracting oil from his desired sea field as soon as possible. To prove at all costs that this was what made the greatest sense.

"There are quite a few other important matters on land,"[340] he heard from one of the biggest millionaires in Baku, and in fact everyone around him thought so.

Zglenicki's expeditions to Poland, London and Paris to gain funds were also fruitless. At the same time, there were envious rumours that he wanted to earn with other people's money. Deprived of the possibility of realising his large investment, Zglenicki launched exploitation of fields in Karachchuri and Chile for smaller expenditures.

---

340 Andrzej Chodubski, *O zasługach inżyniera geologa Witolda Zglenickiego "polskiego Nobla"* in: Maciej Boryń, Andrzej Chodubski, Bartosz Duraj, *Historia polskiego przemysłu naftowego*, (Toruń, 2014), p. 15.

Technology, science and art

From the end of the 19th century, there had been a branch of the Russian Technical Society in Baku, which, in the spirit of the times, organised assemblies, congresses and meetings of the engineering community. In Baku, technical progress in the refining industry was the focus of interest.

From the beginning of his stay in Azerbaijan, Zglenicki had taken an active part in the work of the Society.[341] He did not miss a single meeting and was head of the editorial office of the magazines *"Trudy"* and *"Problemy nafty."* Here he found a friend – Włodzimierz Abramowicz – an enthusiastic metallurgist from the Kielce region, which Zglenicki remembered with great longing despite his professional problems. Neither Pole lost hope in a return to their homeland and working for it. They considered the Polish land to be rich in opportunities for industrial development. They tried to involve as many of their compatriots in Azerbaijan as possible in the structures of the Society.

Zglenicki never used the Society's activities for his own purposes. He paid particular attention to safety at work, taking the example of the Nobel brothers in terms of caring for people. He devoted himself to the organisation of the Society's chemical laboratory. He developed a passion for improving technology. He believed that setting new paths for the development of science, living with no mental limitations, beyond the horizons of everyday life and organising a centre of modern information for industry and science should be the priorities of work for the future.

With a typically Polish, 19th-century respect, proved in the depths of Russia by thousands of deportees, most of whom were Polish elite by birth, class and education, he observed the people with whom he met every day in his Assay Office. He repeatedly postulated the organisation of artisanal schools for the local youth, and in his opinion, the love of beauty and artistic skills shared with those features of Poles led him to believe that it was necessary to give the nations of the Orient a chance to learn in the professions of stonemasons, goldsmiths, jewellers and armourers. It is highly probable that this idea was inspired by the schools of lace-making and artistic crafts created at the end of the 19th century in Polish Podhale on the initiative of Dr. Tytus Chałubiński, who, as a student of the Main School, Zglenicki could not fail to have known in Warsaw and admired during his visits to his homeland.

---

341 Andrzej Chodubski, *Działalność cywilizacyjna Polaków na Kaukazie* [The Civilizational Activity of Poles in the Caucasus], "Cywilizacje w czasie i przesterzeni" 1995, Vol. 1, pp. 65–77.

No happy ending

In 1901, at the age of 51, when Zglenicki, as a consequence of his titanic work, climbed to the heights of popularity as the authority on geology and oil mining, the owner of over a thousand hectares of oil-producing fields and over two hundred hectares of marine fields, a diagnosis came, which at that time meant a death sentence – diabetes. Zglenicki did not give up for a long time. He even worked harder.

Three years later diabetes beat him. In the spring of 1904 he could only lie in bed. On July 3, he summoned the titular counsellor Józef Wojewódzki and three witnesses to dictate his last will, a few versions of which he had prepared earlier. He died three days later, on July 6, 1904.

His body and information about his will was brought to Poland by Maria Winogradov[342] According to the wishes of the deceased, he was buried in his native land, next to his parents and brother Bolesław, in the cemetery in Wola Kiełpińska near Zegrze nad Narwią.

After the legal registration of the will and court approvals, the chancellery of the Governor General of the Kingdom issued a declaration of its validity on 27 January 1905. As the executor of the will, Zglenicki had appointed a friend from his youth, who had visited him in Baku and became familiar with his affairs – Władysław Smoleński, a lawyer from Warsaw. He believed that he would be an honest mediator among the heirs.

A legacy "for everlasting times"

During her stay in Poland, Maria Winogradov is supposed to have told the Zglenicki family that Witold's last will was strange, but in terms of records she never opposed his will, although at that time in Azerbaijan, in the land of unbelievable oil fortune, there were no philanthropic traditions. Zglenicki's testament was as unique as its author. It was a testimony of respect for the most important values, love for the people and places where he had lived, and reverence for science, culture and religion. Zglenicki did not forget about Płock and the lands of the Kingdom of Poland, nor about Baku, where he felt good both among the Polish community and the indigenous people.[343]

Zglenicki donated the proceeds from the sale or exploitation of the indicated oil-bearing plots to Maria, her brothers and son Anatol, the four children of his brother Bolesław, who had died in 1897, his godson, as well as to a large number of former collaborators whom he had employed in private

342 Andrzej Chodubski, O zasługach…, pp. 16–33.

343 Andrzej Szczypiorski, "Bajeczna opowieść o złocie i miłości" [A Fairytale about Gold and Love], *Polityka* 5 July 1969, Vol. 27.

geological work and in the State Assay Office. He established, among other things, a compensation fund for the Office's foremen. He did not forget about the doctors and many people with whom he had come into contact.

His valuable mineralogical collection and his own library with all available publications on geology and a set of Polish classics, on which he never regretted a penny, he gave to the Technical Society of Baku. The oilfields donated to the Society brought huge income in the future.

Significant amounts, together with detailed recommendations, Zglenicki bequested for the purpose of creating artisanal schools in Baku, "where craftsmen of silver and gold products with oriental taste should be educated," as he dictated. He provided funding for the establishment and maintenance of ten such schools in the Kingdom of Poland[344] which, however, for complex reasons, did not come to fruition in the future.

In his will he bequested, among other things, serious sums to the Płock Catholic Society, the Catholic Society in Baku, as well as 30,000 roubles for the construction of a Catholic church in that city, which was built, albeit with time, by historical judgment, it fell into disrepair.

The first paragraph of Witold Zglenicki's first will read:

> "The income from the half of the plot of land given to me and Aleksander Benckendorf on the basis of the Najw. Bay. 14 May 1900 the temporary regulations on the release without auction of plots of treasury land containing kerosene for exploration and extraction of kerosene in a town near the village of Surachany in the Babiński district, mentioned in Section 8, par.1 of the same regulations, I bequesth to the Mianowski Fund in Warsaw, on condition that the fund does not sell its rights to the income, but uses it, as it receives it, for eternity." [345]

Section 8 of the Testament also spoke about the Mianowski Fund:

> "If, after the satisfaction of all the above mentioned legacies, there will remain sums which are free to be disposed of, then I bequeath them to the Mianowski Fund in Warsaw for the creation of an endowment, provided that a percentage of this capital is used to issue prizes, at the discretion of the Management Board of the Fund, for the best work of art, literature and science of the Nobel Prize type."[346]

---

344 Cf. *Źródła wschodnie w Archiwum Koronnym do dziejów orientalizacji smaku artystycznego w Polsce XVI – XVIII w.* [Eastern Sources in the Crown Archive for the History of Orientalisation of Artistic Taste in Poland in the 16th-18th centuries], (Warsaw, 1946).

345 Adamczewski, *Zapis testamentowy Witolda Zglenickiego dla Kasy im. Józefa Mianowskiego, Studia z Dziejów Rosji i Europy Środkowo Wschodniej* [Testamentary recording by Witold Zglenicki for Józef Mianowski Fund, Studies in the History of Russia and Central and Eastern Europe], PAN, XLVIII, p. 190.

346 Adamczewski, *Zapis testamentowy...*, p. 190.

After the legal arrangements made by the Mianowski Fund, it turned out that thanks to these provisions, for the benefit of Polish artists of culture and science, the income from the following areas would consist for "Eternal Times":

- half of the oil-producing field No. 8, located between the villages of Surachany and Amiradzan in the Baku area, with an area of 37.5 tithings.
- from oil field 29 in the Gulf of Bibi-Heybat on the Caspian Sea, together with the family heirs, as the first to be satisfied
- four plots of land with a total area of 13 tithings and 680 fathoms in four towns of the Bakinese Governorate
- from a plot in the Zakatal region with copper resources
- from a parcel in the Szemachi district with Glauber's salt deposits [347]

The "Eternal Times" for the Polish heirs lasted about ten years.

Shafts in the Gulf of Ilyich Lenin

Scoffers, willing to joke about Zglenicki's alleged megalomania, comparing his legacy to Alfred Nobel's foundation, soon had to admit that Zglenicki – a man of science endowed with "sage's glass and eye", indeed – knew with 100 % certainty what value his oil and other fields, rich in mineralogical resources, represented. He did not doubt to his death that extracting oil from the rich seabed was a matter of the near future.

The state project of backfilling a part of the Bibi-Heybat Bay was finally carried out in Soviet Russia. In 1923, after the construction of a drilling tower on the artificial land from a shaft 460 metres deep, a river of crude oil broke out of the Caspian Sea bottom. It happened in a different world, in different realities, after the terrible nationalistic events in Baku, massacres, battles and accidents of the First World War and the bloody revolution. The Polish engineer, Paweł Potocki, the chief constructor of the artificial land, lived to see pioneering fame as an oil extractor from the seabed.[348] Despite the loss of sight, he did not interrupt the construction of infrastructure in the Bay of Bibi-Heybat. Respected by the Russians, his praises were sung by their writers. On Potocki's tomb, erected in a prestigious point of the land of which he was the creator, an inscription was engraved in Russian with the then current patron of the bay:

347 Adamczewski, *Zapis testamentowy...*, p. 191.

348 Andrzej Chodubski, "Paweł Potocki (1879–1932) – pionier eksploatacji ropy naftowej spod dna morskiego" [Paweł Potocki (1879–1932) – A Pioneer of Crude Oil Exploitation from under the Seabed], *Kwartalnik Historii Nauki i Techniki* 1988, Vol. 4, pp. 981–994.

Figure 28: An oil rig on the Caspian Sea. Source: © Shutterstock.com/aquatarkus

> "Eternal glory to a talented engineer – an oilman, the pioneer of the Ilyich Bay backfill."[349]

To the construction of a drilling platform according to Zglenicki's idea, without artificial land, and on the high seas, mankind had to mature for decades until 100 km from Baku in the Caspian Sea, the Neft Daslari platform was built in 1949. In 1954, Americans placed their first platform, Mr. Charlie, in the Gulf of Mexico.

And the oil is still pouring

During World War I, Baku was the front for fierce battles intent on taking over the oil. In Azerbaijan, warfare was waged by Russia, Germany, Great Britain and Turkey. Particularly determined, Germany used the tactic of an alliance with Turkey. They inspired the Muslim element, which was to conquer Russia, leaving areas rich in oil for the Germans.

The last sums from Zglenicki's oil-bearing plots in Azerbaijan reached Poland at the beginning of 1918. During the war they had been coming slowly, but they only stopped under the leadership of Lenin's revolutionaries.

349 Chodubski, *Paweł Potocki...*, p. 68.

Zglenicki's material legacy, the achievements of his geological work, collections and, above all, sea and land fields, were seized for the benefit of Bolshevik Russia.[350]

The attempts by the Zglenicki family and the Mianowski Fund to regain the benefits of the areas of exploitation of natural resources misappropriated by the USSR are a complicated story which, without effect, dragged on through the time of the Socialist People's Republic of Poland to the completely ignored interventions of the post-communist general and president in one person, Jaruzelski. For the Zglenicki family the matter turned out to be particularly painful, and the method of execution of Witold Zglenicki's will, including its management by the Mianowski Fund, almost from the very beginning, unacceptable.

The family heirs of Witold Zglenicki, despite the efforts and trials, received a tiny part of the amounts due to them under his will. The price of the defeat very quickly included even the loss of the family estate of Dębe in the vicinity of Zegrze. The family representative Władysław Smoleński, who during Witold Zglenicki's lifetime seemed to be his most trustworthy friend, took over Dębe as compensation for exceeding the loan repayment deadline. He gave it to one of his family heirs to go to Baku in order to personally discern matters and start his own business, which ended in failure. The fields Witold Zglenicki, the brilliant treasure hunter, turned out to be too money-rich, too gushing of geysers of crude oil, too immeasurable a reservoir of natural gas and many other natural resources, for it to be otherwise possible in Soviet Russia.

At the beginning of the century, even before the revolution of 1917, as a result of which a new Bolshevik state was created, legal representative Wladysław Smoleński began the implementation of Zglenicki's will by starting the sale of plots of land, including those that the testator forbade to sell. In 1908 Maria Winogradov came to Poland with an alarm that Witold's property was being wasted. Smoleński decided to sell the valuable sea field no. 28 to the "Caucasus" Oil Society. A very long and serious dispute arose from the lease of plots of land by the Mianowski Fund through Władysław Smoleński to the Caspian-Black Sea Society, which was a company of the Rothschild millionaires. In the family's opinion, it was more advantageous to lease them to the Polish company of the Rylski family.

There are hints and calculations which indicate that in Poland, both for the benefit of Zglenicki's heirs and the Mianowski Fund, a total of 20 % of the amounts due in the period up to 1918 were received. However, until the Mianowski Fund, although not without perturbation during the war and

350 Chodubski, *O zasługach…*, pp. 30–31.

revolutionary battles, received cash, and so far exceeding its wildest dreams and needs, Smoleński's work was not accepted by it. Until 1918, under foreign rule, the finances of the Fund were conducted half conspiratorially. The Fund's Committee tried not to advertise the state of its finances, as well as the amount of scholarships and the number of scholarship holders and various benefits, explaining this by fear of the Russian authorities, confiscation, liquidation of the Fund and harassment of the beneficiaries.

After 1920, in the period of the Second Polish Republic, in free Poland, the Mianowski Fund continued its activity and its management was taken over by the next generation. The generation of Zglenicki's Main School crumbled and lived its days in the circumstances of the First World War. In the period between the World Wars of the 20th century, a new statute was drawn up for the Fund. The Fund gained a Scientific Council, a printing house, houses of creative work and organised scientific conventions. New donors arrived, including property owners. The bequest, which in a fraction would be equal to Zglenicki's will, never reached its target again.[351]

Meanwhile, after Zglenicki's assets had been taken over by Soviet Russia and intensive exploitation of crude oil and natural gas had started, the amount of raw materials extracted from its fields increased over ten times.

On an archival postcard from 1932, preserved as a reminder by one of the guests of "Mądralin" – as one of the houses of creative work of the Mianowski Fund was named – an appeal for payment was printed, including the words:

> "The aim of the Mianowski Fund is to support Polish creativity by granting benefits to scientists and institutions for research and scientific publications, and in general by organising scientific work. During its 48 years of existence, the Fund has spent about 6,000,000 Polish zlotys and printed more than 1,000 volumes of scientific works. (...) From 1917 the main income from oil fields in the Caucasus ceased to exist. (...) This happened at a time when the needs of Polish science in independent Poland increased. (...) Warsaw, Staszic Palace."[352]

Negotiations with Russia on the restoration of the Caucasus shares were then conducted at the interstate level, but the Fund was still in legal dispute with the Zglenicki family heirs. The Rothschild family lawyer came to Warsaw for the hearing as their joint plenipotent with the Fund. This is probably the reason why for the fact that the name of Zglenicki did not even

351 Z. Szwykowski, *Zarys historii Kasy im. Mianowskiego* [Outline of the history of the Mianowski Fund], in: *Nauka polska i jej potrzeby, organizacja i rozwój*, Vol. 15, (Warsaw, 1932), p. 135.

352 *Kasa Mianowskiego 1881–2011* [The Mianowski Fund 1881–2011], ed. Zasztoft Leszek, (Warsaw, 2011), p. 467.

appear in the context of oil fields on the advertising card of the Fund from those years.

Hitler's campaign in Russia during the Second World War was, among other things, an attempt to capture Azerbaijan's oil fields for Germany. With this intention, Hitler sent selected armoured divisions and alpine rifle units trained for fighting in mountainous terrain like the Caucasus region. The defeat at Stalingrad turned around the German march for the Baku crude oil.

In Azerbaijan, in the early 1990s, the Soviet Union died. Private oil companies from all over the world have come back to the "country of fire", the Eldorado of gas and oil with their immeasurable labyrinth of impressive offshore oil platforms.

This is still in progress. Witold Zglenicki's works, just like the works of Ignacy Łukasiewicz, are still writing new history and determining the fate of the world.

IV The Power of Bóbrka

Political and social life of the Austrian Partition until 1914. The economy of Galicia until 1914. Galician industry. Banking. Transport. Subcarpathian petroleum mining. Drilling, or the breakthrough. From Canada and Pennsylvania. Borysław fever. Entering the 20th century. Innovation, patents and competition. Geology or divining rod. Eruptions, fires, landslides, groundwater. The Subcarpathian refinery industry. Factories of drilling machines and tools in Subcarpathia and Lesser Poland at the turn of the 20th century. Austrian authorities vis-à-vis the Subcarpathian oil industry. Societies and journals. Professional vocational education. MacGarvey and others. Łukasiewicz's heirs in the world. Congresses and exhibitions. Success in spite of obstacles

Political and social life of the Austrian Partition until 1914

Ignacy Łukasiewicz lived and worked in the Subcarpathia region, i.e. in the south-eastern areas of the Republic of Poland. At that time, this area was the northernmost province of the Austrian monarchy, since the end of the 1860s transformed into an Austro-Hungarian monarchy, in which it theoretically remained within the state until 1918 – the end of the First World War and Poland's regaining independence. The area was annexed to the Austrian monarchy at the end of the 18th century and was the part of Poland seized by Austria as a result of the allied aggression of its neighbouring countries – Russia, Prussia and Austria. The Austrians called this province Galicia.

During Łukasiewicz's life, important political events, as well as social and economic transformations of great importance, took place in the Austro-Hungarian Partition, i.e. Galicia. All of them shaped his personality to a great extent. Even a superficial knowledge of the environment and conditions in which Łukasiewicz lived is extremely important for a full understanding of the actions he took both in his youth and maturity. In this context, it is necessary to characterise the socio-political situation of Galicia and its economic development. Only against this background can we describe the history of the oil industry and point to its exceptional significance for the economic development of the country in which Łukasiewicz lived and to which he himself contributed in the first place.

Poland's loss of independence, which took place as a result of three Partition acts in 1772, 1793 and 1795, led to the division of the country between three neighbouring countries – Russia, Prussia and Austria. The Poles did not accept this and that is why they took action several times in order to regain their lost independence. At the beginning of the 19th century, they believed that Napoleon Bonaparte, who waged wars with all the partitioners, could bring them freedom. In 1807 they received a substitute for statehood in the form of the Duchy of Warsaw, which two years later was enlarged by a part of the Austrian Partition. However, Napoleon's army was finally defeated, and the Congress of Vienna of 1815 perpetuated the new division of Polish lands for over 100 years.

Striving to regain independence, Poles established underground organisations and undertook armed struggle. In 1830 they organised the November Uprising, in 1846 the Cracow Uprising, in 1848 they participated in the battles of the Springtime of Nations, and in 1863 in the January Uprising. All these armed uprisings were unsuccessful.

After the defeat of the Cracow Uprising in 1846, the area of Galicia was enlarged by the territory of the so-called Cracow Republic, which, according to the provisions of the Congress of Vienna, was intended to be a free city. At that time the area of Galicia reached 78 thousand km^2. The number of inhabitants of the Province was also growing rapidly. In 1846 it amounted to 4.7 million, in 1880 it was almost 6 million, and in 1910 over 8 million[353]. A characteristic feature of Galician society was its diversity of nationalities and religions. Apart from Poles, traditionally belonging to the Roman Catholic Church (approx. 46 %), Galicia was inhabited by Ruthenians (approx. 42 %) associated with the Orthodox and Uniate churches, and Jews (about 11 %). Moreover, among other nationalities there were Germans, Armenians and other groups[354]. In this form, Galicia survived under Austrian rule until 1918, when it became a part of reborn Poland.

In the political and socioeconomic history of Galicia after 1815, two periods can be clearly distinguished: the period before regaining autonomy and the autonomous period, which began at the end of the 1860s and lasted until 1918. Until the middle of the 19th century, Vienna's central government was not interested in the social and economic development of Galicia, and Viennese officials often called the province "bear country"[355]. They believed that agriculture should be the sole economic activity of the Galician population. The sons of Galician peasants, on the other hand, were to be recruits in the ranks of the increasingly numerous Austrian army.

---

353 Krzysztof Zamorski, *Informator statystyczny do dziejów społeczno-gospodarczych Galicji. Ludność Galicji w latach 1857–1910* [Statistical guide to the socio-economic history of Galicia. Population of Galicia in the years 1857–1910] (Cracow-Warsaw: UJ-PTS, 1989), p. 45.

354 Zamorski, *Informator statystyczny*, p. 71.

355 Stanisław Grodziski, *W Królestwie Galicji i Lodomerii* [In the Kingdom of Galicia and Lodomeria] (Cracow: Wydawnictwo Literackie, 1976), p. 51.

In February 1846 Polish patriots made an attempt to throw off Austrian rule by organising a revolt, later called the Cracow Uprising. This attempt ended in a complete defeat. The Austrian army suppressed the uprising, and the peasants, stirred up by the Austrians, carried out a cruel slaughter of the Galician nobility. It was only as a result of the revolutionary events during the Springtime of Nations that the Viennese authorities carried out quite limited reforms, which also affected Galicia.

One of the most important measures taken by the central administration since the middle of the 19th century, which had a social, but also political and economic consequences, was the enfranchisement of peasants in 1848. This reform was particularly significant in Galicia, where peasants constituted 82 % of the population. However, the manner the reform was carried out, consolidated the unfavourable structure of land ownership in Galicia. According to the agricultural census of 1902, dwarf farms with an area not exceeding 2 ha constituted almost 50 % of all farms, occupying at the same time only slightly more than 9 % of the total cultivated area. Also small farms, with an area between 2 ha and 5 ha, were numerous (about 34 % of all farms), occupying less than 20 % of the cultivated area[356]. Larger farms, with an area exceeding 5 ha, accounted for about 70 % of the cultivated land. They constituted only about 16 % of all farms. A characteristic phenomenon occurring in the Galician countryside was the "chessboard of fields", i.e. the division of land belonging to one farmer into many plots of land that remained at a considerable distance from each other. However, the enfranchisement reform had an extremely important social dimension. The decisions taken under enfranchisement enabled peasants to change their place of residence and thus created the possibility for the free labour force to move from agriculture to industry. This was important for all sectors of the economy, but became particularly important in the case of the emerging oil industry.

Since the mid-19th century, the authorities introduced important changes to stimulate the economy. From 1850, a new customs tariff was in force, abolishing most of the customs duties on goods imported into and exported from Galicia. In the same year, the law on chambers of commerce and industry representing the interests of industry, crafts and commerce entered into force. Three such chambers were established in Galicia – in Lviv, Cracow and Brody. In 1859, an industrial law was passed, abolishing all guild and social restrictions and introducing freedom of production activity[357]. In the

356 *Podręcznik Statystyki Galicji* [Galician Statistical Manual], Vol. 8 (1908), p. 149.
357 Wojciech Saryusz-Zaleski, *Dzieje przemysłu w byłej Galicji 1804-1929* [History of industry in former Galicia 1804–1929] (Cracow: P.A. Zieleniewski i Fitzner Gamper, 1930), pp. 78–79.

following year the restrictions on the industrial activity of Jews were lifted. Ignacy Łukasiewicz started his business activity at exactly that time.

The difficult history of the Austrian monarchy initiated by the outbreak of the Springtime of Nations, and in particular the military defeats suffered by Austria in the late 1850s and mid-1860s forced the Viennese authorities to carry out reforms of the ossified political system. Achieving autonomy in the 1860s was a landmark event in the history of Galicia. It was a process extended over time, covering the political, cultural, social and economic spheres. It was initiated by the so-called October Diploma developed by Agenor Gołuchowski, the Austrian Minister of the Interior. The Emperor signed the Diploma on October 20, 1860. It divided the legislative power in the monarchy into state power exercised by the Viennese Parliament, and the legislative powers of individual provinces, known as the countries of the monarchy. The most controversial element in the Diploma was the division of competences between central and national authorities. The powers of the Viennese authorities were scrupulously listed, with the application of the so-called presumption of competence in relation to local authorities, which meant that all matters that were not reserved for the Viennese Parliament were transferred to their decisions. These decisions were changed in the so-called February Patent, which was created by Anton von Schmerling in 1861. The document specified the powers of the provincial authorities and the presumption of powers was included in the powers of the Viennese authorities. On the basis of this patent and the statute issued together with it, 17 local parliaments called national parliaments were established in individual provinces. Galicia was one of those provinces included in the so-called Cisleithanian countries, i.e. the non-Hungarian part of the Austro-Hungarian monarchy.

The most important centre of the Galician autonomous authorities was the National *Sejm* with its seat in Lviv. The first paragraph of the Galician National Statute of 1861 states:

> "The Kingdom of Galicia and Lodomeria with the Grand Duchy of Cracow is represented in national matters by the National Parliament."

In accordance with paragraph 6 of the Statute, the term of office of the *Sejm* lasted six years[358]. One extremely important issue was to determine the scope of competences of the *Sejm*, which were referred to as "national matters". In accordance with § 18 of the Galician National Statute, several issues were identified as such, of which the notion of "national culture" was

358 "Statut Krajowy Galicyjski z 1861 r." [Galician National Statute of 1861], in: Stanisław Grodziski, *Sejm Krajowy galicyjski 1861-1914. Źródła*, vol. 2 (Warsaw: Wydawnictwo Sejmowe, 1993), pp. 5–6.

of key importance[359]. On the one hand, it referred to issues falling within the broad meaning of the word culture today, but at the same time it also covered economic matters. Initially it was limited to agriculture only, but over time it also covered industry, communications, banking and other areas of economic and social life (e.g. education, healthcare). Further liberalisation of the Austrian political system took place on 21 December 1867, when the emperor signed a law on "representation of the state", also known as the December Constitution. It alluded to the solutions adopted in the October Diploma of 1860. In 1873, with the introduction of direct elections to the Vienna State Council, the process of fundamental political reforms of the monarchy was completed, and at the same time the scope of Galician autonomy was established, which remained in force until the end of the First World War[360].

With the granting of autonomy, there were much greater opportunities for the development of the Provinces. In the sphere of culture, a breakthrough was the replacement of German as the official language with Polish, which also became obligatory in education at all levels. Great progress was made in the area of education. The introduction of the statutory obligation to attend school contributed to the dissemination of elementary schools. Their number increased from 1,670 in 1828 to 6,000 in 1913. While in 1868 more than 156 thousand children attended folk schools, less than half a century later the number was more than 1.1 million[361]. Just before the outbreak of the First World War, spending on elementary education amounted to 31.5 million crowns in the national budget[362]. With the increase in the number of secondary schools – from 19 at the end of the 1860s to 128 in 1914 – the number of students attending them increased rapidly, from about 5.4 thousand to over 40 thousand[363]. Vocational education and higher education in Galicia developed successfully. Although the largest number

359 "Statut Krajowy Galicyjski z 1861 r.", p. 8.

360 Piotr Franaszek, "Walka wyborcza do galicyjskiego Sejmu Krajowego" [Electoral struggle for the Galician National Sejm], in: *Systemy reprezentacji i parlamentaryzm w Europie Środkowej w rozwoju historycznym*, ed. Antoni Barciak (Katowice-Zabrze: PAN, 2016), pp. 213–231.

361 Julian Dybiec, "Galicja na drodze do wielkiej przemiany" [Galicia on the Road to Great Change], in: *Cracow i Galicja wobec przemian cywilizacyjnych (1866-1914). Studia i szkice*, ed. Krzysztof Fiołek and Marian Stala (Cracow: Universitas, 2011), p. 32.

362 Julian Dybiec, *Finansowanie nauki i oświaty w Galicji 1860-1918* [Funding for science and education in Galicia 1860–1918] (Cracow: Wydawnictwo Uniwersytetu Jagiellońskiego, 1979), p. 205.

363 Henryka Kramarz, *Nauczyciele gimnazjalni Galicji 1867-1914. Studium historyczno-socjologiczne* [Teachers of Galician Gymnasiums 1867–1914.

of students were in the law and philosophy faculties, the importance of departments of engineering, mechanics, construction, chemistry, and other technical faculties was growing. The workforce trained in this way contributed to the economic development of Galicia.

The economy of Galicia until 1914

In the second half of the 19th century the face of the Galician economy changed slowly but systematically. Banking and transport, which were the prerequisites for the recovery of individual sectors of the economy, were developing. Agro-technical progress was visible in agriculture. Some industries were also characterised by an increase in production.

However, the oil industry, in particular oil production, was characterised by the highest dynamics. The new political situation that emerged after autonomy had a positive impact on the economy of this backward province of the Austrian monarchy, and favoured discussions on the paths of economic development led by economic and local government activists, politicians, scientists and journalists, but above all by Polish entrepreneurs, including oil entrepreneurs.

They all pointed to the need to industrialise the country as a basic path leading to the recovery of economic life. As far as possible, economic policy was pursued through legislative and organisational measures. Stanisław Szczepanowski was the most famous economic activist and at the same time an entrepreneur also involved in the Galician oil industry. This businessman from Greater Poland indicated, in his opinion, the best way to achieve economic independence for Galicia. In 1888 he published a book in Lviv entitled *"Galician Poverty in Numbers"*[364]. In his work, Szczepanowski addressed many important social and economic issues. Supporting his theses with statistics, he demonstrated the economic backwardness of Galicia. A comparative analysis of the standard of living of the inhabitants of Galicia with the standard of living of the other two Partitions and a few selected European countries showed up definitely against the Austrian Partition. The title of the paper itself indicated the disastrous state of the provincial economy, as documented by the statistical summaries contained in the book. Szczepanowski described the poverty of the Galician countryside,

---

A Historical and Sociological Study] (Cracow: Wydawnictwo Naukowe WSP, 1987), p. 14.

364 Stanisław Szczepanowski, *Nędza Galicji w cyfrach i program energicznego rozwoju gospodarstwa krajowego* [Galician Poverty in Numbers and a Programme for the Energetic Development of the National Economy] (Lviv: Gubrynowicz i Schmidt, 1888).

which limited the development of the internal market, which was one of the factors determining the backwardness of industrial production. He was involved in the activities of the Austrian Parliament, i.e. the Council of State and the Galician National *Sejm*. He popularised the need for work at grass-roots level through his publications, encouraging other politicians, social activists and entrepreneurs in them to actively work for the economic revival of the country[365]. He saw the greatest opportunity to modernise economic and social relations in the development of the oil industry. Unfortunately, his initiatives in coal mining on the basis of loans from the Galician Savings Bank led to the crisis of the Bank, which resulted in the collapse of Szczepanowski's undertakings in the oil industry. The director of the Bank, Franciszek Zima, was sued, and Stanisław Szczepanowski was also involved. These experiences exhausted his health and led to his death[366].

One oil industrialist involved in the country's economic development problems was a landowner by origin, August Gorayski. He was a long-time president of the National Oil Society and a member of the National *Sejm*. He made attempts to revive the textile industry, among others by organising the "*Prządka*" Weaving Society in Krosno. Kazimierz Odrzywolski was also an oilman and friend of Szczepanowski's, and he popularised industrial issues in his daily "*Słowa polskiego*", which was published in Lviv. Representatives of the outstanding Zieleniewski family from Cracow, in particular Edmund Jan Kanty and his brother Leon, were also active in the economic development of Galicia. Scientists such as Tadeusz Pilat, head of the Department of Statistics and Administration Science at the University of Lviv, and Zygmunt Olszewski, secretary of the National Oil Society, co-founder of the Oil Technicians' Society and editor of the oil magazines *Nafta* and *Ropa*, were in favour of the modernization of social and economic life. Bronisław Pawlewski, a chemist, was a member of the National Industrial Council and the National Mining Council, while Stanisław Anczyc was the editor of the *Czasopismo Techniczne*, which was published in Lviv. Juliusz Leo was active as a politician, as a member of the National *Sejm* in Lviv and a member of the Vienna State Council, and finally for many years the mayor of Cracow. Many local government activists, publicists and politicians also spoke on

---

365 Jan Wnęk, *Rozwój ideologii przemysłowej w Galicji 1866-1918* [The Development of the Industrial Ideology in Galicia 1866–1918] (Warsaw: PAN, 2015), pp. 54–55.

366 Piotr Franaszek, *Myśl techniczna w galicyjskim wiertnictwie naftowym w latach 1860-1918* [Technical Thought in Galician Oil Drilling in the Years 1860–1918] (Cracow: Uniwersytet Jagielloński, 1991), pp. 54–55; Alison Fleig Frank, *Oil Empire. Visions of Prosperity in Austrian Galicia* (Cambridge Massachusetts, London England: Harvard University Press, 2005), pp. 82–89.

economic issues, including Włodzimierz Dzieduszycki, Andrzej Lubomirski, Andrzej Wrotnowski, Tadeusz Romanowicz, Tadeusz Rutowski, Teofil Merunowicz, Aleksander Szczepański, Henryk Mianowski, Roger Battaglia, Artur Benis and many others[367].

Galician industry

At the end of the 18th century, the level of industrial development in southern Poland, which soon found itself within the borders of the Austrian state, did not differ significantly from the level in the rest of Poland. In Galicia, however, these seeds could not be transformed into a modern factory industry. On the contrary, under Austrian rule in the early 19th century, a process of "deindustrialisation" began. This was evident in the textile industry, which entered a stage of stagnation and finally destruction. The situation was similar in other branches of industrial production. Fundamental progress in industrialisation can only be seen in the period of autonomy, especially at the turn of the 20th century. Not without significance in this respect was the change in the perception of industrialization as a fundamental factor responsible for the modernization of the province by the then economic and opinion-forming elites. In 1903, the Central Association of Galician Factory Industry was established, headed by Prince Andrzej Lubomirski. Its aim was to care for, advise and represent the interests of the factory industry vis-à-vis the Galician and Austrian authorities and to promote industrialisation[368].

Due to the strong agricultural background, the food industry, using raw materials of agricultural and forestry origin, was of exceptional importance. Brewing and distillation were of great importance, and they became a very important subject of debates both in the National Parliament and in the Vienna Council of State on tax and budgetary policy. Distilleries operating within larger farms played an important role in the development of the spirits industry. The distilleries processed potatoes and rye, supplying the farm with by-products, especially the so-called broth, which was the basic feed for farm animals. The distilleries were conducive to potato cultivation, as high demand stimulated farms supplying this raw material to local distilleries. The technical, technological and organisational changes introduced in distilling favoured technical progress throughout the agricultural economy.

---

367 Wnęk, *Rozwój ideologii przemysłowej*, pp. 50–66.

368 Mariusz Kulczykowski, "'Deindustrializacja' w procesach uprzemysłowienia Galicji w XIX wieku. Problemy badawcze" ['Deindustrialisation' in the industrialisation of Galicia in the 19th century. Research problems], *Zeszyty Naukowe Uniwersytetu Jagiellońskiego. Prace Historyczne*, No. 78 (1986), pp. 75-87.

Distilleries created a local labour market and thus had a positive impact on the economic level and social relations in the Galician countryside. In addition, the commodity nature of distillery production influenced factors such as water resources and quality, the development of local railways and roads, means of transport, telegraph, warehouses and commercial knowledge in the broadest sense of the term. The Galician agricultural spheres attributed an extremely important role to distilleries in the process of intensification of national agriculture, pointing out their influence on the industrialisation of eminently agricultural areas[369]. Spirits distillers operating outside noble granges, e.g. the J.A. Baczewski vodka and liqueur factory in Lviv, created an exceptional brand[370].

The brewing industry with leading companies in Żywiec and Okocim developed very well. The second deserves special attention. The establishment of the Okocim brewery in the mid-1940s and its functioning until the outbreak of World War II is connected with the life and activity of three representatives of the Goetz family. They were in turn: Jan Ewangelista Goetz-Okocimski, his son Jan Albin and his grandson Antoni. The first of them, of German origin, was the founder of the company. He came to Okocim on 26 April 1845 and only two months later the construction of the brewery started. The first brew of beer was obtained as early as 26 February 1846.[371]. Between 1846 and 1884, beer production increased from 4.5 thousand hectolitres to 40.8 hectolitres, i.e. almost tenfold. In 1886, Jan Ewangelista handed over the management of the brewery to his son, Jan Albin, who continued the great plan of investments started by his father. The plant was systematically expanded and modernised. Sixty air-conditioned wagons, standing on a siding built in the 1890s, were used to transport beer over long distances. Just before the outbreak of World War I, the plant employed 620 people. From the time Jan Albin took over until

369 Jerzy Michalewicz, *Przemysł gorzelniany Galicji doby autonomicznej. Między monopolem dworskim a monopolem państwowym* [The Distillery Industry of Autonomous Galicia. Between the Court Monopoly and the State Monopoly] (Cracow: Uniwersytet Jagielloński, 1988), pp. 9, 14-15.

370 Andrzej Chwalba, *Historia Polski 1795-1918* [History of Poland 1795–1918] (Cracow: Wydawnictwo Literackie 2000), pp. 509–510.

371 Jan Maria Włodek, *Goetz-Okocimscy, Kronika rodzinna 1590–2000* [Goetz-Okocimskis, Family Chronicle 1590–2000] (Cracow: Księgarnia Akademicka, 2001), pp. 26-29; Jan Burlikowski, "Z biegiem Uszwicy" [Downstream on the Uszwica], in: *Kronika miasta Brzeska 1385–1944* (Brzesko: ZETO SA, 2005), Vol. 3, p. 29; Burlikowski, "Kto założył browar 'Okocim'" [Who Founded the 'Okocim' Brewery], in: *Kronika miasta Brzeska*, Vol. 4, p. 93; Burlikowski, "Założenie browaru 'Okocim'" [The Foundation of the 'Okocim' Brewery], in: *Kronika miasta Brzeska*, Vol. 1, p. 30.

the outbreak of World War I, beer production increased to over 350,000 hl, and the Okocim brewery became the largest brewery in Galicia and one of the largest of the 1,200 breweries in the entire Austro-Hungarian Empire. Despite its good beginnings, the Galician sugar industry collapsed at the end of the 19th century.

Salt mining had been flourishing in the Polish Subcarpathia region for many centuries. Apart from the old mines in Wieliczka and Bochnia, salinas in the Eastern Subcarpathian region gained in importance. According to estimates, the number of salt mines in Galician times was to reach 500. In some of them salt was mined using mining methods, while in others brine was exploited and evaporated. The two largest mines (Wieliczka and Bochnia) belonged to the government and the others were in private hands. The salt mined was sold in the province, and transported to the Russian Partition as well as to Silesia and Moravia. Despite periodical and relatively small fluctuations, the total production of salt in the Subcarpathia region at the turn of the 20th century grew steadily. At the same time, the proportions between the production of rock salt, brewed salt and industrial salt changed considerably. The production volume of the former decreased significantly, while the production of evaporated salt increased. In 1887, potassium salt was exploited in Kałusz, which was used as a fertiliser in agriculture. The growing demand for salt was caused not so much by its consumption as by its increasing use in breeding and industrial processes[372].

Cracow, Lviv and Sanok became important centres of the metal industry. However, all this was not enough for the Austrian Partition to enter the path of intensive industrialization. As in other countries, a flywheel effect was needed, whose role in the years of the industrial revolution was played by heavy industry, in particular metallurgy and mining, in particular the mining of hard coal, a raw material so important in the production of iron and steel. An opportunity for accelerated economic development of the country could have been the exploitation of energy resources – coal in the west of the province and oil in the east. Each of these centres contained a potential that was able to trigger the process of intensive industrialization, but ultimately none of them fulfilled such a role. On the other hand, oil mining had the greatest

372 Kazimierz Bukowski, Aleksander Jackiewicz, *Sól i saliny polskie* [Polish Salt and Salt Pans] (Warsaw: Ministerstwo Przemysłu i Handlu, 1926), p. 8; Karol Jonca, *Dzieje gospodarcze Polski do 1939 roku* [Economic History of Poland till 1939] (Wroclaw: Wydawnictwo Uniwersytetu Wroclawskiego, 1996), p. 181; Janusz Bar, Piotr Franaszek, *Informator statystyczny do dziejów przemysłu w Galicji. Górnictwo i hutnictwo* [Statistical guide to the history of industry in Galicia. Mining and metallurgy] (Cracow: Uniwersytet Jagielloński 1981), pp. 67-74.

impact on the economic recovery of the whole province, activating local entrepreneurs and economic activists.

Hard coal was mined in the Cracow Basin to the west of Lesser Poland. It covered an area of about 1.2 thousand km^2 and the main centres were the cities of Jaworzno, Chrzanów, Trzebinia, Libiąż and Siersza. The formation of the Cracow Basin at the beginning of the 19th century and its development during this century was connected with the mining of hard coal, zinc and lead ores, and zinc and lead metallurgy, although the beginnings of the mining of lead and silver ores in this region date back to the end of the 13th century. However, the development of zinc ore mining on an industrial scale took place only at the beginning of the 19th century, with the establishment of zinc smelters in Silesia, belonging to Prussia[373].

In the 19th century, hard coal mining gained the greatest importance in the Cracow Basin. Hard coal seams stretching from Upper Silesia reached Krzeszowice and Tenczynek in the east and Żarki and Grojec in the south. Coal was mined on a small scale in the vicinity of Siersza as early as in the 16th century, during the times of the Commonwealth of Poland. In 1792, the Moszyński Coal Mine in Jaworzno started operating, and four years later the government mine. The presence of hard coal in Tenczynek had been known at least since the 17th century. At the beginning of the 19th century, demand for coal increased rapidly, which was associated with its growing industrial use. In addition, coal ash was one of the basic raw materials used in the production of alumina, which was used in the textile industry. For this reason, the exploitation of previously known seams was extended and

373 Danuta Molenda, "Zastosowanie ołowiu na ziemiach polskich od XIV do XVII wieku" [Use of lead on Polish soil from the 14th to the 17th centuries], in: *Metale nieżelazne na ziemiach polskich od XIV do XVIII wieku* (Cracow-Wroclaw: Ossolineum, 1987), pp. 13–14; Danuta Molenda, "Polski ołów na rynkach Europy środkowej w XIII-XVII wieku" [Polish lead in the markets of Central Europe in the 13th-17th centuries], *"Studia i Materiały z Historii Kultury Materialnej"*, Vol. 69 (2001), p. 14; Stanisław Polaczek, *Powiat chrzanowski w W. Kp. Cracowskiem. Monografia historyczno-geograficzna* [Chrzanowski County in W. Kp. Cracow. A Historical-Geographical Monograph] (Cracow: Wydział Rady Powiatowej Chrzanowskiej, 1914), p. 156; Edward Pietraszek, *Ośrodek górniczy Siersza 1804-1861-1961* [Siersza Mining Centre 1804–1861–1961] (Cracow: Wydawnictwo Artystyczno-Graficzne, 1961), pp. 7–8; Witold Kula, *Szkice o manufakturach w Polsce XVIII wieku* [Sketches about manufactories in Poland in the 18th century] (Warsaw: Państwowe Wydawnictwo Naukowe, 1956), pp. 118–135; Krzysztof Zamorski, *Folwark i wieś. Gospodarka dworska i społeczność chłopska Tenczynka w latach 1705-1845* [The Farm and the Village. The Manor Economy and Peasant Community of Tenczynek in 1705–1845] (Cracow-Wroclaw: Ossolineum, 1987), pp. 64–65.

more and more were discovered, with an increasing number of new mines being established. Qualitatively good deposits were found in Jaworzno and Siersza. Smaller companies were located in Libiąż, Moczydło, Myślachowice, Tenczynek, Zwierzyniec, Czarny Bagno, Jelenie, Luszowice, Filipowice and in Góry Luszowskie[374].

In 1804, coal was mined in a mine established on the Kozi Bród stream, on the border of the Siersza and Góry Luszowskie. In 1816 the mine was bought by Princess Izabela Lubomirska of the Czartoryski family. Already earlier, on her behalf, coal deposits were being explored in the region of Siersza. They resulted in the establishment of the mine named "Izabella" in 1808. In the 1820s, the "Zofia" mine was established in Krz, whose production in the years 1823–1824 amounted to about 11 thousand tonnes. There were a dozen or so shafts in the mine, which were open until the middle of the 19th century. In 1843, the "Elżbieta" mine was established. It operated for 25 years. In 1854, on the initiative of Count Adam Potocki, the company "Zakłady Górnicze i Hutnicze w Sierszy" was established. Apart from Potocki, the co-owners were Henryk and Wilhelm Rau and Leonard Erbreich. About a kilometre from the "Izabella" mine, the company built a modern mine called "Nowa Izabella". At the same time it was the beginning of regular coal mining in Siersza[375]. The largest hard coal mining company was "Gwarectwo Jaworznickie" with huge areas of mining rights purchased from the government in 1854. Until 1846, these areas were owned by the Free City of Cracow. In 1871, the Gwarectwo passed into the hands of "Viennese coal barons" – Guttmann, Springer and Oppenheim. In 1914, the Gwarectwo had two mines in Jaworzno – "Fryderyk August" and "Jacek Rudolf", where three shafts were in operation: "Helena", where coal was mined from a depth of 120 m, "Paulina" (160 m) and "Rudolf" (102 m). In 1907 the company "Galicyjskie Zakłady Akcyjne Zakłady Górnicze

---

374 Polaczek, *Powiat chrzanowski*, p. 145; Edward Pietraszek, "Zagłębie Cracowskie w latach 1796–1848. Zarys społeczno-gospodarczy" [The Cracow Basin in the years 1796–1848. A Socio-Economic Outline], *Kwartalnik Historii Kultury Materialnej*, No. 4 (1961), pp. 746–753.

375 Stanisław Orłowski, "Trzebinia i region w latach 1815–1918" [Trzebinia and the District in the Years 1815–1918], in: *Trzebinia. Zarys dziejów miasta i regionu*, ed. Feliks Kiryk (Cracow: Wydawnictwo i Drukarnia Secesja, 1994), pp. 210–218; Piotr Franaszek, "Przemysł Zagłębia Cracowskiego w XIX i pierwszej połowie XX wieku (do 1945 roku)" [The Industry of the Cracow Basin in the 19th and First Half of the 20th Century (until 1945)], in: *Jaworzno interdyscyplinarnie. Przemysł pogranicza śląsko-małopolskiego na przestrzeni wieków*, ed. Adrian Rams (Jaworzno-Częstochowa: Muzeum Miasta Jaworzna, 2016), pp. 13–28.

w Sierszy" was established, which took over all the mining and metallurgical companies, including all the assignments and mining rights previously belonging to Count Andrzej Potocki. The "Galicyjskie Zakłady" employed over 2.3 thousand workers[376]. The "Compagnie Galicienne des Mines" association operated in Libiąż Mały. The works on the "Janina II" shaft, carried out since the end of 1908, resulted in 1911 in the exploitation of seams located at a depth of over 300 metres. The "Société Anonyme Minière et Industrielle de Verviers", based in Strasbourg, owned the "Bory" and "Sobieski" mines in the municipality of Jeleń. The "Sobieski" mine was one of the most modern in the Basin and employed about 1000 miners and the coal mined there was of good quality[377]. In 1906, near Oświęcim, the "Brzeszcze" mine was established. At the beginning of the 19th century, total hard coal mining in the Cracow Basin mines amounted to over 5,000 tonnes, and in 1913 the total production volume reached nearly 2 million tonnes[378].

Although, as far as the conditions of the Austrian Partition were concerned, the Cracow Basin had serious industrial potential, it was at the same time relatively modest in comparison with other industrialised centres of the Austrian monarchy. Only a small part of the coal mined here was of high quality. Most of the seams were difficult to exploit due to the prevailing water conditions. Coal was largely used in local zinc smelters. The extraction of zinc ore, i.e. galvanic ore, began to decrease over time, which also adversely affected the extraction of hard coal. However, the greatest threat to coal mining in the Cracow Basin was the competition for coal from Prussian

376 Polaczek, *Powiat chrzanowski*, pp. 147, 149; Artur Benis, *Zagłębie Cracowskie. Uwagi gospodarcze. Przemysł górniczy i fabryczny* [The Cracow Basin. Economic remarks. The Mining and Manufacturing Industry] (Cracow: Związek górników i hutników polskich w Austryi, 1910), p. 13.

377 Piotr Franaszek, "Zagłębie Cracowskie" [The Cracow Basin], in: *Historia polskich okręgów i regionów przemysłowych*, ed. Łukasz Dwilewicz, Wojciech Morawski (Warsaw: Polskie Towarzystwo Historii Gospodarczej, 2015), Vol. 1, pp. 44–72.

378 Benis, *Zagłębie Cracowskie* [The Cracow Basin], pp. 4–5; Jerzy Jaros, "Górnictwo węgla kamiennego" [Hard Coal Mining] in: *Zarys dziejów górnictwa na ziemiach polskich*, Vol. 2, ed. Jan Pazdur (Katowice: Wydawnictwo Górniczo-Hutnicze, 1961), pp. 65–67; Wacław Długoborski, "Górnictwo i hutnictwo do 1918 r." [Mining and metallurgy until 1918], in: *Uprzemysłowienie ziem polskich w XIX i XX wieku. Studia i materiały*, ed. Irena Pietrzak-Pawłowska, (Cracow-Wroclaw: Ossolineum, 1970), pp. 118–121, 132–135; Andrzej Burzyński, *Robotnicy w przemyśle ciężkim w Galicji w dobie autonomicznej. Struktura zatrudnienia* [Heavy Industry Workers in Galicia in the Autonomous Period. Employment Structure] (Cracow-Wroclaw: Ossolineum, 1985), pp. 17–18.

Silesia. Polish industrial activists drew attention to this many times. This is how the problem was described in 1912:

> "Galicia is facing the threat of a coal monopoly resting in the hands of Prussian barons, flooding our country with their coal and constantly lowering the prices of coal from domestic mines, thus leading to a wasting away. Galicia, the country of inexhaustible coal wealth, imports around fourteen and a half million metric cetnars of Prussian coal every year, paying a total of twenty-five million crowns a year for this extortion"[379].

In this situation, local entrepreneurs fought for their own survival rather than for the development of industrial activities in other areas of the province. As a result, coal mining in the Cracow Basin did not play a significant role in the process of industrialization of the Austrian Partition.

Subcarpathian mining of metal ores and sulphur was relatively underdeveloped, which resulted from the low quality of seams, backward mining methods, lack of capital for investments, as well as growing competition from the Upper Silesian and Austrian industries. Until the middle of the 19th century, the exploitation of iron ore in Galicia was connected with the functioning of a large number of small one-furnace smelters established on the territory of large land holdings. There were about of these smelters, but because they were based on primitive technology most of them collapsed in the 19th century. Larger companies were also closed down, e.g. in 1875 the smelter in Zakopane, in 1886 in Sucha, and in 1905 also in the Węgierska Górka near Żywiec, the largest smelter working mainly on ore imported from Sweden and Russia[380].

Iron ore mines were concentrated in the districts of Chrzanów and Żywiec. Extraction of ore in the Subcarpathia region was subject to frequent and high fluctuations. There were even years when it was completely abandoned. The most ore was mined in 1885 – 28,718 tonnes. Since then, mining declined, with the cessation of forged iron production in Galicia and, in 1905, of cast iron. Galicia was a producer of lead ore and zinc ore. The only Galician mine where lead ore was mined was the "Matylda" mine near Chrzanów. The maximum output was obtained in 1903 and amounted to 7,241 tonnes. But lead ore was not smelted in Galicia and all mining was exported to the

379 Zdzisław Kamiński, "Wywłaszczenie Cracowskiego Zagłębia Węglowego" [Expropriation of the Cracow Coal Basin], *Przegląd Górniczo-Hutniczy*, Vol., 9, No. 10 (1912), p. 359.

380 Zygmunt Jędrkiewicz, "Widoki rozwoju hutnictwa żelaznego w Galicji" [Views of the Development of the Iron and Steel Industry in Galicia], *Prace Biura Statystyki Przemysłowej*, No. 2 (1913), p. 77.

smelters of Upper Silesia[381]. Zinc ore mining was marked by a clear downward trend. In 1892, 14 thousand tonnes of this raw material were obtained, but in 1913 only 1,530 tonnes. Most of the raw material came from the "Matylda" mine. Although there were still seven zinc ore mines in operation at the beginning of the 1890s, most of them were closed down and, as a result, only two mines were in operation after 1905. This was a result of the exhaustion of the seams currently in operation and the rising costs of production from year to year. In contrast to zinc ore mining, zinc metallurgy flourished perfectly in Galicia. Zinc production grew steadily – e.g. in 1867 it amounted to 890 tonnes and in 1913 already 15 thousand tonnes. The smelting took place in two smelters belonging to the Upper Silesian Cartel in Niedzieliska and Trzebinia, and in the Galician Mining Plant in Krz. These smelters processed imported raw material, originating in 30 % from other provinces of the Austro-Hungarian Empire, and in 70 % from Germany. Most of the melted zinc was exported deep into the Austro-Hungarian Empire, to other European countries and even to America. Small quantities of zinc were processed by the sheet metal rolling mills in Oświęcim and Trzebinia[382].

Lignite mining was completely marginal in Galicia. Its output grew until the beginning of the 20th century, when in 1901 the maximum amount of raw material was reached, amounting to 112,784 tonnes. From that moment, mining declined rapidly as a result of the depletion of shallow coal seams. As a result, the Partition's share in the total production of Austrian lignite was very low and in the year of maximum extraction reached only 0.5 %.In 1885, the only sulphur mine in Galicia, located in Swoszowice near Cracow, was closed down. At the end of the 1860s, it produced about 1,000 tonnes of sulphur. However, since the mid-1870s, sulphur production began to decline very quickly, which was associated with the depletion of shallow seams and the unprofitable exploitation of deeper seams. The stone mining and processing industry, glassworks and the ceramic industry played an important role in the economic life of Galicia.[383]

381 Franciszek Bujak, *Galicja. Leśnictwo. Górnictwo. Przemysł* [Galicia. Forestry. Mining. Industry] (Warsaw – Lviv: Księgarnia H. Altenberg, Księgarnia E. Wende i Spółka, 1910), Vol. 2, pp. 210-211.

382 Kazimierz Piesowicz, "Cynk" [Zinc], in: *Encyklopedia gospodarcza Polski do 1945 roku*, ed. A. Mączaka, Vol. 1, (Warsaw: Wiedza Powszechna, 1981), p. 111; Bar, Franaszek, *Informator statystyczny*, pp. 163-172.

383 Bar, Franaszek, *Informator statystyczny*, pp. 124-126.

Banking

Positive changes occurred in Galician banking in the second half of the 19th century. In the country both Galician banks, as well as those coming from outside the province, began to operate. In 1867 the Shareholders Mortgage Bank S.A. was founded in Lviv. The bank granted a mortgage loan secured on real estate in rural and urban areas. It also operated outside of Galicia, establishing representative offices in Cieszyn Silesia, Moravia and Bukowina. However, the attempts to directly finance industrial activity undertaken by this bank did not give the expected results of at the turn of the 20th century[384]. In 1869 in Cracow the Galician Bank for Trade and Industry S.A. was founded. It was the first capital bank in Cracow and the first which provided short-term credit. It repeatedly cooperated with banks in Vienna, including Unionbank and Österreichische Credit-Anstalt. In the same year Bielsko-Biała Trade and Industrial Bank was established, which, however, was too weak to operate independently and therefore in 1891 was transformed into a branch of the Czech Union Bank[385]. In addition to banks aimed at financing industrial projects, banks for agriculture were set up. Such were the Peasants' Credit Facility in Lviv, established in 1868, the Galician Land Credit Facility in Cracow, established in 1872, or the General Agricultural and Credit Facility for Galicia and Bukowina in Lviv, established in 1873.

The most important credit institution in Galicia was the National Bank for the Kingdom of Galicia and Lodomeria with the Grand Duchy of Cracow, which had been operating since 1883. This bank was a public institution and was supervised by the National Parliament. The initial capital of the bank was 1 million guilders and in 1907 the capital was increased to 15 million crowns. The bank not only granted loans, but also financed the construction of railway lines, granted municipal credit and carried out rescue operations for failing private credit institutions. From 1895 it financed the establishment of companies and its own company was a joint stock company for the construction of wagons in Sanok.

At the turn of the 20th century, in the face of positive changes in the economic development of Galicia, local entrepreneurs increasingly demanded the establishment of a credit institution focused exclusively on financing industrial projects. Such was the National Industrial Fund established in 1886. In 1910 the Industrial Bank for the Kingdom of Galicia and Lodomeria with the Grand Duchy of Cracow was established in Lviv. The

---

384 Wojciech Morawski, *Słownik historyczny bankowości polskiej do 1939 roku* [Historical Dictionary of Polish Banking until 1939] (Warsaw: Muza SA, 1998), p. 89.

385 Morawski, *Słownik historyczny bankowości*, pp. 120–121, 176–177.

shareholders in this bank were: the State Department, the municipalities of Lviv and Cracow, and the Viennese bank Niederösterreichische Escompte Gesellschaft. The bank held shares in companies as important for the economic life of Galicia as the Zieleniewski Machine Works in Cracow, the Society for the Exploitation of Potassium Salts in Kałusz and the Power Station in Siersza. In the same year, the Vienna Österreichische Länderbank and Galician oil industrialists established the Galician People's Bank for Agriculture and Trade Ltd. in Lviv. Before the First World War, out of nine stock banks operating in Galicia, only four were based on Austrian capital, with share capital of 27 million crowns[386].

Transport

The development of bulk transport was essential for the proper functioning of the economy. In view of the underdeveloped road network, rail and river transport were of fundamental importance. The first railway line in the Austrian Partition was established in 1847. It led from Cracow to Myślenice. The development of the iron rail network was decisively influenced by the policy of the Viennese government. The railway lines in Galicia were driven by political, military and strategic considerations and, to a much lesser extent, by economics. Significant impact on the delineation of railway routes was caused by difficult geographical and topographical conditions, which significantly increased the costs of investments. Most of the railway connections served to integrate Galicia more closely with the Austro-Hungarian Empire. At the same time, the natural links between this Austrian province and the former Polish lands under Russian rule were deliberately weakened. As a result, there were only three border crossings between Galicia and Russia: in Maczki, Radziwiłłów and Podwołoczyska[387]. In addition to crude oil, railways were used only to a small extent for the transport of industrial goods. Agricultural crops were also transported, but their quantity was subject to large seasonal fluctuations.

---

386 Janusz Kaliński, "Bankowość austriacka na ziemiach polskich", [Austrian Banking on Polish Soil] in: *Między stabilizacją a ekspansją. System finansowy w służbie modernizacji (z warsztatów badawczych historyków gospodarczych)*, ed. Jerzy Łazor, Wojciech Morawski, (Wroclaw: Wydawnictwo Gajt, 2014), pp. 270-271.

387 Stanisław Szuro, *Informator statystyczny do dziejów społeczno-gospodarczych Galicji. Koleje żelazne w Galicji w latach 1874–1914* [Statistical Guide to the Socio-Economic History of Galicia. Galician Railways 1874–1914] (Cracow: UJ, 1997), pp. 1-6.

In this situation, water transport on the Galician section of the Vistula River became particularly important. In the middle of the 19th century, within the borders of the Austrian state, there was a three-hundred-kilometre-long upper course of the Vistula River. Compared to previous centuries, in the 19th century the assortment and quantity of goods transported along this section of the river decreased. The highest regularity was observed in the flow of wood and the transport of coal and limestone. In the case of other articles, statistical information is very random, which indicates the high irregularity of their movements. This was the case with rock salt, agricultural produce (especially grain), iron, oil, bricks, cement, building materials and other goods. From the mid-1870s to 1913, the total weight of all goods registered by customs offices along the Galician section of the Vistula amounted to nearly 4.2 million tonnes[388]. In the 19th century the Vistula, as a communication artery, was used primarily in the trade of goods within Galicia. Goods were also imported and exported to Prussia and the Kingdom of Poland this way. However, this turnover was much smaller in relation to the pre-Partition period of the Commonwealth of Poland, when a much greater quantity of goods and of greater weight were transported from southern Poland to Gdansk.

Economic activists were well aware of this unfavourable situation and that is why they undertook actions conducive to the revival of transport and communication on the Vistula River. The slogan of restoring the economic significance of the Vistula was given in a memorial from 1898, addressed by the Chamber of Commerce and Industry in Cracow to the Polish Circle in the Vienna State Council. It pointed to the need to launch steam navigation on the Vistula and to establish a steam shipping company, modelled on the companies operating on the Elbe and Danube. High land transport costs increased the price of Galician goods, depriving them of the opportunity to compete on European markets. This concerned both agricultural crops and wood, as well as coal mined in the Cracow Basin mines or cement produced in Szczakowa. It was therefore proposed to extend greatly the export by water, which was a much cheaper alternative to rail and road transport. It was also pointed out that it is easier and cheaper to import colonial goods, coke, iron, agricultural and industrial machinery and food products such as fish, fats and oils into Galicia. The possibility of using the Vistula River in transit trade on the North-South and East-West lines, especially from Russia

388 Jan Drużbacki, *Informator statystyczny do dziejów społeczno-gospodarczych Galicji. Towarowy transport wodny w Galicji doby autonomiczne* [Statistical guide to the socio-economic history of Autonomous Galicia. Water Transport in Galicia], (Cracow: UJ, 1986), pp. 93-97.

to Western Europe, was stressed. However, meeting these demands required both the implementation of many organisational and legal activities, as well as serious investments[389].

Subcarpathian petroleum mining

The figure of Ignacy Łukasiewicz in the popular message is combined with the first kerosene lamp, but Łukasiewicz was first and foremost a pioneer of the world oil mining and refining industry. In the history of oil extraction and processing, in the second half of the 19th century he played an unprecedented role not only in the Polish lands of the Subcarpathian region under the rule of the Austro-Hungarian Empire and as such called Galicia, but also in the global arena. Thanks to his pioneering role in the creation of professional oil mining and industrial refining, the crude oil obtained in the Polish Subcarpathia region accounted for 100 % of the entire output of the Austro-Hungarian monarchy until the end of its existence, i.e. until 1918 with the exception of the First World War, which broke out in 1914.

This raw material had been known in the Carpathians and Subcarpathia for a long time. It was collected in hollows of the earth, in places where there were natural leaks. Crude oil was assigned therapeutic properties in case of various diseases (e.g. rheumatism), used to oil wheels and axles, and people even tried, ineffectively, to use it to light rooms instead of wax candles, vegetable and animal oils. Replacement of bituminous oils, called photogen, or hydrocarbon oil obtained by roasting lignite and bituminous shale, which after mixing with turpentine and alcohol were called Camphine was also tried. The oils used in properly constructed lamps were a safe and odourless product, but expensive. Since the end of the 1880s, gas obtained from hard coal was also used for lighting.

Distillation of crude oil was already being attempted at the beginning of the 19th century. In the years 1810–1817 in Drohobych, the Czech Józef Hecker, an Austrian saline official, was interested in it. Even peasants and merchants undertook to do this. Scientific research on oil distillation was carried out in 1837 by Filip Walter, a Pole, and Joseph Pelletier, a Frenchman, but with no practical effect.

Only a few barrels of smear liquid from the village of Borysław near Drohobycz in the area of Lviv – the capital of Galicia – delivered by local traders to the pharmacy of Piotr Mikolasch in Lviv turned out to be an important link in learning about this intriguing substance. The liquid offered

389 Tomasz Kargol, *Izba Przemysłowo-Handlowa w Cracowie w latach 1850–1939* [Chamber of Commerce and Industry in Cracow in the Years 1850–1939] (Cracow: Historia Iagellonica, 2003), pp. 206–207.

to the Lviv pharmacy was oil obtained from natural leaks and cleaned by Borysław peasants in order to obtain lubricants. Its enterprising suppliers counted on the interest in this peasant "distillate" from the owner of the pharmacy known for his openness to innovations and the first producer of soda water in Galicia, drawing in front of him the mirages of obtaining spirit.

Mikolasch bought a large amount of oil from them and ordered two of his employees – Ignacy Łukasiewicz and Jan Zeh – to investigate it. Difficult and tedious attempts to eliminate the lightest fractions in the pharmacy laboratory made the obtained preparation safe thanks to the increase in the temperature of self-ignition. Oil devoid of heavy components did not smoke during combustion and was 30 % cheaper than the *hydrocarbure* imported from Germany[390].

The pharmacist did not live to see both the spirit and the expected profits from the sale of kerosene. There were also no attractive pharmaceutical products for which Mikolasch cared, which caused him to lose interest in crude oil. However, Ignacy Łukasiewicz decided to exploit the properties of oil distillate and commissioned Adam Bratkowski, a Lviv tinsmith, to make a lamp in which kerosene became a fuel and gave very good lighting. Contrary to Bratkowski's doubts, it was successful.

As early as at the end of 1853, Ignacy Łukasiewicz moved to Western Subcarpathia to work in a pharmacy in Gorlice. Less than a year later he was found there by Tytus Trzecieski – an educated Polish landowner fascinated for a long time both by the spontaneous exploitation of crude oil by the population of the surrounding areas, and by oil itself as a chemical compound, before which he saw a great future. Felicjan Laskowski, vice-president of the Land Credit Society in Lviv, and Karol de Lance, secretary of the Chamber of Commerce, recommended it to Trzecieski as the only one who would be able to distil oil and explain and demonstrate the process. This fact seems to determine, on an equal footing with the further fate of Jan Zeh, which of the Mikolasch pharmacists in this duo was the brain and the main executor of a successful petroleum refining experiment. Since Lviv's financiers and administrators of chambers of commerce knew Łukasiewicz's competences, they couldn't fail to know either Jan Zeh or Mikolasch.

Łukasiewicz turned out to be the person Trzecieski was looking for. Not only did he present him with the innovative lamp, but he immediately went for a vision of the oil-bearing areas of Bóbrka and Polanka, owned by

390 Stanisław Brzozowski, "Ignacy Łukasiewicz", in: *Historia polskiego przemysłu naftowego*, ed. Ryszard Wolwowicz (Brzozów-Cracow: Muzeum regionalne PTTK im. Adama Fastnachta w Brzozowie, 1994), Vol. 1, pp. 16–17.

Trzecieski himself. He also chose Gorlice, where Łukasiewicz ran a pharmacy, probably because it is located on the most oil-bearing route leading from the Dukla Pass in the Carpathian Mountains north to the city of Rzeszow. Having found a common language, the two enthusiasts accelerated the realisation of the vision of creating the oil industry and carrying out broad exploration and production works, with Łukasiewicz being the specialist chemist – the experimenter and mentor of the project. The excavation of the first shaft, a timbered well in the forest area of the Bóbrka estate, which was owned by a landowner named Klobassa, took place in 1855. This first crude oil mine in history was named "Franek". The shaft is still in operation and oil is still being extracted from it today.

Such shafts began to multiply in the Subcarpathia region at a rapid pace. They were dug by hand using shovels and pickaxes, hence they were popularly called *"kopanki"*. Hard levels were demolished with dynamite. A tripod consisting of three beams was placed above the shaft, and on its top a rope with a bucket hung on it was led through the block. The bucket was used for extracting the excavated material and for lowering the digger to the bottom of the hole being drilled. The inner walls of the shaft were lined with boards protecting the well from slipping ground. Fresh air was pumped to the bottom of the shaft by means of a so-called lute, i.e. a manually driven fan. The aim was to eliminate excessive concentrations of extracted gas and supply fresh air to the digger working at the bottom of the hole. Initially, mining specialists from Silesia and Hungary were brought to the mine in Bóbrka. They taught local mining workers and then returned home.

Encouraged by the effects of these innovations, in 1856 Karol Klobassa, the owner of the area in Bóbrka, initiated the establishment of the first Oil Company, which included Ignacy Łukasiewicz and Tytus Trzecieski together with him. At the same time, Klobassa applied to the mining authorities in the *Starosty* of Cracow for a mining permit to carry out official large-scale mining work. This was certainly another starting point for the large oil consortia of today.

Twenty years later there were 61 productive shafts in Bóbrka. The owners of the mines became Galician entrepreneurs and newcomers, the most famous of whom was Stanisław Szczepanowski from Greater Poland. The businessmen were professionals with technical education, among others: Wacław Wolski, Kazimierz Odrzywolski, Stanisław Jurski, Leon Syroczyński, Leon Mikucki, Stefan Bartoszewicz, Bolesław Łodziński, Władysław Długosz and others. Oil fever overcame small savers investing their savings in oil mines. There were also foreigners coming to Galicia who soon played an extremely important role in the development of the oil industry, including Albert Fauck, a German from Słupsk, William Henry

MacGarvey and John S. Bergheim, who came to Galicia from Canada via Germany

Drilling, or the breakthrough

The breakthrough in crude oil mining came with the use of the auger. Initially, drilling for oil covered only the western part of the Subcarpathia region. The first trials were made under the supervision of the industrialist Robert Doms from Lviv in Borysław and Sanok. On April 13, 1861, in the "Wojciech" Bóbrka mine shaft, an auger penetrated to the depth of 22 metres through the layers of thick sandstone and an impressive outflow of crude oil in the amount of about 5,000 kg per day was achieved. History knows the names of the creators of this success. Under the direction and supervision of Henryk Walter, this was done by a worker, already a professional miner, Tomasz Jabłecki. The drill was made on site, in the Bóbrka mine, under the direction of Jan Horytiak. Thanks to the use of drilling methods, the excavated shafts were deepened with great success.

Henryk Walter, who was employed at the Bóbrka mine in 1862 and introduced methods of manual percussion drilling using Fabian's free-fall jars, was an extremely deserving person for raising the level of oil mining. Until the outbreak of the January Uprising in 1863, in which he took part, he was in charge of mining work in Bóbrka, and after his fall he returned to Bóbrka and cooperated with Łukasiewicz, but as a former insurgent he could not occupy an official, mining position controlled by the government.

The "Małgorzata" shaft, opened in Bóbrka in 1862, was already giving a value of approx. 3,800 litres of crude oil per day and decided on the profitability and further development of the mine and the the dynamic arrival of several dozen more wells.

Until the mid-1860s, the auger was set in motion in a drilling derrick and other drilling work was carried out by hand. Significant technical progress in this area was the use of a steam machine to drive the drilling derrick. In the Subcarpathia region it happened in the second half of the 1860s.

Local designers and inventors developed modern systems and technical solutions of global importance, making Lesser Poland one of the leading regions in the field of scientific and technical progress in oil mining. The oil industry had an impact on other spheres of economic and social life. Łukasiewicz was perfectly able to read the importance of the changes taking place, engaging not only in economic undertakings, but also in social and organisational activities, co-creating the National Oil Society and engaging in economic policy at the forum of the National *Sejm*.

From Canada and Pennsylvania

Meanwhile, the fame of oil went around the world. In mid-1859, Captain Edwin Drake inaugurated oil exploration by drilling on the Pennsylvanian oil fields near Titusville, USA. The drilled well turned out to be a rich source of crude oil. The great commercial success contributed to the popularization of his method, called the "Pennsylvanian" drilling method in Europe.

At that time, the Canadian drilling method appeared in the oil fields of the Subcarpathia region, which in a short time completely reigned there. The oil exploration streak did not bypass Canada either. The drilling method used in Canada was based on impact drilling principles. From a technical point of view, it differed from the free-fall method of using automatic shears and the drilling derrick was of different construction. During the first drilling using the Canadian method, it was intended to reach seams at depths of 300–500 m, but it turned out that the new method allowed oil seams to be reached below 1000 metres.

In the middle of 1883, i.e. 29 years after the excavation of the first shaft "Wojciech" in Bóbrka, the Canadian method was applied for the first time in the Subcarpathian drilling industry. Ignacy Łukasiewicz had been dead for a year then. The emergence of the Canadian drilling method there was closely related to the entrepreneurship of foreign oil workers, who, in pursuit of the new, promising raw material, were lucky in the oil-bearing basin of the Polish Subcarpathia region.

William Henry MacGarvey turned out to be a special individual among them. Shortly after his arrival, he became one of the most important entrepreneurs and constructors, and thus a leading figure of the Subcarpathian oil industry until the end of the 19th century and the personification of the American dream from a shoeshine boy to a millionaire. It was this immigrant's company that presented for the first time the Canadian drilling method in the Uherce mine in Subcarpathia, acquired by the Berlin Society – a company of which he was a shareholder[391]. The efficiency of the Canadian system aroused great interest among local entrepreneurs, and its popularization was of breakthrough importance for the Podkarpackie oil mining industry and contributed to the reduction of the costs of the work carried out.

Local oil workers soon began to innovate in the construction of the device called an "old derrick", with which MacGarvey came to Galicia. The most important improvements were introduced by such oil entrepreneurs as Wacław Wolski, Kazimierz Odrzywolski, Felicjan Łodziński, Julian

391 August Gorayski, "Ignacy Łukasiewicz", *Górnik,* Vol. 2 (1882), pp. 166–167.

Timoftiewicz, and others. Soon there was no longer talk about the Canadian system, but about the Galician-Canadian or Polish-Canadian system[392].

The introduction of the effective Canadian drilling method initially established the dominance of the Western Subcarpathian mines, mainly in the Gorlice and Krosno regions. There was drilling, among other places, in Bóbrka, Wietrzno, Potok, on the extension of the oil line in Libusza, Lipinki, Wójtowa, Kobylanka, Siara and Sękowa. Thanks to this method, deep oil deposits were finally discovered in the Eastern Subcarpathia region, which proved to be the most productive in the whole province. Therefore, the main centres of the Subcarpathian oil mining industry moved from the region of Krosno, Jasło and Gorlice in the Eastern Subcarpathians. The oil-bearing areas of Eastern Subcarpathia during the Austro-Hungarian monarchy called Eastern Galicia, after Poland regained independence in 1918, returned to the borders of the Republic of Poland. After 1945 – the end of the Second World War, with the Yalta Agreement's verdict, these lands found themselves within the borders of the Ukrainian Republic of the Soviet Union.

Borysław fever

Oil had also been known in the Eastern Subcarpathia region for years. The first place where its rich deposits were found was the village of Sloboda Rungurska. In the mid-1860s, several hand-dug wells with timbering were built there. In 1879 manual drilling was used here for the first time, activating the "Jadwiga" shaft.

In the same year, Stanisław Szczepanowski established an oil mine in Sloboda. Thanks to him, for several years this town was the leading centre of oil production in the Eastern Subcarpathia region. Szczepanowski was also responsible for the discovery at the end of the 19th century at Skhidnytsya, another important oil mining centre in the Eastern Subcarpathia region, where his company also began drilling in 1890. The wells drilled produced large quantities of oil. The "Jakub" shaft drilled in 1896 turned out to be a particularly efficient shaft with a daily production of more than 500 tonnes of crude oil. Unfortunately, just before the discovery of these fields Szczepanowski sold his mine.

392 Antoni Klebert, "Rozwój żurawia kanadyjskiego wiercenia w Galicji" [The Development of the Canadian Drilling Rig in Galicia], *Przegląd Techniczny Naftowy*, dodatek do czasopisma *Nafta*, R. 1907, p. 17–20, 25–27, 29–31, 37–41, 45–46; Zygmunt Bielski, "Wybór metody wiercenia. Referat na Zjeździe Naftowym we Lwowie 26 czerwca 1927 r." [Choice of Drilling Maethod. Presentation at the Oil Congress in Lviv on June 26, 1927.], *Przemysł Naftowy*, Vol. 2, No. 21 (1927), pp. 385–391.

In the early 1890s, Bergheim and MacGarvey started deep drilling in Borysław. This work was managed by a Pole, Władysław Długosz. It was thanks to his persistence that in 1896, after exceeding the depth of 500 metres, oil spurted to a volume of up to 40 tonnes per day, unprecedented in a single shaft in the oil fields of the Subarpathians. It was the beginning of the "Borysław oil fever".

Ultimately, the dominance of the Eastern Galicia Subcarpathian region in terms of oil production was sealed by the discovery of oil fields in the Tustanovice region. This was done in 1903 by Dr. Władysław Szujski, the son of a well-known historian from Cracow. The "Litwa" shaft he drilled produced over 1,000 tonnes of oil per day. Shortly afterwards, further shafts produced similar capacity, and the "Oil City" shaft with a capacity of up to 2.5 thousand tonnes of oil per day went down in history. Thus, the Eastern Subcarpathian region gained a dominant position in oil production. This dominance was also maintained in the interwar period, despite the enormous destruction of oil installations by the Soviet army. The Russian forces, having driven Austria-Hungary out of this area, were forced by the Polish army to withdraw from the former Austrian Galicia for the city of Kiev.

Figure 29: The oil field in Borysław. Source: The National Digital Archives.

Entering the 20th century. Innovation, patents and competition

With time, three oil mining districts in the vicinity of Stanisławow, Drohobycz and Jasło developed in the territory of the Polish Subcarpathia region. The first two districts were located in the eastern part of Subcarpathia, the third in the western part[393]. The number of oil mines in Subcarpathia at the turn of the 20th century was estimated at over 250, and at the beginning of the 1910s this number reached 400[394]. Most of them were small, with one shaft, others were serious companies with over a dozen shafts belonging to joint stock oil companies.

The production of Subcarpathian crude oil at the turn of the 20th century was characterised by extraordinary dynamics. Oil production statistics show an almost fifty-fold increase in production – from about 40,000 tonnes in 1885 to 2 million tonnes in 1909, which, representing 5.2 % of world production, gave the third place among oil producers after the USA and Russia, where the Basin near Baku had been experiencing a climax of oil mining development for several years. Such a rapid increase in production was connected with the process of setting up of new mines. The number of employees in the mines of the Polish Subcarpathia region increased from 2917 in 1886 to 7268 persons in 1913.[395]

However, the Pennsylvanian method of rope drilling did not work in these geologically difficult seams. The Canadian method was still practised in the Subcarpathia region in parallel with the tendency to replace it with more modern systems. Domestic constructions, e.g. those proposed by Albert Fauck, were tested. Attempts were made to disseminate wash boring apparatuses. Wash drilling, in which cleaning of the bore with the use of a so-called water scrubber pressed into it operated simultaneously with the work of the auger, was a significant step forward in relation to the "dry" methods – free fall and Canadian. Drilling with diamond bores, an expensive

---

393 Stanisław Olszewski, "Przemysł naftowy w Galicji w latach 1884–1901" [The Oil Industry in Glaicia in the Years 1884–1901], *Nafta*, Vol. 10 (1902), pp. 133–137; Stanisław Olszewski, "Z dawnych lat przemysłu naftowego" [Olden Times in the Oil Industry], *Przemysł Naftowy*, Vol. 4, No. 13 (1929), pp. 406–409; Władysław Długosz, "Wspomnienia z Borysławia" [Memoirs of Borysław], *Przemysł Naftowy*, Vol. 4, No. 9 (1929), pp. 329–332; Franaszek, *Myśl techniczna*, pp. 91–93.

394 Bar, Franaszek, *Informator statystyczny*, pp. 130–133; Burzyński, *Robotnicy w przemyśle ciężkim*, p. 99.

395 Piotr Franaszek, "Zatrudnienie w przemyśle naftowym Galicji (do roku 1914)" [Employment in the Galician oil industry (until 1914)], *Studia Historyczne*, Vol. 33, No. 1 (1990), pp. 37-46.

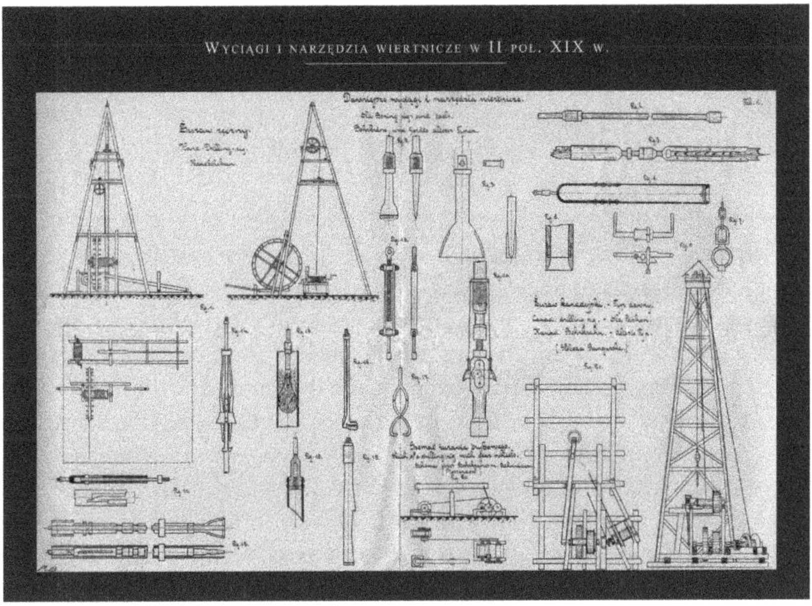

Figure 30: Drilling rigs and tools from the 2nd half of the 19th century. Source: Subcarpathian Museum in Krosno

version of wash drilling, was tested without much success[396]. Józef Howarth, Władysław Pruszkowski, Stanisław Janiszewski and Walery Siterski[397] also worked on wash drilling constructions in the Podkarpackie mines.

A number of very successful, homegrown projects for this kind of apparatus appeared[398]. An innovation in this method was the use in Borysław

396 Marcin Maślanka, *Zarys kopalnictwa naftowego. Podręcznik dla urzędników kopalni naftowych* [An Outline of Oil Mining. A Handbook for Oilfield Officials] (Stanisławów: Wysoki Wydział Krajowy, 1885), pp. 19, 23.

397 Franaszek, *Myśl techniczna*, pp. 123–140.

398 Albert Fauck, "Ein Drittel Jahrhundert der Entwicklung der Tiefbohrtechnik in Galzien", *Naphta*, Vol. 8 (1900), pp. 99–102; Zygmunt Bie1ski, "Żuraw wiertniczy Express" [Express drilling rig], *Nafta*, Vol. 12 (1904), pp.129–134; Wacław Wo1ski, "O nowych systemach wiertniczych" [On New Drilling Systems], *Nafta*, Vol. 9 (1901), pp. 1–4, 17–21; Wiktor Petit, "Nowy system wiertniczy" [A New Drilling System], *Ropa*, Vol. 1 (1911), pp. 278–284; Wiktor Petit, "Kombinowany przyrząd wiertniczy systemu Petita" [Petit System Combined Drilling Tool], *Ropa*, Vol. 4 (1914), pp. 278–282, 297–300; Wacław Wolski, "O taranie wiertniczym" [On the Drilling *Taran*], *Nafta*, Vol. 10 (1902), pp. 120–126, 137–142; Wacław Pruszkowski, Józef Horwath, "Hydraulisch betriebener Tiefbohrer", *Naphta*, Vol. 8 (1900), pp. 65–67; Stanisław Czerwiński,

of a derrick of his own construction by Leon Mikucki in 1901. Mikucki managed to drill a shaft with a depth of more than 510 metres, which was a sensation for those times. In 1909 Wiktor Petit, with the use of a scrubber of his own design, exceeded the results of the Canadian method by 50 %. Unfortunately, neither Mikucki's nor Petit's constructions were accepted on a larger scale.

The peak achievement and absolute innovation of the world's early 20th-century drilling technology was the prototype of Wacław Wolski with the meaningful name *"Taran"* ["battering ram"], for which the creator received the highest award during the international exhibition in Liége. It was a derrick for wash drilling using a water hammer as the result of sudden stopping of the accelerated stream of pumped water under high pressure. This solution was a complete novelty. In 1902, the companies Deutsche Tietbohr-Aktien Gesellschaft in Nordhausen and Internationale Tietbohrgesellschaft in Erkelenz acquired the right to the patent of *"Taran"* in all European countries for the dizzying sum of 600,000 marks, with the exception of Galicia as a Polish province under the annexation of Austria-Hungary. However, it was a ploy to destroy competition. Formally, these companies committed themselves to improving the system and advertising it. In fact, they aimed to prevent its use and dissemination in Europe.

In 1913, the company "Galicyjskie Karpackie Naftowe Towarzystwo Akcyjne, dawniej Bergheim i MacGarvey" (GKNTA) [Galician Carpathian Oil Joint Stock Company, formerly Bergheim and MacGarvey] undertook to adapt the rotary drilling system to the conditions of the Subcarpathia region. As a result of this work, a rotary drilling structure called the "Karpath Rotary" with a Sharp and Hughes patented auger was developed. In May 1913, in Tustanovice this device was used to drill the "Wageman III" shaft, the first in the history of Subcarpathian oil drilling with the rotary washing method. The obtained results of the drilling confirmed the great possibilities of the rotary drilling method, however, the outbreak of World War I stopped further experiments in this area[399]. The rotating scrubber system is now widely used to make oil seams available.

"Taran p. Wolskiego" [Mr. Wolski's *Taran*], *Nafta*, Vol. 11 (1903), pp. 145–146; Jan J. Cząstka, "Wacław Wolski", *Nafta*, Vol. 11 (1955), pp. 152–153; Antoni Plutyński, "Inżynier Wacław Wolski" [Engineer Wacław Wolski], *Kwartalnik Historii Nauki i Techniki*, Vol. 11, No. 4 (1966), pp. 397–416; Albert Fauck, "Użycie płuczki na nowych terenach" [Use of the Flusher in New Areas], *Ropa*, Vol. 1 (1911), pp. 421–423.

399 "Wiercenie rotacyjne w Galicji 'Parker-Karpath-Rotary' – świdry 'Sharp and Hughes' " [Rotary Drilling in Galicia, with the Parker-Karpath-Rotary – Sharp and Hughes Auger], *Ropa*, Vol. 3 (1913), pp. 218–221; "System rotacyjny

Geology or divining rod

Among many important factors in the development of oil mining in the Subcarpathia region, the specific geological conditions determining both the type of drilling techniques and the threats to the oil industry and the natural environment were of crucial importance. The Subcarpathian deposits are characterised by an unfavourable and complicated system of geological layers, diversity, frequent and steep arrangement and a high degree of flowability of the seams. These circumstances make it extremely difficult to carry out both oil exploration and production.

In 1878, the National *Sejm* in Lviv passed a resolution ordering the commencement by the National Department of the action of collecting geological information on deep oil deposits. Thanks to this, drilling works by selected companies were given the chance of subsidy, and drilling engineers received detailed information about the methods of graphical presentation of stratigraphy and petrography of the drilled geological layers, methods of preliminary description of water and oil samples taken, which they were obliged to send for further research in Lviv. On the one hand, this action was of a scientific nature, and on the other hand, it aimed at attracting interest in the above-mentioned problems from oil companies. In 1880, the National Parliament passed a resolution to allocate 50,000 guilders for the study of oil deposits. Three years later, 13 thousand guilders were allocated for the same purpose, as it was stated – "directly for geological research and 500 guilders for the Physiographic Commission of the Academy of Arts and Sciences in connection with the Geological Atlas of Galicia which it is developing". In the 1890s, the National Department began a campaign of action of rewarding geological work providing hitherto unknown information about the geology of Galicia.

In 1912, on the initiative of a Polish geologist, professor of the Jagiellonian University Józef Grzybowski, a Geological Station was established in Borysław, whose task was to conduct research and training in the field of oil geology. Bolesław Kropaczka was appointed head of the station[400].

---

wiertniczy Parkera" [Parker's Rotary Drilling System], *Nafta*, Vol. 21 (1913), pp. 168–170.

400 "Geologiczne badania głębszych warstw ziemi – odczyt Leona Syroczyńskiego, inżyniera Wydziału Krajowego we Lwowie, miany na IV Zjeździe Polskich Przyrodników i Lekarzy w Poznaniu dn. 22 czerwca 1884 r." [Geological studies of deeper layers of the earth – lecture by Leon Syroczyński, Engineer of the National Department in Lviv, delivered at the 4th Congress of Polish Naturalists and Doctors in Poznań on 22 June 1884.], *Górnik*, Vol. 3 (1884), pp. 84–86; Józef Grzybowski, "O potrzebie stacji geologicznej w Borysławiu" [On the need for a geological station in Boryslaw], *Ropa*, Vol. 2 (1912), pp. 73,

A series of lectures entitled "Geology applied to oil mining in the Carpathians", conducted by Rudolf Zuber at the University of Lviv since 1893, also served to promote geological knowledge among oil companies. A great undertaking and at the same time an achievement was the work on the "Geological Atlas of Galicia" undertaken by the Physiographic Commission of the Academy of Arts and Sciences in Cracow. Numerous geological maps of various regions with the distribution of oil areas were published. A great number of scientific papers on geological conditions in the Subcarpathia region, and in particular its oil areas, were written[401]. Popularization of geology with petroleum mining was supported by numerous articles published in the professional press, which had been in existence for years. Exhibitions presented the achievements of the oil industry, geological specimens and drawings of geological cross-sections of the areas.

Unfortunately, theoretical knowledge was not sufficiently appreciated and used in business practice. The designation of areas for future drilling was mainly based on spontaneous surface oil spills or by joining existing oil wells. Stanisław Olszewski, writing in 1910 – perhaps with some exaggeration – stated:

> "A characteristic feature of those years was that entrepreneurs in Borysław, Mraźnica, Dźwiniacz and Staruń adhered to the principle, which I call peasant geology, to look for oil where it shows itself and to put the shafts as close to each other as possible, even at a distance of 2 metres. Hence the number of shafts in those towns where relatively small spaces (about 142 ha) were dug (according to statistics from 1881 3,616 shafts). Such areas with shafts were called 'peasants' sieves."[402]

242–244; "Program wykładów z geologii naftowej" [Programme of Lectures on Oil Geology], *Nafta*, Vol. 20 (1912), pp. 364–365.

401 Rudolf Zuber, Nowa strefa ropna w Galicji [A New Oil Zone in Galicia], *Górnik*, Vol. 3 (1884), pp. 29–30; Rudolf Zuber, *Geologia pokładów naftowych w Karpatach galicyjskich* [Geology of Oil Fields in the Galician Carpathians] (Lviv: author's print, 1899); Rudolf Zuber, *Geologiczne podstawy występowania oleju skalnego* [The Geological Basis for the Occurrence of Rock Oil] (Lviv: author's print, 1916); Józef Grzybowski, *Przegląd obszarów naftowych Karpat polskich* [Review of Oil Areas in the Polish Carpathians] (Cracow: Tow. TEWUGE, 1919); Klaudiusz Angerman, *Ogólna geologia naftowa. Podstawy do badań terenów naftowych* [General oil geology. Basis for oilfield exploration] (Lviv: author's print, 1902).

402 Stanisław Olszewski, "Galicyjski przemysł naftowy jego dotychczasowe położenie i widoki na przyszłość" [The Galician Oil Industry's Current Location and Future Prospects], *Nafta*, Vol. 18 (1910), p. 349.

It happened that fraudsters pretending to be diviners, indicating the sources of oil occurrence, were involved in designating the locations of oil mines. This is how Olszewski described it:

> "Herr Meissel und Frau Pumpe, as it was said, are the experts they used to proudly point to as the only ones geologists believe in. And although it is difficult to deny that only on this basis of forced, often blindly undertaken drilling the now powerful Carpathian Society grew up, we allow ourselves to claim that hundreds of kilometres drilled, with the skilful use of geology could have led to results much sooner and to a greater extent."[403]

However, it must be stated that similar, non-scientific practices were not a local feature of Subcarpathia. For example, similar remarks are known from the lips of the most eminent geologist of Baku at the turn of the 20th century, Witold Zglenicki, a Pole. This forefather of seabed oil extraction, pioneer of drilling platforms and creator of precise geological maps of Azerbaijan and the Gulf of Bibi-Heybat of the Caspian Sea sometimes mocked the magical, irrational practices used by various nations of entrepreneurs brought to the Baku El Dorado from all over the world.

Drilling works undertaken in such a way, almost blindly, often led to serious financial losses, bankruptcies of enterprises and at the same time senselessly degraded the environment. In 1912, therefore, by decree of the Drohobycz Mining Authority and pursuant to § 37 of the Petroleum Act of 1908, all mines were obliged to keep logbooks and to store samples of drilled geological layers.

Eruptions, fires, landslides, groundwater

The complex geological system of Galician oil deposits and the accompanying gas were often the cause of sudden and uncontrolled oil eruptions. The powerful pressure of pushing oil and gas ejected the drill bit with the drilling line from the borehole. Such an eruption was a very serious threat to the workers. A characteristic plume of gushing oil formed a few or even a dozen metres above the surface of the ground, which flooded the fields and flowed into the hollows of the terrain. Although it could sometimes be collected in tanks, most of it flowed down into streams and rivers, poisoning the fauna and flora.

A large number of such explosive shafts were found in mines located near Wietrzno. The mixture of oil, gas and air often ignited. This fire may have

403 Stanisław Olszewski, "Rozwój przemysłu naftowego w Rumunii oraz w innych krajach" [The Development of the Oil Industry in Romania and Other Countries], *Nafta*, Vol. 4 (1896), p. 128.

been caused both by natural spontaneous ignition, as well as by the fault of workers or by sparks from operating equipment. The most spectacular event of this type was the fire of the richest shaft in terms of oil production called "Oil City" in Tustanovice. In the summer of 1908, the year in which the magnificently oil-bearing well was drilled, the shaft started to burn as a result of a lightning strike. It took several months to extinguish it, and all the oil in it burned out[404].

Petroleum shaft fires were also intentionally ignited. In 1915, during the First World War, the retreating Russian army in 1915 not only systematically pillaged and destroyed the equipment of oil mines, but also set great fires in shafts. It took the Poles many weeks to extinguish the 322 shafts set on fire by Russians at that time.

A serious danger for the drilling works was the collapse of the walls of the drilled hole. Therefore, each section of the shaft had to be secured systematically. Initially, in order to protect the hole against collapsing, rolled pipes made of sheet metal 2–3 mm thick were used. These were so-called lost pipes, because there was practically no possibility of their reuse. Another danger that occurred during drilling were natural groundwater veins, flooding the drilled shaft and, as a result, the inability to exploit it. A solution to this threat was found when deep deposits of oil were discovered in Borysław. It turned out to be the introduction of sheet metal pipes, hermetic, especially steel, connected by a thread, lowered directly behind the auger. By protecting the hole from collapse, these pipes simultaneously protected it from flooding.

Their first application in Subcarpathian oil mining took place in Bóbrka in 1874. These were pipes imported from the United States[405]. With the introduction of the Canadian drilling method, they proved essential. A movable column of pipes led directly behind the auger was created. If such a column, having reached a certain depth, could not be moved further, a column of smaller diameter pipes was introduced centrifugally into it. The use of the

---

404 Wit Sulimirski, "O rozwoju kopalnictwa naftowego w Galicji" [On the Development of Oil Mining in Galicia], *Ropa*, Vol. 3, No. 5 (1913), pp. 479–485, 502–507; Stefan Bartoszewicz, "Pogląd na dzieje przemysłu naftowego i środki jego sanacji" [View of the History of the Oil Industry and the Means of Its Remediation], *Nafta*, Vol. 16 (1908), pp. 367–377.

405 Jan J. Cząstka, "U źródeł powstania polskiego przemysłu naftowego. (Z okazji 120-rocznicy powstania tego przemysłu)" [At the Source of the Emergence of the Polish Oil Industry. (On the Occasion of the 120th Anniversary of the Establishment of this Industry)], *Nafta*, Vol. 29 (1973), pp. 337–344; Roman Mierzecki, "Przemysł naftowy w Polsce w XIX i XX wieku" [The Oil Industry in Poland in the 19th and 20th Centuries, *Analecta*, Vol. 8/2 (16), No. 55–71 (1999), pp. 55–71.

eccentric auger, patented by MacGarvey, which makes holes with a diameter larger than its width, prompted drillers to pipe sections several hundred metres long with a column of pipes. In 1897, the Galician mining authorities imposed an obligation to close groundwater with hermetic pipes. Removal of the pipes was possible only with their consent, after appropriate protection of the opening with clay or cement mortar[406].

The Subcarpathian refinery industry

Ignacy Łukasiewicz not only started industrial production of crude oil, but also its distillation in the refineries he founded in Ulaszowice, Polanka and Chorkówka. Soon, other entrepreneurs, usually in the vicinity of oil mines, started to set up similar businesses. The first refineries were not large. The refinery in Stanisławów employed 8 people, in two refineries in Stary Sambor 10.

Walerian Stawiarski, who took over the refinery in Chorkówka after Łukasiewicz's fire, founded the refinery in Krosno in 1905 and equipped it with the most modern equipment for oil processing. Over time, he also purchased installations for the production of white paraffin and gasoline rectification.

One of the most modern plants was the refinery in Glinik near Gorlice, founded around 1883 by Bergheim and MacGarvey, which from 1895 became the property of the Galician Carpathian Oil Joint Stock Company. The refinery covered an area of 18 ha and had a railway connection to the national railway network. In the early 1890s it was equipped with two distillation boilers with a capacity of 51 m^3 and three with a capacity of 71 m^3. Over time, the plant's processing capacity increased to 100,000 tonnes per year.

In 1894, during the General National Exhibition in Lviv, samples of petroleum distillation products from the Glinik Marijampole refinery, from gasoline, through several types of kerosene called "imperial, saloon, farming kerosene", to lubricants, were presented in the naphtha distillation section.

In 1878 Count Adam Skrzyński established a refinery in Libusza, which after World War I was transferred to Standard-Nobel. At the turn of the 20th century, the Jedlicze refinery was established, equipped with 10 distillation boilers from 1908 to 1910. The factory employed 80–100 people. In

406 Piotr Franaszek, "Eksploatacja ropy w Galicja a zagrożenia dla środowiska naturalnego" [Oil Production in Galicia and Environmental Dangers], in: *Od regaliów po dobro narodowe. Ochrona i wykorzystanie zasobów środowiska narodowego na ziemiach polskich – aspekt historyczny*, ed. Tomasz Głowiński, Marek Zawadka, (Wroclaw: Wydawnictwo Gajt, 2016), pp. 39–54.

1888 the construction of a refinery in the village of Niegłowice, then adjacent to Jasło, began. The construction was completed in 1895. It was the largest plant in the area and employed about 300 people. It belonged to the Dutch concern Petinag A.G. Gartenberg Schreier et-Amsterdam[407].

About 30 distilleries and refineries operated in the Borysław-Drohobycz district. Among them, there was the refinery "Galicia" in Drohobycz, which initially produced ceresin. At the end of the 19th century, the refinery was taken over by English capital and a joint stock company, The Anglo-Galician Comp. Ltd., was formed, which first processed oil from Schodnica and then Borysław in the amount of about 60 thousand tonnes per year. After the First World War, the refinery was expanded in free Poland, thanks to which it obtained a processing capacity of about 140,000 tonnes of crude oil per year. Drohobycz also had a refinery owned by the "Nafta" Limited Company, with an annual processing capacity of 42 thousand tonnes of crude oil and the refinery "Dros" with a processing capacity of 30 thousand tonnes.

The following refineries operated in Borysław: Galicyjska Spółka dla Spożytkowania Oleju Ziemnego with an annual processing capacity of 1,300 tonnes, the "Gazolina" S.A. refinery and the Mineral Oils inż. Baltuch i Tow. refinery, processing annually about 5,300 tonnes of crude oil. The refinery of the Borysławska Spółka Przetwórów Oleju Skalnego "Borysław" had a similar processing capacity. In the Borysław-Drohobycz region there were also a number of smaller companies: the Schutzmann, Zuckerberg and Ska refinery, the H. Altman and J. Gottlieb refinery, the Salomon Kreppel refinery, the Hoffman Brothers and Ska refinery, the "Iriag" refinery in Drohobycz and many others[408].

Far from the oil fields there was an oil refinery in Trzebinia, located west of Cracow. Count Andrzej Potocki's application for a building permit for it was approved on June 6, 1895 by the outstanding Polish Count, who at that time became the Prime Minister of Austria, Kazimierz Badeni. Count Potocki's refinery already in the following year processed about 400 tonnes of crude oil per month. In the following year, the refinery's production capacity increased to 1,000 tonnes. The refinery produced several types of lighting kerosene. After ownership reorganisations and equipping it with a modern installation, in 1905 5,000 tonnes of crude oil per month were distilled there. In the following years, a paraffin factory was established in the refinery, producing 12 thousand tonnes of this product annually. Gasoline was recovered and heavy fractions left over from oil distillation were processed. In 1907,

407 Kachlik, *"Rafinerie nafty w Polsce"*, pp. 431–432, 437–438, 442–443.
408 Kachlik, *"Rafinerie nafty w Polsce"*, pp. 408–411, 418–446.

the refinery obtained all final products derived from processed crude oil. In 1913 the production of toluene, used for the production of explosives, began. During the First World War, the construction of new tanks allowed the amount of processed oil to be increased to 60 thousand tonnes per year. The factory had its own power plant of 690 kW. Just before the outbreak of World War I about 300 people worked there[409].

Refineries in Czechowice and Limanowa were also located far from the raw material base. The first one was founded in May 1905 by the Vacuum Oil Company Ltd. owned by the American Rockefeller Vacuum Oil Company. Its processing capacity amounted to 12,000 tonnes of crude oil per year, and the production was intended for sale to neighbouring Silesia. The refinery in Sowliny-Limanowa was built just before the outbreak of World War I by the "Limanowa" Oil Society, supported by French capital. The refinery had an annual processing capacity of 35,000 tonnes of crude oil and employed approximately 200 people[410].

Nevertheless, the development of the refining industry in the Subcarpathia region did not keep up with oil extraction. In the mid-1880s, nearly 60 oil production plants were registered in Galicia. These were small enterprises, using simple technology.

There is no doubt that this state of affairs was linked to the effect of the longstanding discriminatory policy of the Austrian authorities towards the economic initiatives of the indigenous peoples. As a result of confiscation, post-Uprising repressions and foreign colonisation, Poles did not have the capital capable of competing not only with financial powers from abroad, but also with other fellow citizens of the Austro-Hungarian monarchy. Poles encountered an incomparably greater number of the bureaucratic barriers for which the Austrian administration was famous. From the beginnings of the oil industry, from the times of the first borehole in Bóbrka and the first distillery organised by Ignacy Łukasiewicz, while priority, inventiveness, knowledge, intelligence, logistic abilities, innovativeness and diligence decided, Poles pulled Subcarpathia up to the leading position in this industry. When a great deal of capital was needed for large projects, the situation changed.

Meanwhile, at the end of the 19th century, it turned out that the rapid increase in oil on the market caused an overproduction crisis in Galicia. The first wave of this crisis occurred around 1896. The "Ropa" Association of Galician Oil Producers established in the same year[411] undertook to regulate

---

409 Orłowski, *"Trzebinia i region w latach 1815-1918"*, pp. 222–223.

410 Kachlik, *"Rafinerie nafty w Polsce"*, pp. 452–453.

411 Stanisław Olszewski, "Kryzys naftowy w Galicji i zgromadzenie producentów naftowych we Lwowie" [The Oil Crisis in Galicia and the Assembly of Oil

Figure 31: The oilfield personnel in Bóbrka 1908. Source: The Ignacy Łukasiewicz Oil and Gas Industry Museum in Bóbrka

and control trade in the raw material and, until 1901, managed to maintain a profitable price for it for the producers. However, further improvements in drilling technology and a surge in production, especially in the Boryslav and Tustanowice region, led to another crisis in overproduction and a collapse in oil prices. While in 1901 prices were 5.1 crowns per 100 kg and in 1902 – 2.84, in the following year only 2.5 crowns[412]. Therefore, in the situation when drilling technology flourished in the Subcarpathia region, oil producers, and with them the whole Galician province, suffered increasing losses.

Producers in Lviv], *Sprawozdania Krajowego Towarzystwa Naftowego we Lwowie*, Vol. 2 (1895), pp. 1–4.

412 Bar, Franaszek, *Informator statystyczny*, pp. 132–133; August Gorayski, "Przemysł naftowy" [The Oil Industry], *Przegląd Polski*, Vol. 37 (1903), pp. 519–532; Stefan Bartoszewicz, "Pogląd na dzieje przemysłu naftowego i środki jego sanacji" [View of the History of the Oil Industry and the Means of its Remediation], *Nafta*, Vol. 16 (1908), pp. 367–377; Stanisław Olszewski, "Przemysł naftowy w Galicji w latach 1884–1901" [The Oil Industry in Galicia in the Years 1884–1901], *Nafta*, Vol. 10 (1902), pp. 133–137.

In these circumstances, in Vienna, in 1903, the central joint stock company "Petrolea" was founded, bringing together oil producers and refiners. The Society granted advance payments for oil and undertook to build new reservoirs of raw material. However, it soon turned out that the "Petrolea" Society acted primarily in the interests of the refineries, ignoring the interests of oil producers. The Society's activity did not therefore help the situation in the market, which is why in 1908 Subcarpathian entrepreneurs established the Association of Oil Producers in Lviv on their own. It was only the work of this Lviv-based association that allowed the situation to be brought under control and a price of oil to be obtained which guaranteed the profitability of production[413].

Cyclical crises of oil overproduction indicated the need to build a strong and modern domestic refinery industry, for which, however, low and uncertain prices were certainly not a stimulating factor. It is also a fact that at the beginning of the 20th century the market was saturated with oil and its products. The role of petroleum lighting was systematically decreasing due to the spread of gas and electric lighting. Demand for oil in the chemical and automotive industries was still low at the beginning of the 20th century. Looking for markets, Galician oil activists pointed out the benefits of using oil as a fuel. At that time, for example, salinas in Stebnik and Drohobycz were heated with oil.

In 1906 the magistrate of Lviv decided to heat the city's bathhouse with oil. Distillery owners were encouraged to do the same, based on the positive examples of the Lviv power plant and the sugar plant in Przeworsk. The constructions of furnaces and burners for home oil firing were also developed. However, for heating purposes, only a small part of Subcarpathian crude oil was used (in 1906/07, barely 4,000 tanks)[414]. High hopes were placed on the use of oil for fuelling locomotives on Austrian railways, which would allow for the creation of a constant, stable and quite capacious market

413 "Układy z Petroleą" [Agreements with Petrolea], *Nafta*, Vol. 14 (1906), p. 149–152; Stefan Bartoszewicz, "Nowa organizacja naszego przemysłu" [A New Organization for Our Industry], Vol. 14 (1906), pp. 277–280; Stefan Bartoszewicz "Embarras de richesse", *Nafta*, Vol. 15 (1907), pp. 237–239; Sulimirski, "O rozwoju kopalnictwa naftowego", pp. 479–485, 502–507; Franciszek Bujak, "Przemysł naftowy w Galicji i obecne jego przesilenie" ["The Oil industry in Galicia and its current solstice"], *Ekonomista*, Vol. 9 (1909), pp. 111–156.

414 Stefan Bartoszewicz, "Opalanie ropą" ["Heating with crude oil"], *Nafta*, Vol. 15 (1907), p. 329; "Nowy palnik ropy inż. Zygmunta Rodakowskiego" [The new oil burner by engineer Zygmunt Rodakowski"], *Nafta*, Vol. 15 (1907), pp. 171–172.

for oil. Negotiations on this matter between oil companies and the authorities represented by the Ministry of Railways and the directorates of state railways in Cracow, Lviv and Stanisławów began as early as the mid-1890s. In February 1895, the first tests with oil-fired locomotives were carried out on the Rzeszow-Dębica route. Further trials were conducted in 1897, but the talks broke down on the price of fuel. In 1902, locomotives fired with so-called *ropał*, i.e. a mixture of crude oil and heavy oils, were tested on the Lviv – Przemyśl, Przemyśl – Podwołczyska, Przemyśl – Chyrów routes. The *ropał* was supplied by refineries in Ustrzyki, Lviv and Zniesienie. The results were promising, as 1 kg of fuel allowed 12 litres of water to evaporate, while 1 kg of coal was less than 5 litres. These results still did not encourage the Austrian state railway authorities to take a deeper interest in this problem, even though in 1908 five steam locomotives fired with Silesian coal, which at that time was foreign, were running between Borysław and Drohobycz. This was a characteristic feature of Austrian policy towards the economy of the Polish province, whose excessively intensive growth was often suppressed in an artificial, mostly bureaucratic way. At the same time, 500 oil-fired locomotives were running in neighbouring Romania.

It was not until 1908 that a contract was signed for the Subcarpathian oil industry to supply in the years 1909–1914 30 thousand cisterns of oil per year to fuel locomotives, in the furnaces of which the oil was sprayed with the Holden system burner. This solution made it possible to move away from coal-fired locomotives, which contributed to the stabilisation of the oil market and a gradual increase in oil prices, which reached 5 crowns per 100 kg in 1912[415].

In 1910, the "Ropa" association conducted negotiations with the Standard Oil Company on the sale of Subcarpathian oil to this powerful American monopolistic group. The talks, which were already close to signing the agreement, were interrupted as a result of direct intervention by the Austrian government, fearing that the oil mining industry in the occupied Galician province would be taken over by American capital[416].

415 "O zastosowaniu ropy do celów opałowych" ["On the use of oil for the heating purposes"] *Nafta*, Vol. 13 (1905), pp. 356–359; "Potrzeba organizacji" ["The need for the organisation"], *Nafta*, Vol. 16 (1908), pp. 121–122; "Opalanie lokomotyw ropą" ["Fueling locomotives with the oil"], *Nafta*, Vol. 17 (1909), pp. 243–247; Stefan Bartoszewicz, "Projekt większej assocjacji w naszym przemyśle" [A Project for Greater Association in Our Industry], *Nafta*, Vol. 20 (1912), pp. 97–98.

416 "Pertraktacje z towarzystwem Standard Oil Company" [Talks with the Standard Oil Company], *Nafta*, Vol. 17 (1909), pp. 145–147; "Układ z towarzystwem Standard Oil Company" [An Arrangement with the National Oil Company],

However, pressure from the Polish scientific and industrial circles, lobbying on behalf of autonomous oil workers, the Polish authorities of the Galician province in order to increase demand for Subcarpathian oil and its products did not stop. The National Petroleum Society founded in Lviv made a special contribution to the switch of the Austrian railways to liquid fuel. On its initiative, in 1910 a state refinery in Drohobycz was established, a modern refinery called a "decontamination plant". The refinery was initially focused on the production of fuel oil.

Stanisław Pilat, an outstanding Polish economic activist, became the director of this plant. The supervision over the plant was taken up by the Ministry of Public Works. Thanks to the personal merits of Director Stanisław Pilat, within two years the company was expanded by adding an installation for the rectification of gasoline and kerosene. In 1914, the refinery was equipped with a large research laboratory and a paraffin factory was built. After the outbreak of the war, in 1915 Russian troops invaded Drohobycz. The Russians unsuccessfully tried to run the facility. Pushed out by the Polish army, the oil reservoirs were set on fire, which led to serious damage to the plant. After Poland regained independence in 1918, the plant was taken over by the government of the Republic of Poland. The refinery under the name Państwowa Fabryka Olejów Mineralnych "Polmin" had a processing capacity of about 200,000 tonnes of crude oil per year and employed 820 people in 1927[417].

Refineries in Czechowice and Limanowa were also located far from the raw material base. The first one was founded in May 1905 by the Vacuum Oil Company Ltd. owned by the American Rockefeller Vacuum Oil Company. Its processing capacity amounted to 12,000 tonnes of crude oil per year, and the production was intended for sale to neighbouring Silesia. The refinery in Sowliny-Limanowa was built just before the outbreak of World War I by the "Limanowa" Oil Society, supported by French capital. The refinery had an annual processing capacity of 35,000 tonnes of crude oil and employed approximately 200 people[418].

Nafta, Vol. 17 (1909), pp. 161–163; Frank, *Oil Empire. Visions of Prosperity*, pp. 168–169.

417 Stanisław Olszewski, "Zapotrzebowanie nafty w Austro-Węgrzech" [Oil Requirements in Austria-Hungary], *Nafta*, Vol. 6 (1898), p. 45; "Statystyka przemysłu naftowego w Galicji w roku 1900" [Statistics on the Oil Industry in Galicia in 1900], *Nafta*, Vol. 9 (1901), p. 67; Aleksander Szczepański, *Stan wytwórczości przemysłowej i górniczej w Galicji w roku 1910* [Industrial and Mining Production in Galicia in 1910] (Lviv: Nakładem Wydziału Krajowego, 1912), pp. 54, 104–107, 142–144; Kachlik, "Rafinerie nafty w Polsce", pp. 409, 433–437.

418 Kachlik, *"Rafinerie nafty w Polsce"*, pp. 452–453.

Figure 32: Drilling tools from a historic mine in Bóbrka. Source: Anna Kozicka-Kołaczkowska

Factories of drilling machines and tools in Subcarpathia and Lesser Poland at the turn of the 20th century

Along with the dynamically developing oil mining industry, there was a demand for the production of equipment and tools necessary for drilling and oil extraction. Therefore, from the very beginning, mechanical workshops were set up in the oilfields, which were often transformed into factories of drilling equipment, producers of oil pumps, oil tanks, machines and propulsion equipment, as well as various types of tools used for work on oilfields. Simple repairs and sharpening of augers were carried out in forges next to the mines.

Until the 1880s, in the Polish lands of the Austrian Partition, the only plant producing machinery and equipment for the mining industry was the Zieleniewski and Deskura factory in Cracow, producing drills and some other drilling tools. Throughout the entire 19th century, from the beginning of the Austrian colonization, there had been a policy of suppressing the development of metallurgy in the Galician province. Beyond the nearby border with the Russian Partition, on Polish lands in the area of the Świętokrzyskie Mountains and Kielce, iron ore mining and metallurgy had been developing since time immemorial. The territory of Poland occupied by the Austro-Hungarian Empire for the whole of the 19th century did not include strictly mining areas, hence the lack of such traditions and needs for mining machinery. This market opened with the discovery of oil in the Subcarpathia region, and Ignacy Łukasiewicz and the first oil workers were forced to use imports from abroad, including from the USA. Until the 1880s, mining equipment was imported, among others, from J. Schenk's factory in Messendorf and from the Österreichisch Alpine Montangesellschaft in Vienna.

The first specialist machinery factory for oil mining in Subcarpathia was founded by Stanisław Szczepanowski in Słoboda Rungurska in 1880. Initially, these were workshops which, thanks to the efforts of their manager, Felicjan Łodziński and Mikołaj Fedorowicz, the then director of the Szczepanowski mines, were equipped with a steam machine, driving a lathe and a modern steam hammer. Łódzkiński replaced the existing wood-fired boiler with a gas-fired boiler, using natural gas from the nearby "Zygmunt" shaft. In this respect, he benefited from his previous experience in the United States, where he observed the use of natural gas for heating and lighting purposes in the town of Bratforth for companies and residential buildings. In S. Szczepanowski's workshop in Słoboda Rungurska, one of the first attempts to build a Canadian derrick in the Subcarpathian region was made. Also here, in the second half of the 1880s, the first trial drilling with a diamond auger was carried out. The plant played an important role in the development of drilling technology in Galicia. Słoboda Rungurska was not only an important centre of oil extraction and production of drilling equipment, but also a place of intensive training for oil workers in the Subcarpathia region[419].

Szczepanowski also contributed to the development of Schodnica, which also became one of the most important centres of technical progress in the Subcarpathian oil mining industry.

Kazimierz Odrzywolski and Wacław Wolski also established a drilling equipment factory in this town. In their factory, a number of inventions and improvements in drilling technology were developed – a new type of derrick drilling rig constructed by Wolski and Łodziński and a design and construction company Wolski Weydlich, Korsak and Co. In 1897, Wolski

419 "Spostrzeżenia nad 'Spostrzeżeniami nad systemem Raky'ego' " [Observations on 'Observations on the Raky System'], *Nafta*, Vol. 9 (1901), pp. 58–61; Jan J. Cząstka, "Z dziejów górnictwa naftowego w Bieszczadach" [History of Oil Mining in the Bieszczady Mountains], *Nafta*, Vol. 17 (1961), pp. 171–174; Jan J. Cząstka, *Wiertnictwo* [Drilling], (Katowice: Wydawnictwo Śląsk), 1975, p. 19.

and Odrzywolski's company obtained a patent for their own type of eccentric auger in the Ministry of Commerce. The plant tested drilling equipment, tools and materials used in their production. The results of these tests were the basis for numerous theoretical dissertations by Wolski. The Szczepanowski factory collapsed in 1899 with invaluable damage to the innovativeness of the oil industry, pledged as debt security during the trial of Franciszek Zimy and Stanisław Szczepanowski, which was a consequence of the crisis of the Galician Savings Bank[420].

The largest enterprise producing equipment and machines for oil mining was the Factory of Drilling Tools and Machines in Glinik Mariampolski near Gorlice, founded in 1883[421]. The establishment of this workshop in the early 1880s, which was transformed into a Factory of Drilling Tools and Machines, was directly connected with the introduction of the Canadian drilling method in the Subcarpathia region. The mechanical workshop established at the refinery in Glinik Mariampolski was owned by the company "Bergheim and MacGarvey", which subsequently took the form of a joint stock company and changed its name to the Galician Carpathian Oil Joint Stock Company (GKNTA) and became the largest oil company in Galicia[422]. In a short period of time this workshop became one of the most important enterprises in the machinery industry in Subcarpathia. Initially, it provided repair services as well as tools and machines for its own mines and the refinery in Glinik. With time, the manufactured products started to be sold. In the face of an increasing number of orders, they enriched their production assortment, which led to the transformation of the service plant into a factory. The company grew spatially and organisationally. New production departments were created – foundry, forge, a machine tool hall, a boiler room and an electric workshop. At the beginning of the 20th century it opened a branch of the factory in Borysław. The products of the Glinik factory were known not only in the Subcarpathia region, but were also exported to many countries.

Important manufacturers of equipment for the oil industry were: the Lipiński Factory in Sanok and the MacIntosh & Perkins Drilling Machines

420 Franaszek, *Myśl techniczna*, pp. 55, 92, 101–104.

421 Ryszard Wolwowicz, "Maszyny, urządzenia i narzędzia dla górnictwa naftowego" [Machinery, Equipment and Tools for Oil Mining], in: *Historia polskiego przemysłu naftowego*, Vol. 1, p. 357.

422 Piotr Franaszek, "Fabryka Maszyn i Narzędzi Wiertniczych w Gliniku Mariampolskim (do 1918 roku)" [Factory of Drilling Tools and Machines in Glinik Mariampolski (until 1918)], in: *Z dziejów przemysłu przed 1945 rokiem*, ed. Jędrzej Chumiński i Marek Zawadka (Wroclaw: Wydawnictwo Gajt, 2012), pp. 161–175.

and Tools Factory in Stryj, also running workshops in Borysław and Tustanowice and offering derricks and steam engines[423].

The Roman Gierszyński Co Factory of Drilling Tools and Instruments also operated in Tustanowice, which at the beginning of the 20th century produced drills for impact drilling, shears, extensions and rescue tools. An important role in the production of drilling equipment was played by other workshops, which in many cases did not differ from the existing factories in terms of production profile. Among them were the Mechanical Workshops of F. Dudziak and C. Mermon, the Borysław Mechanical Workshops of Józef Dawidowicz and Jerzy Meszaros, and the TARAN Mechanical Workshops in Tustanowice[424].

Austrian authorities vis-à-vis the Subcarpathian oil industry

The development of the oil industry was strongly influenced by the economic policy of the central Viennese authorities, all the more so as the actions taken did not keep up with the pace of development of oil mining. They were characterised by slowness or even reluctance towards its real needs. In 1907, when Galician mining had been in operation for over 50 years, Stefan Bartoszewicz wrote:

> "Thus far, the government's care of this industry is almost non-existent, and the country's care is still too small and not yet based on an understanding of the importance of this industry."[425]

This was the case with the legal regulation of matters relating to the industry. The Austrian and Galician laws on this matter were not passed until 1884, 30 years after the beginning of the Galician oil industry. Previously, rules regulating legal problems were issued on an *ad hoc* basis. The most important ones include the definition of ownership rights to oil seams. The aim was to determine whether, like other minerals, oil was to be treated as a reserved mineral, i.e. belonging to the state (the so-called *regalia*) or whether the right to exploit it was to be associated with the ownership of the land. This issue was the subject of tenders and contradictory decisions by various authorities. In October 1810, by the decision of the General Court

423 Franaszek, "Fabryka Maszyn i Narzędzi Wiertniczych w Gliniku Mariampolskim", pp. 161–175.

424 Wolwowicz, "Maszyny, urządzenia i narzędzia dla górnictwa naftowego", pp. 358–359.

425 "O kongresie naftowym w Bukareszcie, odczyt dr Stefana Bartoszewicza wygłoszony w Towarzystwie Politechnicznym we Lwowie, dn. 16 bm." [On the Oil Congress in Bucharest, Lecture by Dr. Stefan Bartoszewicz Delivered at the Polytechnic Society in Lviv, 16 July.], *Nafta*, Vol. 15 (1907), p. 303.

Commission, oil flowing spontaneously to the surface of the land belonged to the owner of the land. In 1838 and 1840 the same office confirmed the reserved nature of crude oil and earth wax, although in 1841 the owner of the land was allowed to use the oil seams in his area, unless they were present in the presence of coal seams, earth wax and asphalt exploited by mining methods. The State Mining Act of 1854 was intended by its creators to unambiguously regulate all mining problems in the area of the monarchy. In accordance with the provisions, oil was recognised as *regalia*, which was further confirmed by the regulation Treasury Minister from 1860. However, as early as January 1862, the emperor issued a decree excluding oil from the *regalia*, provided that it was used exclusively for the production of kerosene for lighting purposes. The following years were marked by a constant struggle between the autonomous national authorities to explicitly exclude oil from the group of reserved minerals and the central authorities to confirm its reserved nature.

In a draft petroleum law passed by the Galician *Sejm* in 1878, bituminous materials were excluded from the group of reserved minerals. Although this project did not receive imperial sanction, it contributed to the resignation of the central government from projects to subordinate oil to the mining *regalia*. As a result, a state law passed by the Council of State recognised the exploitation of bituminous raw materials as a so-called land culture and excluded oil and earth wax in Galicia from the group of reserved minerals. The oil law of December 1884, based on the Galician state law, linked the right to exploit oil and earth wax with the ownership of the land. Thus, the owner of the land directly entered into an agreement with the prospecting entrepreneur, surrendering the right to drill and exploit oil. As a rule, in addition to the fees related to the lease of land, a rule of remunerating the owner of the land with a certain amount of oil extracted on his land (the so-called gross percentage) was established. Sometimes the "payment" in this form reached a high level, e.g. in 1913 in the region of Borysław-Tustanowice it reached 20 % of the quantity of extracted oil[426].

---

426 "Olej i wosk ziemny wobec rządowego projektu prawa górniczego" [Oil and Earth Wax in View of the Government's Mining Law Bill], *Górnik*, Vol. 3 (1884), pp. 2–5; Stanisław Olszewski, "Uwagi do rządowego projektu prawa górniczego" [Comments on the Government's Draft Mining Bill], *Górnik*, Vol. 3 (1884), pp. 103–109;"Przemysł naftowy i woskowy. Szkic na 50 letni jubileusz" [The Oil and Wax Industry. A Sketch for a 50-Year Jubilee], *Nafta*, Vol. 11 (1903), pp. 109–112; "Ustawodawstwo naftowe wobec powszechnej ustawy górniczej", [Oil Legislation on the General Mining Act], *Nafta*, Vol. 12 (1904), pp. 226–230; "Historia i stan ekonomiczny przemysłu naftowego w Galicji"

The Acts came at a time when the Galician oil industry was entering a new, extremely important stage related to the adaptation of the Canadian drilling system. It turned out that the introduced regulations were not adjusted to the changed situation, in particular to the dynamically developing drilling traffic (e.g. regulations on the manner of exercising supervision by mining authorities, statutory establishment of the so-called oil fields and registers). More importantly, however, the granting of rights to oil seams to the owner of the land, while at the same time maintaining the system of payment in the form of "gross percentages" undercut the profitability of oil exploitation. Entrepreneurs often complained about this:

> Every fifth or tenth kerosene barrel extracted must be handed over to the owner of the land, because only under this condition will an industrialist be granted the right to exploit, in addition to 200 crowns per drilling shaft, per morga of occupied land 160 to 400 crowns per year, 200 crowns per road to a shed, which he will have to construct provisionally, and before works commence from several hundred to several thousand when drawing up the contract. When counted together, the entrepreneur has to pay the value of a quarter or even a third of the product to he who did not put in any capital, did nothing, and who did not deserve it at all. Therefore, the entrepreneur barely vegetates, because this part is often his pure income, and individuals fatten up who do not know how to do anything in the industry. Inept individuals are lining their pockets and industry, and the country suffers it, so the well-being of individuals goes against the common good, quite the opposite as it should be in healthily organised societies.[427]

But drilling practice and technical progress developed independently of regulations, although the lack of appropriate legal solutions undoubtedly acted as a brake. Temporary mining and police regulations issued for oil mines in the years 1898, 1904, 1911 were to provide immediate help in solving problems. The amendment of the Petroleum Act of 1907 also served this purpose.[428]. In March 1908, a new national law was passed, which was in force until the outbreak of World War I. In points significant for the development of drilling traffic, the minimum surface area of oil mines (12,000 m^2), the minimum distance between the mine boundary and the drilled shaft (30 m), as well as the statutory qualifications of oil mine managers were

---

[History and economic state of the oil industry in Galicia], *Nafta*, Vol. 14 (1906), pp. 293–299.

427 Klaudiusz Angerman, "Krótkowidze" [Short-sightedness], *Nafta*, Vol. 9 (1901), p. 28.

428 "Petycja Krajowego Towarzystwa Naftowego w sprawie zmiany ustawy naftowej" [Petition by the National Petroleum Society on an Amendment to the Petroleum Act], *Nafta*, R. 1904, pp. 312–314; Leopold Caro, "Projekt reformy ustawodawstwa naftowego" [Draft Reform of Oil Legislation], *Nafta*, R. 1901, pp. 75–76, pp. 86–89.

clearly defined. However, in the mining and police regulations for Galicia of 1911, the technical conditions which had to be met by appropriate drilling structures were specified in a very detailed manner.[429]

Societies and journals

The oil industry of Lesser Poland and Subcarpathia, integrating scientific and economic circles, also played a significant cultural and social role. The stakeholders quickly realised how beneficial and necessary professional organisations and oil workers' organisations were, as well as a professional press as a method of economic lobbying and a forum for the exchange of scientific ideas and information. In December 1873, during an assembly of oil entrepreneurs in Gorlice, a resolution was passed for the first time on the necessity to establish an oil workers' association. Three years later, during a convention in Jasło, an informal committee was established to promote the interests of the oil industry. Ignacy Łukasiewicz was unanimously elected president.

The following year in Lviv, during the oil congress, the establishment of an "association for the development and care of the oil industry" was discussed. The statute of the Society, prepared by Wojciech Biechoński, was adopted. In June 1879, the Governor's Office approved the statute of the Galician Oil Society, which in 1881 adopted the name of the National Oil Society (KTN). The seat of the Society was Gorlice, and then Lviv. The Society operated until 1939. Its first president was Ignacy Łukasiewicz, and from 1881 until 1915 the Society was headed by August Gorayski.

Among the leading activists of the KTN one can meet all outstanding representatives of the Galician oil industry: Stanisław Szczepanowski, Wacław Wolski, Leon Syroczyński, Tytus Trzecieski, William Henry MacGarvey, Leon Mikucki, Stefan Bartoszewicz, Bolesław Łodziński, Władysław Długosz and others. At the end of 1884, the Kołomyskie Oil Society, founded by oil

---

429 "Zmiana ustawy krajowej" [Amendment of the National Law], *Nafta*, R. 1906, pp. 23–27; Stefan Bartoszewicz,"Ankieta rządowa w sprawie ustawy naftowej" [Government Survey on the Petroleum Act], *Nafta*, R. 1907, pp. 33–37; id.,"Wniesienie do Sejmu ustawy naftowej" [Submitting the Petroleum Act to the Sejm], ibid., pp. 65–67; "Krajowa ustawa naftowa" [National Petroleum Act], *Nafta*, R. 1908, pp. 91–92; "Przepisy górniczo-policyjne dla kopalń oleju ziemnego w Galicji z 24.06.1911 roku ogłoszone 18.08.1911 w Dzienniku Ustaw i Rozporządzeń Krajowych Nr 101" [Mining and Police Regulations for Galician Oil Mines of 24.06.1911 Announced on 18.08.1911 in the Journal of Laws and National Regulations No. 101.], *Nafta*, R. 1911, pp. 273–280.

entrepreneurs from the Eastern Subcarpathian region, was incorporated into the KTN.

The KTN became the organisation representing the interests of oil mining to both central and national authorities. KTN representatives entered national and state economic institutions (e.g. the Lviv Chamber of Commerce, the Agricultural and Industrial Council, the State Railways Council), gaining influence on the decisions of these bodies, especially in matters relating to the oil industry. The KTN was the spokesman, initiator and organiser of various actions aimed at transforming oil mining into a modern branch of production, acting as a lever for the economic development of the country[430].

The KTN was the most important, but not the only organisation integrating the oil industry. In the years 1893–1897 the Society of Petroleum Technicians (TTN) operated. The fact of its existence and the objectives it set itself, and in particular the establishment of the journal *Nafta*, testify to the conviction that joint action was necessary to modernise the domestic oil mining industry.

Among the co-creators and leading activists of TTN we meet people of the measure of Antoni Błażowski, Wacław Wolski, Rudolf Zuber, Felicjan Łodziński, Tytus Trzecieski, Kazimierz Odrzywolski, Zenon Suszycki, and thus the most active activists, constructors and rationalisers of the national

430 "Kongres naftowy w Przemyślu 1882" [Oil Congress in Przemyśl 1882], *Górnik*, Vol. 1 (1882), pp. 307–315; *Górnik*, Vol. 2 (1883), p. 80; *Górnik*, Vol. 3 (1884), p. 162; "Nadzwyczajne walne zgromadzenie Krajowego Towarzystwa Naftowego" [Extraordinary General Meeting of the National Petroleum Society], *Nafta*, Vol. 7 (1899), pp. 2–4; "Protokół posiedzenia Wydziału Krajowego Towarzystwa Naftowego z dnia 9 grudnia 1903 r." [Minutes of the Meeting of the National Petroleum Society Department of 9 December 1903.], *Nafta*, Vol. 12 (1904), pp. 13–14; "Z historii Krajowego Towarzystwa Naftowego" [From the History of the National Petroleum Society], *Nafta*, Vol. 12 (1904), pp. 375–377; "Wnioski Krajowego Towarzystwa Naftowego na Państwową Radę Kolejową" [Applications of the National Petroleum Society to the National Railway Council], *Nafta*, Vol. 21 (1913), p. 333; Olszewski, "Z dawnych lat przemysłu naftowego", pp. 406–409; Brzozowski, *Ignacy Łukasiewicz*, pp. 151–153; Wojciech Biechoński, "Pierwsze lata działalności Krajowego Towarzystwa Naftowego" [The First Years of Activity of the National Petroleum Society], *Przemysł Naftowy*, Vol. 4 (1929), pp. 300–303; Stanisław Olszewski, "Ważne momenty w działalności Krajowego Towarzystwa Naftowego w czasie od 15 sierpnia 1881 r. do maja 1900 r." [Important moments in the activity of the National Petroleum Society from 15 August 1881 to May 1900], *Przemysł Naftowy*, Vol. 4 (1929), pp. 303–304; Stefan Bartoszewicz, "Przemysł naftowy i działalność Krajowego Towarzystwa Naftowego w latach 1902–1914" [The Oil Industry and Activities of the National Petroleum Society in the Years 1902–1914], *Przemysł Naftowy*, Vol. 4 (1929), pp. 304–308.

oil drilling industry. According to the statutory assumptions, the Society's activities were to focus on the problems of geology, drilling, petroleum technology, wax mining and commercial matters related to petroleum products. Another objective was to maintain contacts between technicians employed in the oil industry and wax mining in order to jointly improve their qualifications. Discussions, oil competitions, professional journals, joint study tours and professional libraries were planned[431].

In 1905 the Union of Drilling Technicians (ZTW) was established. Its founders included: Władysław Dunka de Sajo, Stanisław Glazor, Jan Longchamps de Berier, Mieczysław Longchamps de Berier, Paweł Setkowicz and others. The association consisted mainly of managers of oil mines and managers of industrial plants associated with the drilling industry. They represented the middle-ranking personnel of the businesses and were not as well-known in the field as the activists of the KTN, but, thanks to direct contact with drilling works and drilling technology, they understood no less well the need to modernise equipment and tools and rationalise drilling methods. Apart from social and living protection of mine managers, they strongly emphasised the active involvement of this professional group in the course of activities for the advancement of drilling technology[432].

An exceptional role in the history of domestic oil mining was played by the professional oil press. In view of the territorially dispersed nature of mining, with communication difficulties, journals became the basic glue binding the community. The professional press was one of the main drivers of the efforts to modernise the oil industry. It was a basic source of information about the transformations taking place also in oil mining in other countries and a special place for presenting scientific achievements. The oil press became the most important platform for the exchange of experience between local and foreign oil companies. It was often an arena for heated discussions and polemics.

The first oil magazine *"Górnik"* published in 1882–1886, played a very important role in this respect. It became an excellent platform for discussion on modern drilling methods. *"Górnik"* contained reprints of articles from various technical journals, and information was provided on events in the domestic and other countries' oil mining industry. In a tribute to the

431 "Protokół z I Walnego Zgromadzenia członków Towarzystwa Techników Naftowych odbytego dnia 11 czerwca 1893 w Jaśle" [Minutes of the first General Meeting of members of the Society of Petroleum Technicians held on 11 June 1893 in Jasło], *Nafta*, Vol. 1 (1893), p. 2.

432 "Polski Kalendarz Naftowy" [Polish Oil Calendar], *Nafta*, Vol. 16 (1908), p. 4.

German-speaking citizens of the empire, they were informed about the situation in the oil mining industry.

The leading role was played by *Nafta*, which was published continuously since 1893. Initially, its pages were dominated by technical issues. In the face of many problems affecting the oil industry, a few years later these topics took the background, so a group of experts from the Union of Drilling Technicians decided to create their own body, the *Przegląd Techniczny Naftowy*, devoted exclusively to the issues of technology in oil mining. The same objectives guided the activists of the Union of Drilling Technicians when they started to publish their own organ, *Ropa* in 1911. There were fierce polemics concerning such important issues as, for example, the use of drilling with water in oil mining[433].

It should be emphasised that people who gave direction to changes in the oil mining industry were characterised by a high level of education. They formed a group of people with profound professional knowledge gained through technical or mining studies at the Austrian Academy of Mining in Leoben, the mining college that was most easily accessible to Poles from the Austrian Partition. Among others, Henryk Walter, Franciszek Zamojski, Julian Fabiański, Kazimierz Gąsiorowski, Stanisław Jurski and Ludwik Zdanowicz studied there. Wacław Wolski graduated from the mechanical engineering department of the Vienna Politechnikum. Other graduates of this college were Stanisław Szczepanowski, Leon Mikucki and Antoni Błażowski. Władysław Dunka de Sajo, Zygmunt Bielski and Władysław Szaynok studied at the Faculty of Mechanical Engineering of Lviv Polytechnic. Stanisław Olszewski graduated in geology, mineralogy and chemistry from the Jagiellonian University. Stefan Bartoszewicz graduated in technical chemistry from the Karlsruhe Politechnikum, Ludwik Stocker from the Lviv Polytechnic and Zurich, Zenon Suszycki from the École des Minès Superieure in Paris and Leon Syroc from the mining department of the University of Liège. Władysław Długosz completed his technical studies in Prague, Adolf Jabłoński, a graduate of the Agronomic Institute in Marymont, Warsaw, during his one and a half year stay in the United States, studied geology, mining and chemistry at the University of Virginia. Kazimierz Odrzywolski graduated from the Faculty of Chemistry and Technology of the Academy of Industry and Technology in Cracow. Stanisław Szczepanowski, after graduating from the Vienna Politechnikum, studied metallurgy and chemical technology in Paris and London[434]. All of them sought to create a system of professional, vocational oil education in the Austrian Partition.

---

433 Franaszek, *Myśl techniczna*, pp. 45–46.

434 "Antoni Błażowski", *Nafta*, Vol. 2 (1894), pp. 34–35, 51–53; "Stanisław Jurski", *Nafta*, Vol. 13 (1905), p. 156; *Nafta*, Vol. 16 (1908), p. 62; "Inż. Leon

Professional vocational education

As early as in 1875, Ignacy Łukasiewicz proposed that a lower mining school be established in Bóbrka to train oil miners. Three years later, the head of the District Mining Office in Lviv, Eng. Henryk Wachtel, approached the National Department and the National School Council on this matter. However, the establishment of a school in Bóbrka not happen[435].

In February 1881, the National Mining Council of the National Faculty unanimously approved a motion to establish an oil mining school. Engineer Wachtel, who had been involved in the subject for years, was commissioned to draw up a statute, regulations and an estimate of the school's annual expenses. The school was to improve the qualifications of people employed in mines and refineries. In October 1882, the National *Sejm*, composed of Polish MPs of autonomous authorities, passed a resolution calling on the government to establish a mining division at the Academy of Technology and Industry in Cracow with the main task of educating managers of oil mines. However, the letters of the Speaker of the *Sejm* and the National Department did not receive an answer from the Austrian central authorities in Vienna.

Vienna's lack of reaction forced other solutions to be sought. It was therefore proposed to set up a training facility for Canadian derricks[436] and,

---

Korwin Mikucki", *Przegląd Naftowy*, Vol. 1 (1921), pp. 65–66; *Czasopismo Techniczne*, Vol. 43, No. 14 (1925), p. 252; Stefan Margold, "W rocznicę śmierci śp. Stanisława Szczepanowskiego" [On the anniversary of the death of the late Stanisław Szczepanowski], *Dwutygodnik Naftowy*, Vol. 32 (1925), pp. 1–2; "45-lecie pracy Władysława Długosza" [45th Anniversary of the Work of Władysław Długosz], *Nafta*, Vol. 49 (1931), pp. 269–272; "Prezes Władysław Długosz" [Chairman Władysław Długosz], *Przemysł Naftowy*, Vol. 6 (1931), pp. 526–527; "Śp. inż. Władysław Dunka ds Sajo" [The late Eng. Władysław Dunka ds Sajo], *Przemysł Naftowy*, Vol. 7 (1932), pp. 50–51; "Inż. Ludwik Stocker", *Przemysł Naftowy*, Vol. 7 (1932), p. 105; Ferdynand Jastrzębski, "Fragmenty z pamiętników starych nafciarzy" [Fragments from the Memoirs of Former Oil Workers], *Przemysł Naftowy*, Vol. 8 (1933), pp. 337–342, 370–372, 396–398; "Kronika żałobna śp. inż. Kazimierz Gąsiorowski" [Funeral Chronicle of the late Kazimierz Gąsiorowski], *Życie Techniczne*, Vol. 13 (1936), p. 150; "Zygmunt Bielski (Saryusz)", *Przemysł Naftowy*, Vol. 11 (1936), pp. 20–21; "Śp. dr inż. Stanisław Olszewski" [The late Dr. Eng. Stanisław Olszewski], *Przemysł Naftowy*, Vol. 14 (1939), pp. 47–48; Jan J. Cząstka, "Henryk Walter", *Nafta*, Vol. 17 (1961), pp. 316–319; Jan J. Cząstka, "Wacław Wolski", *Nafta*, Vol. 11 (1955), pp. 152–153.

435 Tomasz Kapała, "Zawodowe szkolnictwo naftowe" [Vocational oil education], in: *Historia polskiego przemysłu naftowego*, Vol. 2, pp. 303–304.

436 "Wiadomości bieżące" [Breaking News], *Górnik*, Vol. 2 (1883), pp. 42, 67–69, 71–77, 79–80, 137; "W sprawie szkoły górniczej w Galicji" [Concerning the

in October 1885, a Practical Canadian Drilling School was established in Ropianka near Dukla. Students received theoretical knowledge of Canadian drilling and acquired practical skills. At the same time they learned how to organise and conduct drilling works. After three months of study they took an exam. In the following year the course was extended to 6 months. The number of pupils increased from year to year. In the first year of activity there were 4 of them, in the next 10, and soon the limited capacity of the school exceeded the number of eager applicants[437]. The school-leavers were granted the right to supervise the traffic and to drill using the Canadian method.

Zenon Suszycki became the organiser and director of the school. Practical exercises in the field of impact drilling on steam-powered rods were carried out on a complete Canadian derrick, which was purchased by Władysław Fibich, the owner of the mine in Ropianka. The purchase was possible thanks to a subsidy of 16,000 crowns from the National Department. In 1888, the Drilling School was moved to Wietrzno, which was associated with a decrease in drilling traffic in the Ropianka mine and the development of exploration works in the area of Wietrzna near Krosno. The works of Adolf Jabłoński, *Kopalnictwo naftowe* and Marcin Maślanka, *Zarys kopalnictwa naftowego*[438] were used as basic textbooks.

In 1896 the school was moved to Borysław, where it was connected with the existing mining school, which educated workers of earth wax mines, and especially foremen for ozokerite mines in Borysław and in the vicinity of Stanisławów.

As a result of the merger of the National Mining School for overseers of earth wax mines in Borysław with the Practical Canadian Drilling School in Wietrzno, the National Mining and Drilling School in Borysław was established. Its director was Kazimierz Gąsiorowski until 1903, in the years 1903–1913 Kazimierz Szumsk, and then Kazimierz Miński. The school's programme included three courses – a preparatory course and two professional courses: mining and drilling. Each course lasted one year and students with higher education could be exempted from the preparatory course. This course consisted of subjects such as Polish language, stylistics, arithmetic, geometry, physics, technical mechanics, and geometric and handwritten

---

Mining School in Galicia], *Górnik*, Vol. 3 (1884) pp. 10–12, 17–18; "Wiadomości bieżące" [Breaking News],*Górnik*, Vol. 4 (1885), pp. 5–6, 43–44.

437 Leon Syroczyński, "Sprawozdanie o egzaminie uczniów praktycznej szkoły kanadyjskiego wiercenia w Ropiance, odbytym dnia 22 grudnia 1885 r." [Report on the Canadian Drilling School Practical Examination in Ropianka, 22 December 1885.], *Górnik*, Vol. 5 (1886), pp. 14–15.

438 Kapała, "Zawodowe szkolnictwo naftowe", pp. 304–305.

drawing. In the mining course the students learned about mineralogy, geognosy (geology), technical drawing, mining science, mining machinery, electrical engineering, accounting principles and legal regulations. During the drilling course, a number of subjects from the mining course were repeated, extending the programme to include theoretical mechanics, learning about boilers and steam engines, drilling and deep drilling. The lessons took place between 8 a.m. and 1 p.m. and in the afternoon the students worked in the surrounding mines. Men were enrolled in the school from the age of 20, after completing at least two grades of secondary school and with at least one year's practice in an oil or earth wax mine. By the outbreak of World War I, 218 graduates had completed the school. During the holidays, the school organised specialist courses combined with internship in mines for students of technical universities planning work in oil mining. The school ceased to operate in 1914, and warfare caused it to be closed until 1920.[439].

The national oil industry also took steps to educate professionals with higher technical education in the field of oil mining. In the future, they were to be responsible for introducing improvements, designing new structures and developing theoretical issues in the field of oil mining. However, the measures taken immediately met with strong resistance from the Viennese government to the increase in the number of Polish faculties in higher education institutions. Vienna also did not agree to the establishment of the Mining Academy in Cracow. The creation of a national higher education in the field of oil mining was very important because the mining academies in Leoben and Przybram did not take sufficient account of the problems of oil mining and did not meet the needs or ambitions of the Polish oil workers' community.

In 1884, the Mining Council of the National Faculty postulated the launch of lectures on petroleum mining and chemical processing of crude oil and earth wax at the Lviv Polytechnic. As a result, the Senate of the Polytechnic decided to organise a two-year preparatory course in the mining industry in the broadest sense. Bronisław Pawlewski had already been lecturing on oil and wax technology at the Polytechnic School in Lviv since 1882. In 1886 Pawlewski organised a national testing station for the oil industry, of which the management and lectures were later handed over to reader Roman Załoziecki. In 1886, the Polytechnic School launched a two-year preparatory course for candidates for studying mining at the Faculty of Mechanical Engineering. From the academic year 1891/92, the authorities allowed the introduction of lectures on petroleum mining and chemical technology of petroleum products to the study programme of Lviv Polytechnic.

439 Kapała, "Zawodowe szkolnictwo naftowe", pp. 306–307.

In 1898, the general mining course was converted into an oil mining course and its completion gave the right to manage oil mines. Initially, it was a three-year course, and from 1909 – two-year course. It authorised graduates to be admitted automatically to the final year of studies in Leoben and Przybrama. A total of more than 200 oil and earth wax mine managers were educated there. In 1896, the Ministry of Religion and Enlightenment established an Extraordinary Chair for the Encyclopedia of Oil Mining and Deep Drilling at the Polytechnic School, which until 1917 was headed by Leon Syroczyński, followed by Julian Fabiański[440].

However, the main objective pursued by oil miners was to establish a separate mining department, which would not only educate personnel, but also conduct theoretical and implementation research. The greatest hopes were connected with the commencement of a series of lectures on drilling and petroleum mining at the Academy of Mining in Cracow. Stefan Bartoszewicz spoke about this problem in 1913:

> "I suppose that when the Mining Academy in Cracow is opened, when a drilling and oil course is opened there, it will contribute to strengthening the stronger scientific and oil traffic; and our oil industry must, of itself, deliver the first scientific forces for this first mining university."[441]

The opening of the academy was planned for October 1, 1914. The outbreak of World War I annihilated these plans.

MacGarvey and others

The community of Polish oil workers was open to both new technical ideas and newcomers, who always came from the world in crowds to the newly discovered oil Eldorado.

Twelve years after the opening of the first oil well in Bóbrka, in 1867, Albert Fauck, of German origin, born in the now Polish town of Słupsk in Pomerania, was brought to Lesser Poland. He was already an experienced oil miner, as he had worked on Pennsylvania oil fields in the USA. In 1870 he became Ignacy Łukasiewicz's advisor and co-worker in Bóbrka itself, and

440 "O wyższe studia górnicze w kraju" [On Higher Mining Education in the Country], *Nafta*, Vol. 20 (1912), pp. 55–56; "W sprawie wyższych studiów górniczych w kraju" [On Higher Education in Mining in the Country], *Nafta*, Vol. 20 (1912), pp. 116–117; Karol Jahoda, "Wkład Czytelni Polskich Akademików Górniczych w Leoben do rozwoju polskiego przemysłu naftowego" [Contribution of the Reading Room of Polish Mining Academics in Leoben to the Development of the Polish Oil Industry], *Nafta*, Vol. 32 (1976), pp. 428–431.

441 "Walne Zgromadzenie Krajowego Towarzystwa Naftowego" [General Meeting of the National Petroleum Society], *Nafta*, Vol. 21 (1913), p. 179.

it is to him that the transplantation of the Siberian drilling method to Polish soil was owed. Fauck worked on improving the free-fall drilling method, as well as on mud systems with the use of his own original construction of drilling derricks. He developed and built two drilling derricks, which he named "Rapid" and "Express", but which did not find a wider practical application. Fauck made an enormous contribution to the theoretical achievements of oil mining through participation in international oil congresses and organisations, and prepared an oil mining handbook[442].

A Canadian of Irish origin, William Henry MacGarvey, turned out to be a special figure among those who came from around the world. Born in 1843 in Huntington, Canada, to a family of Irish emigrants, he died on Polish soil in 1914. At the age of 14, he began drilling in Wyoming, Ontario. A few years later he was already a mine entrepreneur. His business in hard coal mining forced him to constantly travel around Canada. Still in Canada, he got married, and his choice was Helena Wesołowska, the daughter of Polish emigrants[443]. In 1881, he left Canada and, as manager of Continental Oil Company, began drilling for oil in the Hanover region of Germany. There he met John S. Bergheim, a Viennese banker whose capital enabled him to establish a company under the name "Bergheim and MacGarvey". The company was additionally cemented by the marriage of MacGarvey's son and Bergheim's daughter[444].

The exact date of arrival of this company in Subcarpathia is not known. It was certainly there already in 1883. Its seat was in Marijampol Glinik, today a district of the city of Gorlice.

Bergheim, although in first place in the company's name, did not play an active role in the company – he was the provider of capital. The main manager, organiser and engineer of the company's activities was William MacGarvey. He was probably attracted to the Subcarpathia region by Stanisław Szczepanowski, who, after the establishment of the mine in Słoboda Rungurska, was looking for a drilling system that would give better results than the free-fall system. He planned to commission MacGarvey to carry out drilling work using the Canadian method. However, he set excessive financial conditions and the agreement was not concluded.

In a short time MacGarvey became the leader in drilling in the Austrian Partition, and his company began to buy its own oil fields, becoming one of

442 Jan J. Cząstka, "Doradcy Ignacego Łukasiewicza" [Ignacy Łukasiewicz's Advisors], *Wiadomości Naftowe*, Vol. 3 (1973), pp. 271–272.
443 "Ustawodawstwo naftowe wobec powszechnej ustawy górniczej" [Oil Legislation in Terms of the General Mining Act], *Nafta*, Vol. 12 (1904), pp. 226–230.
444 Józef Grzybowski, "O potrzebie stacji geologicznej w Borysławiu" [On the Need for a Geological Station in Borysław], *Ropa*, Vol. 2 (1912), p. 104.

the largest producers of oil and earth wax. In this way, he violated the general rule in Galicia that large estates could only be the property of the aristocracy. His practical sense and commitment to business forced MacGarvey to purchase "unscrupulously" huge areas of land, which he turned into oil fields and reached a serious state of ownership. He quickly found a common language with Polish engineers and technicians involved in the oil industry[445]. From the mid-1880s until the First World War, his company's rigs pumped oil in both the centres of Western Subcarpathia (Biecz, Gorlice Jasło and others) and Eastern Subcarpathia (Borysław, Schodnica, Tustanowice, Drohobycz). He was the founder of one of the largest refineries in the area and one of the first drilling equipment workshops, which in a short time transformed into the Drilling Tools Factory. Both plants were built in Glinik Marijampolski[446].

MacGarvey introduced many technical innovations. He was one of the first to use pole-tool method in a Canadian rig, for which he used a manila rope, because the small diameter of the rig's drum made it impossible to use a steel rope. In 1899 he patented an eccentric drill bit of his own design. In 1894, during the General National Exhibition in Lviv, a device popularly known as the MacGarvey *pumpryga* system was exhibited, enabling oil to be pumped from three wells at the same time. The factory presented its products at other exhibitions and fairs, including the National Industry Review Exhibition in Gorlice in August 1903. These rapid successes led to a huge accumulation of capital, which was not wasted, but used in further business ventures, and above all to the transformation in 1895 of the company "Bergheim and MacGarvey" into the joint stock company GKNTA. The Society was the next phase of capital concentration, taking over all the firm's companies. Thanks to this MacGarvey became a leading economic activist in the country. He also invested in new projects outside the borders of Galicia.

In 1905 he organised the large "Apollo" refinery near Vienna and the "OPIAG" oil company in Vienna. He purchased oil mines in Russia and two refineries in Hungary[447]. His attitude was appreciated by his contemporaries. The correspondent of the professional magazine *Nafta* wrote about him

---

445 Stanisław Czerwiński, "Taran p. Wolskiego", *Nafta*, Vol. 11 (1903) p. 141.

446 "Odczyt dr Stefana Bartoszewicza wygłoszony w Towarzystwie Politechnicznym we Lwowie dnia 16 bm." [Dr. Stefan Bartoszewicz's lecture given at the Polytechnic Society in Lviv on the 16th of this month], *Nafta*, Vol. 15 (1907), pp. 172.

447 Stanisław Brzozowski, "Mac Garvey William (1843–1914)", in: *Historia polskiego przemysłu naftowego*, Vol. 2, p. 411; Frank, *Oil Empire. Visions of Prosperity*, pp. 90–98.

as "a brave soldier who does not soon leave the battlefield."[448] Other oil activists characterised him as a tireless man with unwavering will, great diligence and excellent administrative skills[449]. Stanisław Szczepanowski, a pioneer of modern economic development, linked his organisational talents with his belonging to the "powerful Anglo-Saxon race". He stated:

> ... he was a master above all masters, who paid homage to the Anglo-Saxon principle that small things are perfection and perfection is a not small things[450]. MacGarvey assimilated easily in the Galician community. The recognition he commonly enjoyed in the circles of Galician oilmen is emphatically confirmed by the following words: There is no oilman in Galicia who does not know Mr. MacGarvey, and if he does not know him personally, he should know who he is, what he has done and what position he has taken in our oil industry. Also far and wide beyond the sphere of oil industry, the name of Mr. MacGarvey became known and famous. They popularly call him the "oil king", so apparently in this man and his career must be something more than a happy coincidence, if he could be a stranger and an unknown stranger from 20 years ago from far away to become the centre and in a way the director of the fate of our oil industry. It has to be admitted that he has not exploited his dominant position to our detriment; on the contrary, Galician oilmen have in him a sincere spokesman and friend, and the country has gained a righteous citizen, taking due regard for his qualities and showing gratitude and sympathy in this practical understanding of the matter, which is characteristic of Anglo-Saxon people's blood.[451]

Other foreigners who served oil mining in the Subcarpathia region included Cyrus and Jacob Perkins, G. Adams, Elgin Scott, D. Issemann, George MacIntosh and others.

Łukasiewicz's heirs in the world

Excellent mastery of drilling craftsmanship, directional education and level of expertise in the field of oil mining made the oil workers, experienced in the oil mines of Subcarpathia a sought-after professional force in many countries of the world. Almost everywhere where oil deposits were discovered, one could meet Poles[452]. They went to neighbouring oil countries such

---

448 Jan Zeitleben, "Nowa Polonia w dziewiczych lasach Południowej Ameryki" [New Polonia in the Virgin Forests of South America], *Nafta*, Vol. 3 (1895), pp. 25

449 Zygmunt Bie1ski, "Żuraw wiertniczy Express" [The Express Drilling Rig], *Nafta*, Vol. 12 (1904), p. 178

450 Franciszek Szeliga-Żychliński, "Moje wspomnienia" [My Recollections], *Przemysł Naftowy*, Vol. 6 (1931), p. 537.

451 Zygmunt Bie1ski, "Żuraw wiertniczy Express", *Nafta*, Vol. 12 (1904), pp. 178–179.

452 "I Zjazd polskich techników wiertniczych we Lwowie" [First Congress of Polish Drilling Technicians in Lviv], *Nafta*, Vol. 17 (1909), pp. 273–276.

as Romania and Russia, but they also worked in oil mines on both American continents.

Among many anonymous experts, leading Polish oil activists were also leaving. And so, in 1872–1873 Adolf Jabłoński, a collaborator of Ignacy Łukasiewicz and later the technical manager of the mine in Bóbrka, stayed in the United States. In the USA, he got acquainted with the American oil industry and studied geology, mining, chemistry and mineralogy at the University of Virginia. In 1880 Felicjan Łodziński went to the United States. In the Pennsylvanian mines, he learned the secrets of the rope drilling method. From 1895 to 1898, Stanisław Jurski worked in a gold mine in Colorado and then drilled for oil in the Los Angeles region.

In the 1880s and 1890s, Poles searched for oil deposits in Argentina using Canadian derricks they brought from Poland. After Rudolf Zuber's research work on oilfields in the province of Mendoza, in 1886 a team of engineers, drillers and workers arrived there, including Kazimierz Odrzywolski, Antoni Błażowski and Michał Rieger. Their drilling work was fully successful.

In 1888, Poles again went to Argentina, where they conducted exploration activities for the Compania Medocina de Petrólea. Two years later, Błażowski started drilling on his own in the south of Argentina. In the same year Odrzywolski, together with Stanisław Czerwiński and Jan Zeitleben, organised a new expedition to South America. This time, too, their activity was successful and they called the new mine "Nueva Polonia" – "New Poland"[453].

Polish oil workers recruited by foreign companies took up work in Java, Sumatra and even in New Zealand[454]. Thanks to this, Polish miners made a significant contribution to the development of the oil industry in many countries. For oil engineers and technicians, such trips were a valuable experience. They were used to learn about the conditions of oil occurrence in other areas and to test the equipment they used. The experience and knowledge gained there paid off in the Polish oil mining industry.

453 Rudolf Zuber, "Antoni Błażowski", *Nafta*, Vol. 2 (1894),pp. 34–35, 51–53; *Nafta*, Vol. 1 (1893), p. 14; Jan Zeitleben, "Nowa Polonia w dziewiczych lasach Południowej Ameryki", *Nafta*, Vol. 3 (1895), pp. 5–8, 15–16, 22–24, 30–32.

454 Stanisław Liebelt, "Nasz Mazur na obczyźnie. Szkice z Sumatry, Łebak" [Our Masuria abroad. Sketches from Sumatra, an Oil Carrier], *Nafta*, Vol. 14 (1906), p. 133; Jan Fedorowicz, "System kanadyjsko-polski w Nowej Zelandii" [The Canadian Polish System in New Zealand], *Ropa*, Vol. 3 (1913), pp. 401–404.

Congresses and exhibitions

The development of the oil industry, like few other sectors of the economy, gave many opportunities for direct international contacts and exchange of scientific ideas.

International oil congresses created a unique opportunity to present theoretical achievements and practical technical solutions in the oil industry. The first such congress was held in 1900 in Paris, on the occasion of a general exhibition, the second in 1905 in Liège, and the third in 1907 was convened in Bucharest.

The organising committee of the Bucharest congress included representatives of the Subcarpathian oilmen – W. Wolski, Roman Załoziecki and W.H. MacGarvey, who prepared themselves especially carefully for the congress. During the congress they distributed a special issue of the *Nafta* journal in German. Their participation was considered an extremely successful and valuable contribution to the organisation of the congress.

The presentation and exchange of technical ideas in the field of oil mining also took place during the annual international congresses of engineers and drilling technicians. They were held three times in Lviv: the eighth congress in 1894, the fourteenth congress in 1900 and the twenty-second in 1908.[455]

Another form of presenting the achievements of the Polish oil mining industry were national industrial exhibitions. It was one of the ways of arousing broader interest in the oil industry among entrepreneurs from other branches of production and representatives of the Galician and Austrian authorities. For oilmen, technicians and engineers, the exhibitions were not only an opportunity to present their own achievements, but also a form of collation of ideas and concepts and their implementation. They were a source of inspiration and encouragement for further action on the road to

455 "Odczyt dr Stefana Bartoszewicza wygłoszony w Towarzystwie Politechnicznym we Lwowie dnia 16 bm." [Dr. Stefan Bartoszewicz's lecture given at the Polytechnic Society in Lviv on the 16th of this month], *Nafta*, Vol. 15 (1907), pp. 301–305; "Sprawozdanie z działalności Krajowego Towarzystwa Naftowego za rok 1908" [Report on the Activities of the National Petroleum Society for the Year 1908], *Nafta*, Vol. 17 (1909), p. 33; "Wystawa Paryska w roku 1900" [The Paris Exhibition of 1900], *Nafta*, Vol. 6 (1898), R. 1898, p. 22; "The First International Crude Oil Congress in Paris", *Naphta*, Vol. 8 (1900), pp. 297–299; *"Earth drilling in oil production. Lecture by Engineer A. Fauck at the International Oil Congress in Paris 1900"*, *Naphta*, Vol. 8 (1900), pp. 338–341. "Udział Polaków w II międzynarodowym kongresie naftowym w czerwcu 1905 r." [Participation of Poles in the Second International Oil Congress in June 1905], *Nafta*, Vol. 13 (1905), p. 209–212.

technical progress. They were also an excellent form of advertising of their own ideas, constructions and produced mining equipment.

Due to the wide audience, the best opportunity to present the achievements of the entire oil industry were the periodically organised national agro-industrial exhibitions. At the exhibition in Przemyśl in 1882, apart from samples of various types of crude oil and distillation products and samples of earth wax, a significant part of the crude oil exhibition was devoted to the achievements of technical thought in the field of oil mining. Figures or models presented new solutions and structures developed by oil engineers (e.g. Adolf Jabłoński's device for shutting off water, Stanisław Jurski's idea for a device for keeping the rope in constant tension during rope drilling, a device for draining gases from the well, used in the Harklowa mine, A. Fauck's device for grabbing the drilling pipes models of derricks for free-fall drilling, made by Piotr Brzozowski and a model for rope drilling). The entire exhibition was enriched by statistical reports giving information about the volume of oil production in individual mines[456].

In 1887, at the Third National Exhibition in Cracow, the oil industry was exhibited in the "Mining and Metallurgy" section. One of the judges of the exhibition was the well-known oil activist Stanisław Olszewski[457]. The richest exhibition of the oil industry took place in 1894 during the General National Exhibition in Lviv. It was prepared for with particular care, since – as it was written – oilmen

> "... together will show their own and strangers what they can do and that kerosene is not a toy or a blind game, but a large and serious national industry"[458]

The exposition of the oil industry was to reflect as faithfully as possible the conditions and landscape of a typical oil mine. The main pavilion of the oil section was built in the shape of a drilling rig with adjacent buildings. A drilling tools warehouse next to the mine, a forge, a boiler room and even flats

---

456 Stanisław Olszewski, "Galicyjski przemysł naftowy i wosku ziemnego na wystawie krajowej we Lwowie" [The Galician Oil and Earth Wax Industry at the National Exhibition in Lviv], *Sprawozdania Krajowego Towarzystwa Naftowego w Galicji*, Vol. 1 (1894), pp. 3–6; P.O., "Przemysł naftowy na wystawie rolno-przemysłowej w Przemyślu" [The Oil Industry at the Agro-Industrial Exhibition in Przemyśl], *Górnik*, Vol. 1 (1882), pp. 267–274, 283–289.

457 "Grupa IX. Wystawa przemysłu naftowego i wosku ziemnego urządzona przez członków Krajowego Towarzystwa Naftowego w Galicji" [Group IX. Exhibition of the Oil and Earth Wax Industry Organized by Members of the National Petroleum Society of Galicia], in: *Katalog Powszechnej Wystawy Krajowej we Lwowie 1894* (Lviv: Piller i Spółka, 1894), p. 28.

458 "Z działu naftowego na przyszłorocznej wystawie krajowej" [From the Oil Department at Next Year's National Exhibition], *Nafta*, Vol. 1 (1893), p. 74.

for workers were erected. These were barracks made of paper pulp, lightweight and easy to build. They were produced by the Szeliga and Łyszkiewicz factory in Lviv. The pavilion of the Galician Carpathian Oil Joint Stock Company (GKNTA) was modelled on the building of the so-called pumping booth from the mine in Bóbrka. It featured a W.H. MacGarvey designed pump, which enabled simultaneous pumping of oil from three wells. The main pavilion contained drawings depicting plans of mines, cross-sections of oil wells, drawings illustrating activities performed during drilling with the Canadian method. The drawings of geological sections of single shafts and mines in Bóbrka, Wietrzno, Równe, Lipniki and Iwonicz, made by Klaudiusz Angerman on the basis of geodetic measurements and drilling notes, were of particular scientific value. The tools used in oil mining, pipes, samples of various types of oil and statistical summaries relating to oil mining were all presented.

Samples of petroleum distillation products were collected – from gasoline through several types of kerosene, e.g. "imperial", "saloon" and "household", to several types of lubricants. They came from the Bergheim and MacGarvey refineries in Glinik Mariampolski, Sholto Douglas in Klęczany, Mikołaj Fedorowicz in *Ropa*, Józef Stawiarski in Chorkówka, Władysław Fibich and Straszewska in Lipniki, Gartenberg, Lauterbach, Goldhammer and Wagemann in Drohobych, Gartenberg and Schreier in Jasło and Kołomyja, Maria Lubomirska in Schodnica, Adam Skrzyński in Libusz, the First Galician Joint Stock Company for the oil industry (formerly S. Szczepanowski and co.) in Peczeniżyn[459]. In this way, the activists of the KTN wanted to show the technical changes in mining, especially in drilling technology, and thus to present to the general public the progress that had been made in this field since the beginning of the Subcarpathian oil mining industry. For this purpose, the exhibition gathered machines and equipment used in mining in the initial period of its development, i.e. in the years 1860–1883. The central point of the exhibition was a derrick for manual free-fall drilling with the use of Fabian shears. In contrast, and to show the major breakthrough in drilling in the early 1880s, a work demonstration using the Canadian method was held at the exhibition site. This very attractive project was made possible thanks to the financial support of the national authorities, the Lviv City Council and private individuals. After these drillings some scientific effects were expected, as well as the discovery of water springs for the city and mineral deposits.

459 "Grupa IX. Wystawa przemysłu naftowego i wosku ziemnego", p. 28; Rudolf Zuber, "Dział naftowy na Wystawie Krajowej" ["The Oil Department at National Exhibition"], *Nafta*, Vol. 2 (1894), p. 51.

In connection with the exhibition, a competition was announced for the best construction of steam locomobiles for deep drilling works. Apart from two machine factories, L. Zieleniewski in Cracow and K. Lipiński in Sanok, two companies from Poznań entered the competition: Urbanowski, Romocki and co. and H. Cegielski, whose construction won the first prize[460]. During the exhibition, the efforts made in favour of technical progress were rewarded. The gold medal was awarded to A. Fauck's drilling rig factory in Vienna, the silver medal of the Ministry of Commerce to engineers from Łodziński and Wolski for their adaptation of the rope bail to a Canadian derrick, and an honorary diploma from this ministry was awarded to Bergheim and MacGarvey's company. The honorary diploma of the Exhibition Committee was awarded to "The First Galician Wagon Construction Plant and Machinery Factory of Kazimierz Lipiński" in Sanok for the produced drilling tools[461].

Success in spite of obstacles

Oil mining was of exceptional importance for both the Polish economy of the Galician autonomy and the entire Austro-Hungarian Empire. It produced nearly 100 % of the oil produced in the empire, although the policy of the central government did not favour the development of the Subcarpathian oil industry, which is clearly indicated by the analysis of legal regulations related to the oil industry in the form of state and national oil laws.

This new field of economic activity triggered a huge potential hidden in Polish entrepreneurs. It was conducive to the formation of capital- and organisationally strong enterprises, taking on the character of companies, including joint stock companies[462]. The dynamic development of the oil industry was conducive to the emergence of workshops and factories for drilling equipment and equipment for oil exploitation[463].

The oil industry stimulated the development of other branches of industry – the metal and engineering industries, refinery and chemical industry plants.

---

460 *Nafta*, Vol. 1 (1893), p. 34; "Próba kotłów na Wystawie Krajowej" [Boiler Test at the National Exhibition], *Nafta*, Vol. 2 (1894), p. 133.

461 "Nagrody przyznane przez prezydium jury na Powszechnej Wystawie Krajowej we Lwowie 1894 wystawcom przemysłu naftowego i wosku ziemnego" [Awards granted by the presidium of the jury at the Lviv General National Exhibition 1894 to exhibitors of the oil and earth wax industry], *Sprawozdania Krajowego Towarzystwa Naftowego w Galicji*, Vol. 1 (1894), pp. 1–3.

462 "Wiadomości bieżące" [Breaking News], *Górnik*, Vol. 2 (1883), pp. 166–167.

463 Franaszek, "Fabryka Maszyn i Narzędzi Wiertniczych w Gliniku Mariampolskim", pp. 161–175.

The wood industry received a new impulse, as most of the buildings of oil mines, including drilling rigs, were built of wood. The network of railway lines to the mines became more and more dense. The rolling stock necessary for the transport of oil was produced.

At the turn of the 20th century, oil mining accounted for about 60 % of the value of industrial production in Galicia and employed 34.5 % of all domestic industrial workers, compared to 33.1 % employed in hard coal mining, 16.7 % in salt mining, 5.4 % in zinc and lead metallurgy, 4.4 % in earth wax mining, 3 % in zinc and lead mining, 2.2 % in lignite mining and 0.7 % in iron ore mining.

Technological progress brought international contacts, implied self-organisation and the creation of industrial associations as a forum for the exchange of scientific and technical ideas. It animated the publishing of a professional oil press and the organisation of national and international oil meetings and congresses.

The Subcarpathian oil industry, although it encountered many obstacles such as the weakness of domestic capital, the fight against foreign unfair competition, insufficient support from central authorities, played the role of an extremely important economic stimulator and fostered positive civilisational changes far beyond the borders of Galicia. It is a magnificent example of the enormous progress of the economy in the second half of the 19th and early 20th century on Polish lands, initiated and elevated by Polish forces on international and world forums.

The death of Ignacy Łukasiewicz in the early 1880s, in the face of the unprecedentedly dynamic march of the trend he initiated and the huge legacy of his thoughts and actions, the ranks of successors, scientists, experts and oil enthusiasts that he left behind did not stop this process. The end of the 19th century and the beginning of the new were a continuous march towards more and more spectacular economic, scientific, cultural and civilisational successes achieved thanks to oil.

And the awareness of the importance of the oil industry and the role that oil played already then in the history of civilisation was to constantly increase. The efforts of Poles over the course of more than half a century from the first shaft in Bóbrka to World War I – that era of the Polish pioneer Ignacy Łukasiewicz and his students, co-workers, competitors and successors – gave a wonderful start to the new century of motorisation, aviation and space conquest.

In October 1907, Stefan Bartoszewicz spoke to the activists of the Polytechnic Society in Lviv:

> "And we are a small country, and we are a nation that cannot think of any kind of colonial expansion, so we should use all its energy more eagerly for internal work,

Figure 33: The antique oilfield in Bóbrka – contemporary view. Source: The Ignacy Łukasiewicz Oil and Gas Industry Museum in Bóbrka

> for the exploitation of the riches that nature has given us. And nature happily gave us oil, so it gave us a product of common and mass use, which is of interest and must be of interest to the whole world..."[464]

At the threshold of the new century, these were wise and prophetic words.

464 "O kongresie naftowym w Bukareszcie, odczyt dr Stefana Bartoszewicza", p. 303

List of Figures

Figure 1: Memorial obelisk in the Museum in Bóbrka. Source: Anna Kozicka-Kołaczkowska 15
Figure 2: Collection of kerosene lamps from The Ignacy Łukasiewicz Oil and Gas Industry Museum in Bóbrka. Source: Anna Kozicka-Kołaczkowska 20
Figure 3: Map of Polish lands after 1815. Source: Authors. 22
Figure 4: Ignacy Łukasiewicz. Source: The Ignacy Łukasiewicz Oil and Gas Industry Museum in Bóbrka 36
Figure 5: Pharmacy. Source: Museum of Pharmacy, Jagiellonian University Medical College 52
Figure 6: Łukasiewicz's manuscript. Source: Museum of Pharmacy, Jagiellonian University Medical College 54
Figure 7: Łukasiewicz's lamp – a replica of the prototype. Source: Subcarpathian Museum in Krosno 70
Figure 8: Łukasiewicz's lamp. Source: Subcarpathian Museum in Krosno 72
Figure 9: A crude oil distillery from the 19th century. Source: Museum of Pharmacy, Jagiellonian University Medical College 80
Figure 10: Leaking crude oil. Source: Subcarpathian Museum in Krosno 81
Figure 11: "Franek" manually dug well. Source: The Ignacy Łukasiewicz Oil and Gas Industry Museum in Bóbrka 83
Figure 12: Ignacy Łukasiewicz. Source: The Ignacy Łukasiewicz Oil and Gas Industry Museum in Bóbrka 85
Figure 13: Diploma from the exhibition of 1858. Source: Museum of Pharmacy, Jagiellonian University Medical College 86
Figure 14: Honorata Łukasiewicz. Source: The Ignacy Łukasiewicz Oil and Gas Industry Museum in Bóbrka. 89
Figure 15: Transport of steam boilers to Ignacy Łukasiewicz's distillery in Chorkówka (1865–1870). Source: Subcarpathian Museum in Krosno 94
Figure 16: Installation of steam boilers at the distillery in Chorkowka. Source: Subcarpathian Museum in Krosno 94
Figure 17: Vehicle for measuring the fluid extracted from a crude oil well from the 19th century. Source: Anna Kozicka-Kołaczkowska. 96
Figure 18: Oil barrels, exposition of Bóbrka. Source: Anna Kozicka-Kołaczkowska 97

Figure 19: Tanks for oil, exhibition in the historic mine in Bóbrka. Source: Anna Kozicka-Kołaczkowska 106
Figure 20: Plan of the mine at Bóbrka. Source: The Ignacy Łukasiewicz Oil and Gas Industry Museum in Bóbrka 108
Figure 21: Ignacy Łukasiewicz's mansion in Chorkówka. Source: The National Digital Archives 136
Figure 22: Łukasiewicz's style. Source: Anna Kozicka-Kołaczkowska ... 139
Figure 23: Medal for the 25th anniversary – head + Medal for the 25th anniversary – tail. Source: Subcarpathian Museum in Krosno 144
Figure 24: Ignacy Łukasiewicz. Source: Subcarpathian Museum in Krosno 147
Figure 25: The oil field in Bóbrka 1911. Source: The Ignacy Łukasiewicz Oil and Gas Industry Museum in Bóbrka 153
Figure 26: Wooden smithy of Łukasiewicz in Bóbrka. Source: Anna Kozicka-Kołaczkowska 157
Figure 27: Witold Zglenicki. Source: The archive of the Płock Scientific Society 177
Figure 28: An oil rig on the Caspian Sea. Source: © Shutterstock.com/aquatarkus 191
Figure 29: The oil field in Borysław. Source: The National Digital Archives 219
Figure 30: Drilling rigs and tools from the 2nd half of the 19th century. Source: Subcarpathian Museum in Krosno 221
Figure 31: The oilfield personnel in Bóbrka 1908. Source: The Ignacy Łukasiewicz Oil and Gas Industry Museum in Bóbrka 230
Figure 32: Drilling tools from a historic mine in Bóbrka. Source: Anna Kozicka-Kołaczkowska 234
Figure 33: The antique oilfield in Bóbrka – contemporary view. Source: The Ignacy Łukasiewicz Oil and Gas Industry Museum in Bóbrka 257

Bibliography

Books

Aeschylus. *Prometheus Bound*, trans. Joel Agee, New York: NYRB Classics, 2002.

Andrusiewicz, Andrzej. *Carowie i cesarze Rosji - szkice biograficzne [Tsars and Emperors of Russia – Biographical Sketches]*, Warsaw: Wydawnictwo Fakty, 2001.

Angerman, Klaudiusz. *Ogólna geologia naftowa. Podstawy do badań terenów naftowych [General Oil Geology. Basis for Oilfield Exploration]*, Lviv: edition by the author, 1902.

Bańkowski, Andrzej. *Etymologiczny słownik języka polskiego [Etymological Dictionary of the Polish Language]*, vol. 2, Warsaw: Wydawnictwo Naukowe PWN, 2000.

Bar Janusz, Franaszek Piotr. *Informator statystyczny do dziejów przemysłu w Galicji. Górnictwo i hutnictwo [Statistical Guide to the History of Industry in Galicia. Mining and Metallurgy]*, Cracow: Uniwersytet Jagielloński, 1981.

Baranowski, Bohdan, Baranowski K. *Polaków kaukaskich drogi [Caucasian Polish Roads]*, Lodz: KAW, 1985.

Bartkiewicz, Magdalena. *Polski ubiór do 1864 roku [Polish Clothing until 1864]*, Wrocław: Zakład Narodowy im. Ossolińskich, 1979.

Benis Artur. *Zagłębie Cracowskie. Uwagi gospodarcze. Przemysł górniczy i fabryczny [The Cracow Basin. Economic Remarks. The Mining and Manufacturing Industry]*, Cracow: Związek górników i hutników polskich w Austryi, 1910.

Bieńkowska, Danuta. *Lekarz starej Warszawy [Doctor of Old Warsaw]*, Warsaw: Czytelnik, 1977.

Birkenmajer, Aleksander. *Astrologia i astronomia w wiekach średnich [Astrology and Astronomy in the Middle Ages]*, Katowice: Instytut Śląski w Katowicach, 1937.

Bonusiak, Włodzimierz. *Życie i działalność Ignacego Łukasiewicza [The Life and Activity of Ignacy Łukasiewicz]*, Rzeszow: Wydawnictwo Towarzystwa Naukowego w Rzeszowie, 1985.

Bonusiak, Włodzimierz. *Szejk z Galicji. Ignacy Łukasiewicz 1822–1882 [The Sheikh of Galicia. Ignacy Łukasiewicz 1822–1882]*, Rzeszow: Libra, 2007.

Brzozowski Stanisław, *Ignacy Łukasiewicz [Ignacy Lukasiewicz]*, Warsaw: Interpress, 1974.

Bujak, Franciszek. *Galicja. Leśnictwo. Górnictwo. Przemysł [Galicia. Forestry. Mining. Industry]*, Warsaw – Lviv: Księgarnia Altenberg H., Księgarnia E. Wende i Spółka, 1910.

Bukowski, Kazimierz, Jackiewicz, Aleksander. *Sól i saliny polskie [Polish Salt and Salt Pans]*, Warsaw: Ministerstwo Przemysłu i Handlu, 1926.

Burchardt, Jerzy. *List Witelona do Ludwika we Lwówku Śląskim: problematyka teoriopoznawcza, kosmologiczna i medyczna [Vitello's letter to Ludwik in Lvivek Śląski: Theoriological, Cosmological and Medical Issues]*, Wrocław: 1979, Vitelo, *Witelonis perspectivae liber (...) = books of perspectivae liber*, Wrocław: Ossolineum, 1983.

Burlikowski, Jan. "Z biegiem Uszwicy" *["Downstream on the Uszwica"]*, in: *Kronika miasta Brzeska 1385–1944*, Brzesko: ZETO SA, 2005.

Burzyński, Andrzej. *Robotnicy w przemyśle ciężkim w Galicji w dobie autonomicznej. Struktura zatrudnienia [Heavy Industry Workers in Galicia in the Autonomous Period. Employment Structure]*, Cracow-Wrocław: Ossolineum, 1985.

Chłędowski, Kazimierz. *Pamiętniki. Tom I. Galicja (1843–1880) [Memoirs. Volume 1. Galicia (1834–1880)]*, Cracow: Wydawnictwo Literackie, 1957.

Chodubski, Andrzej. *Witold Zglenicki, Polski Nobel 1850–1904 [Witold Zglenicki, the Polish Nobel 1850–1904]*, Płock: Towarzystwo Naukowe Płockie, 1984.

Chodubski, Andrzej. *Aktywność kulturalna Polaków w Azerbejdżanie w XIX i na początku XX wieku [Cultural Activity of Poles in Azerbaijan in the 19th and Early 20th century]*, Zeszyty Naukowe, Gdańsk: Uniwersytet Gdański, 1986.

Chodubski, Andrzej. *O świątyni czcicieli ognia "Ateszgiah". Z relacji polskich zesłańców i podróżników [On the "Atesgias" Temple of the Worshippers of Fire. From the Relations of Polish Exiles and Travellers]*, Elbląg: Towarzystwo Inicjatyw Twórczych, 1991.

Chwalba, Andrzej. *Historia Polski 1795–1918 [History of Poland 1795–1918]*, Cracow: Wydawnictwo Literackie, 2000.

Chylak, Karol. *Ubezpieczenia społeczne i zaopatrzenia emerytalne w II Rzeczypospolitej [Social Insurance and Pension Provision in the Second Republic of Poland]*, Warsaw: Wydawnictwo Polskie IHOO, 2017.

Cząstka, Jan J. *Wiertnictwo [Drilling]*, Katowice: Wydawnictwo Śląsk, 1975.

Cząstka, Jan Józef. *Kopalnia ropy naftowej w Bóbrce, najstarsza na ziemiach polskich [Bóbrka Crude Oil Mine, the Oldest on Polish Soil]*, Cracow: Akademia Górniczo-Hutnicza, 1989.

Czuma, Mieczysław, Mazan, Leszek. *Austriackie gadanie czyli encyklopedia galicyjska [Austrian Chats or the Galician Encyclopaedia]*, Cracow: Oficyna Wydawniczo-Handlowa Anabasis, 1998.

Czuma, Mieczysław, Mazan, Leszek. *Pępek świata nazywa się Cracow [The Centre of the World is Called Cracow]*. Cracow: Oficyna Wydawnicza Anabasis, 2000.

Davies, Norman. *Heart of Europe*, London: Oxford University Press, 1995.

Dębski, Jan. *Ignacy Łukasiewicz. Narodziny przemysłu naftowego*, Warsaw: Ludowa Spółdzielnia Wydawnicza, 1955.

Drużbacki, Jan. *Informator statystyczny do dziejów społeczno-gospodarczych Galicji. Towarowy transport wodny w Galicji doby autonomiczne* [*Statistical Guide to the Socio-Economic History of Autonomous Galicia. Water Transport in Galicia*], Cracow: UJ, 1986.

Dybiec, Julian. *Finansowanie nauki i oświaty w Galicji 1860–1918*, Cracow: Wydawnictwo Uniwersytetu Jagiellońskiego, 1979.

Figurowsky, Nikolai Alexandrovich. *Dmitrij Iwanowicz Mendelejew*, Moscow, Izd-vo Akademii nauk SSSR, 1961.

Fita, Stanisław. *Pokolenie Szkoły Głównej [The Generation of the Main School]*, Warsaw: Państwowy Instytut Wydawniczy, 1980.

Franaszek, Piotr. *Myśl techniczna w galicyjskim wiertnictwie naftowym w latach 1860–1918 [Technical Thought in Galician Oil Drilling in the Years 1860–1918]*, Cracow: Uniwersytet Jagielloński, 1991.

Frank, Alison Fleig. *Oil Empire. Visions of Prosperity in Austrian Galicia*, Cambridge Massachusetts, London England: Harvard University Press, 2005.

Gąsiorowska – Grabowska, Natalia. *Z dziejów przemysłu w królestwie Polskim 1815–1918 [From the History of Industry in the Kingdom of Poland 1815–1918]*, Warsaw: PWN, 1985.

Galicia: A Multicultured Land, ed. Christopher Hann, Paul Robert Magocsi, Toronto-Buffalo-London: University of Toronto Press, 2005.

Grodziski, Stanisław. *W Królestwie Galicji i Lodomerii [In the Kingdom of Galicia and Lodomeria]*, Cracow: Wydawnictwo Literackie, 1976.

Grzybowski, Józef. *Przegląd obszarów naftowych Karpat polskich* [*Review of Oil Areas in the Polish Carpathians]*, Cracow: Tow. TEWUGE, 1919.

Himka, John-Paul. *Galician Villagers and the Ukrainian National Movement in the Nineteenth Century*, London: MacMillan Press, 1988.

Jabłoński, Adolf. *Kopalnictwo nafty [Oil Mining]*, Cracow: Wł. Anczyc i Spółka, 1885.

Janion, M. *Powstrzymać Prometeusza [To Stop Prometheus]*, Chotomów: w: Wobec zła, 1989.

Jarry, Alfred. *Ubu Roi*, trans. Barbara Wright, New York: New Directions Paperbook, 1993.

Jonca, Karol. *Dzieje gospodarcze Polski do 1939 roku [Economic History of Poland till 1939]*, Wrocław: Wydawnictwo Uniwersytetu Wroclawskiego, 1996.

Kargol, Tomasz. *Izba Przemysłowo-Handlowa w Cracowie w latach 1850–1939 [Chamber of Commerce and Industry in Cracow in the Years 1850–1939]*, Cracow: Historia Iagellonica, 2003.

Kasa Mianowskiego 1881–2011 [The Mianowski Fund 1881–2011], ed. Zasztowt, Leszek. Warsaw: Oficyna Wydawnicza Aspra-JR, 2011.

Kieniewicz, Stefan. *Historia Polski 1795–1918 [History of Poland 1795–1918]*, Warsaw: Państwowe Wydawnictwo Naukowe, 1968.

Krajewski, Andrzej. *Krew cywilizacji. Biografia ropy naftowej [The Blood of Civilization. A Biography of Crude Oil]*, Cracow: Wydawnictwo WAM, 2018.

Kramarz, Henryka. *Nauczyciele gimnazjalni Galicji 1867–1914. Studium historyczno-socjologiczne [Teachers of Galician Gymnasiums 1867–1914. A Historical and Sociological Study]*, Cracow: Wydawnictwo Naukowe WSP, 1987.

Krasiński, Zygmunt. *Psalmy przyszłości [Psalms of the Future]*, Cracow: Universitas, 2002.

Krośnieńskie kopalnictwo naftowe. Wspomnienia i sylwetki [*Krosno Oil Mining. Memories and Profiles]*, ed. Zuzak, Józef. Krosno: Oficyna Wydawnicza Apla, 2005.

Kula, Witold. *Szkice o manufakturach w Polsce XVIII wieku [Sketches about Manufactories in Poland in the 18th Century]*, Warsaw: Państwowe Wydawnictwo Naukowe, 1956.

Łada, Ładicz, Mancz – polski herb szlachecki z Mazowsza according to T.Gajl, *Herbarz polski [Polish Emblems]*, Gdańsk: Gdański Kantor Wydawniczy, 2007.

Laskowski, Andrzej. *Związki Ignacego Łukasiewicza z Jasłem w świetle materiałów archiwalnych, dawnej prasy i literatury [Relations between Ignacy Łukasiewicz and Jasło in the Light of Archival Materials, Old Press and Literature]*, Cracow: 2005.

Maślanka, Marcin. *Zarys kopalnictwa naftowego. Podręcznik dla urzędników kopalni naftowych [An Outline of Oil Mining. A Handbook for Oilfield Officials]*, Stanisławów: Wysoki Wydział Krajowy, 1885.

Michalewicz, Jerzy. *Przemysł gorzelniany Galicji doby autonomicznej. Między monopolem dworskim a monopolem państwowym [The Distillery Industry of Autonomous Galicia. Between the Court Monopoly and the State Monopoly]*, Cracow: Uniwersytet Jagielloński, 1988.

Michałowski, Witold. *Płonący Kaukaz [Burning Caucasus]*, Wrocław: Alta 2, 2000.

Morawski, Wojciech. *Dzieje gospodarcze Polski [Economic History of Poland]*, Warsaw: Diffin, 2010.

Nicholl, Charles. *Leonardo da Vinci, Lot wyobraźni [Leonardo da Vinci. Flight of Imagination]*, Warsaw: Wydawnictwo WAB, 2006.

Norwid, Cyprian, *Pisma wybrane [Selected Writings] vol. 1*, Warsaw: Państwowy Instytut Wydawniczy, 1968.

Norwid, Cyprian. *Promethidion*, Warsaw: Państwowy Instytut Wydawniczy, 1989.

Orzeszkowa, Eliza. *O sobie [About Myself]*, Warsaw: Czytelnik, 1974.

Pawlikowski, Józef. *Czy Polacy wybić się mogą na niepodległość [Can Poles Break Out to Independence?]*, Warsaw: Wydawnictwo Ministerstwa Obrony Narodowej, 1967.

Pietraszek, Edward. *Ośrodek górniczy Siersza 1804–1861–1961 [Siersza Mining Centre 1804–1861–1961]*, Cracow: Wydawnictwo Artystyczno-Graficzne, 1961.

Polaczek, Stanisław. *Powiat chrzanowski w W. Kp. Cracowskiem. Monografia historyczno-geograficzna [Chrzanowski County in W. Kp. Cracow. A Historical-Geographical Monograph]*, Cracow: Wydział Rady Powiatowej Chrzanowskiej, 1914.

Pratzer, Barbara. *A pieniądze zamieniał na ideały ideały [And He Turned Money into Ideals]*, Warsaw: BMW, 2004.

Prus, Bolesław. *Lalka [The Doll]*, Cracow: Wydawnictwo Greg, 2017.

Roeske, Władysław. *Ignacy Łukasiewicz 1822–1882 [Ignacy Lukasiewicz 1822–1882]*, Warsaw: Państwowy Zakład Wydawnictw Lekarskich, Warsaw 1974.

Rutkowski, Jan. *Historia gospodarcza Polski, vol. II, Czasy porozbiorowe do 1918 roku [An Economic History of Poland, Vol. II, Post-Partition to 1918]*, Poznań: Księgarnia Akademicka, 1950.

Samuel, M., Linde, Bogumił. *Słownik języka polskiego* [*Dictionary of the Polish Language*], vol. III., Warsaw: PIW, 1951.

Saryusz-Zaleski, Wojciech. *Dzieje przemysłu w byłej Galicji 1804–1929 [History of industry in former Galicia 1804–1929]*, Cracow: P.A. Zieleniewski i Fitzner Gamper, 1930.

Shanes, Joshua. *Diaspora Nationalism and Jewish Identity in Habsburg Galicia*, Cambridge University Press, New York 2012.

Sikora, Jerzy. *Ignacy Łukasiewicz [Ignacy Lukasiewicz]*, Katowice: Wydawnictwo "Śląsk", 1978.

Słowacki, Juliusz. *Wyjazd z Neapolu [Departure from Naples]*, w: *Wiersze i poematy* Wroclaw: PIW, 1959.

Staszic, Stanisław. *O ziemiorodztwie Karpatów i innych gór i równin Polski [On the Carpathian Mountains and Other Mountains and Plains of Poland]*, Warsaw: Wydawnictwo Geologiczne, 1955.

Stauter-Halsted, Keely. *The Nation in the Village: The Genesis of Peasant National Identity in Austrian Poland 1848–1914*, Cornell University Press, Ithaca-New York 2001.

Sulmierski, Filip., Walewski Władyslaw. *Słownik geograficzny Królestwa Polskiego i innych krajów słowiańskich*, [*Geographical Dictionary of the Kingdom of Poland and Other Slavic countries]* vol. 10, vol. 15, pt .2, Warsaw: Wydawnictwa Artystyczne i Filmowe, 1975.

Szajnocha, Władysław. *Górnictwo naftowe w Galicji wobec ustawodawstwa górniczego [Oil Mining in Galicia versus Mining Legislation]*, Cracow: Wł. Anczyc i Spółka, 1881.

Szczepanowski, Stanisław. *Nędza Galicji w cyfrach i program energicznego rozwoju gospodarstwa krajowego [Galician Poverty in Numbers and a Programme for the Energetic Development of the National Economy]*, Lviv: Gubrynowicz i Schmidt, 1888.

Szczepański, Aleksander. *„Stan wytwórczości przemysłowej i górniczej w Galicji w roku 1910" [Industrial and Mining Production in Galicia in 1910]*, Lviv: Nakładem Wydziału Krajowego, 1912, pp. 54, 104–107, 142–144.

Szubert, Tomasz. *Jak(ó)b Szela 14 (15) lipca 1787 – 21 kwietnia 1860 [Jak(ó)b Szela 14 (15) July 1787–21 April 1860]*, Warsaw: Wydawnictwo DiG, 2014.

Szuro, Stanisław. *Informator statystyczny do dziejów społeczno-gospodarczych Galicji. Koleje żelazne w Galicji w latach 1874–1914 [Statistical Guide to the Socio-Economic History of Galicia. Galician Railways 1874–1914]*, Cracow: UJ, 1997.

Teleżyński, *Olej skalny, jego zastosowanie w przemyśle i życiu codziennym [Rock Oil, Its Use in Industry and Everyday Life]*, Lviv: Kornel Piller, 1870.

The Limits of Loyalty: Imperial Symbolism, Popular Allegiances, and State Patriotism in the Late Habsburg Monarchy, ed. Laurence Cole, Daniel L. Unowsky, New York: Berghahn Books, 2009.

Tomanek, Ludwik. *Ignacy Łukasiewicz Twórca przemysłu naftowego w Polsce. Wielki inicjator – wielki jałmużnik jałmużnik [Ignacy Łukasiewicz, Founder of the Oil Industry in Poland. A Great Initiator – A Great Almsgiver]*, Miejsce Piastowe: Komitet Uczczenia Pamięci Ignacego Łukasiewicza, 1928.

Tuszyńska, Agata. *Rosjanie w Warszawie [Russians in Warsaw]*, Paryż: Instytut Wydawniczy, 1990.

Wachowicz, Barbara. *Nazwę cię Kościuszko [I Name You Kościuszko]*, Warsaw: Oficyna Wydawnicza RYTM, 2000.

Windakiewicz, Edward. *Olej i wosk ziemny w Galicji [Oil and Earth Wax in Galicia]*, Lwów: E. Winiarz, 1875.

Włodek, Jan Maria. *Goetz-Okocimscy, Kronika rodzinna 1590–2000 [Goetz-Okocimskis, Family Chronicle 1590–2000]*, Cracow: Księgarnia Akademicka, 2001.

Wnęk, Jan. *Rozwój ideologii przemysłowej w Galicji 1866–1918 1918 [The Development of the Industrial Ideology in Galicia 1866–1918]*, Warsaw: PAN, 2015.

Wolff, Larry. *The Idea of Galicia: History and Fantasy in Habsburg Political Culture*, Stanford: Stanford University Press, 2010.

Wyspiański, Stanisław. *Wesele [The Wedding]*, Cracow: Społeczny Instytut Wydawniczy, 2016.

Zamorski, Krzysztof. *Folwark i wieś. Gospodarka dworska i społeczność chłopska Tenczynka w latach 1705–1845 [The Farm and the Village. The Manor Economy and Peasant Community of Tenczynek in 1705–1845]*, Cracow-Wrocław: Ossolineum, 1987.

Zamorski, Krzysztof. *Informator statystyczny do dziejów społeczno-gospodarczych Galicji. Ludność Galicji w latach 1857–1910 [Statistical Guide to the Socio-Economic History of Galicia. Population of Galicia in the Years 1857–1910]*, Cracow-Warsaw: UJ-PTS, 1989.

Żeromski, Stefan. *Przedwiośnie [The Spring to Come]*, Cracow: Wydawnictwo Zielona Sowa,

Zieliński, Jan. *Staropolskie Zagłębie Przemysłowe [The Old Polish Industrial Area]*, Wroclaw – Warsaw – Cracow: Ossolineum, 1965.

Znamirowski, Stanisław. *Przegląd stanu kopalń nafty i wosku ziemnego w Galicji w drugim półroczu* 1881 *[Overview of the Oil and Earth Wax Mines in Galicia in the Second Half of 1881]*, Tarnów: Józef Pisz, 1882.

Zuber, Rudolf. *Nafta i wosk ziemny w Galicyi [Crude Oil and Earth Wax]*, Warsaw: Drukarnia J. Bergera, 1883.

Zuber, Rudolf. *Geologia pokładów naftowych w Karpatach galicyjskich [Geology of Oil Fields in the Galician Carpathians]*, Lviv: published by the author, 1899.

Zuber, Rudolf. *Geologiczne podstawy występowania oleju skalnego [The Geological Basis for the Occurrence of Rock Oil]*, Lviv: published by the author, 1916.

Żeromski, Stefan. *Przedwiośnie [The Spring to Come]*, Cracow: Wydawnictwo Zielona Sowa, 2009.

Articles in periodicals

"45-lecie pracy Władysława Długosza" ["45th Anniversary of the Work of Władysław Długosz"]. *Nafta*, Vol. 49, 1931, pp. 269–272.

"Antoni Błażowski". *Nafta*, Vol. 2, 1894, pp. 34–35, 51–53.

"Geologiczne badania głębszych warstw ziemi – odczyt Leona Syroczyńskiego, inżyniera Wydziału Krajowego we Lwowie, miany na IV Zjeździe Polskich Przyrodników i Lekarzy w Poznaniu dn. 22 czerwca 1884 r." ["Geological studies of deeper layers of the earth – lecture by

Leon Syroczyński, Engineer of the National Department in Lviv, delivered at the 4th Congress of Polish Naturalists and Doctors in Poznań on 22 June 1884."]. *Górnik*, Vol. 3, 1884, pp. 84–86.

"Historia i stan ekonomiczny przemysłu naftowego w Galicji" ["History and economic state of the oil industry in Galicia"]. *Nafta*, Vol. 14, 1906, pp. 293–299.

"Honorata ze Stacherskich Łukasiewiczowa w Pamiętniku Anny z Działyńskich Potockiej" ["Honorata Stacherska Łukasiewiczowa in the Diary of Anna Działyńska Potocka"]. *Wiek Nafty* no. 4, 2005, pp. 47–48.

"I Zjazd polskich techników wiertniczych we Lwowie" ["First Congress of Polish Drilling Technicians in Lviv"]. *Nafta*, Vol. 17, 1909, pp. 273–276.

"Inż. Leon Korwin Mikucki". *Przegląd Naftowy*, Vol. 1, 1921, pp. 65–66.

"Inż. Ludwik Stocker". *Przemysł Naftowy*, Vol. 7, 1932, p. 105.

"Kongres naftowy w Przemyślu 1882" ["Oil Congress in Przemyśl 1882"]. *Górnik*, Vol. 1, 1882, pp. 307–315.

"Kronika żałobna śp. inż. Kazimierz Gąsiorowski" ["Funeral Chronicle of the late Kazimierz Gąsiorowski"]. *Życie Techniczne*, Vol. 13, 1936, p. 150.

"Nadzwyczajne walne zgromadzenie Krajowego Towarzystwa Naftowego" ["Extraordinary General Meeting of the National Petroleum Society"]. *Nafta*, Vol. 7 1899, pp. 2–4.

"Nagrody przyznane przez prezydium jury na Powszechnej Wystawie Krajowej we Lwowie 1894 wystawcom przemysłu naftowego i wosku ziemnego" ["Awards granted by the presidium of the jury at the Lviv General National Exhibition 1894 to exhibitors of the oil and earth wax industry]". *Sprawozdania Krajowego Towarzystwa Naftowego w Galicji*, Vol. 1, 1894, pp. 1–3.

"Nowy palnik ropy inż. Zygmunta Rodakowskiego" ["The new oil burner by engineer Zygmunt Rodakowski"]. *Nafta*, Vol. 15, 1907, pp. 171–172.

"O kongresie naftowym w Bukareszcie, odczyt dr Stefana Bartoszewicza wygłoszony w Towarzystwie Politechnicznym we Lwowie, dn. 16 bm." ["On the Oil Congress in Bucharest, Lecture by Dr. Stefan Bartoszewicz Delivered at the Polytechnic Society in Lviv, 16 July"]. *Nafta*, Vol. 15, 1907, p. 303.

"O wyższe studia górnicze w kraju" ["On Higher Mining Education in the Country"]. *Nafta*, Vol. 20, 1912, pp. 55–56.

"O zastosowaniu ropy do celów opałowych" ["On the use of oil for the heating purposes"] *Nafta*, Vol. 13, 1905, pp. 356–359.

"Odczyt dr Stefana Bartoszewicza wygłoszony w Towarzystwie Politechnicznym we Lwowie dnia 16 bm." ["Dr. Stefan Bartoszewicz's lecture given at the Polytechnic Society in Lviv on the 16th of this month"]. *Nafta*, Vol. 15, 1907, pp. 172, 301–305.

"Olej i wosk ziemny wobec rządowego projektu prawa górniczego" ["Oil and Earth Wax in View of the Government's Mining Law Bill"]. *Górnik*, Vol. 3, 1884, pp. 2–5.

"Opalanie lokomotyw ropą" ["Fueling locomotives with the oil"]. *Nafta*, Vol. 17, 1909, pp. 243–247.

"O wystawie w Jaśle r. 1858" ["On the Exhibition in Jasło in 1858"]. *Tygodnik Rolniczo-Przemysłowy* no. 45, 1858, pp. 356–357.

"Pertraktacje z towarzystwem Standard Oil Company" ["Talks with the Standard Oil Company"]. *Nafta*, Vol. 17, 1909, pp. 145–147.

"Petycja Krajowego Towarzystwa Naftowego w sprawie zmiany ustawy naftowej" ["Petition by the National Petroleum Society on an Amendment to the Petroleum Act"]. *Nafta*, Vol. 12, 1904, pp. 312–314.

"Polska uczciła pamięć pioniera nafty pomnikiem" ["Poland honoured the memory of the kerosene pioneer with a monument"]. *Ilustrowany Kurier Codzienny* no. 296, 1932, p. 3

"Polski Kalendarz Naftowy" ["Polish Oil Calendar"]. *Nafta*, Vol. 16, 1908, p. 4.

"Potrzeba organizacji" ["The need for the organisation"]. *Nafta*, Vol. 16, 1908, pp. 121–122.

"Poświęcenie pomnika Ignacego Łukasiewicza w Krośnie" ["The consecration of the monument to Ignacy Łukasiewicz in Krosno"]. *Gazeta Lwowska* no. 245, 1932, p. 3.

"Prezes Władysław Długosz" ["Chairman Władysław Długosz"]. *Przemysł Naftowy*, Vol. 6, 1931, pp. 526–527.

"Program wykładów z geologii naftowej" ["Programme of Lectures on Oil Geology"]. *Nafta*, Vol. 20, 1912, pp. 364–365.

"Protokół posiedzenia Wydziału Krajowego Towarzystwa Naftowego z dnia 9 grudnia 1903 r." ["Minutes of the Meeting of the National Petroleum Society Department of 9 December 1903."]. *Nafta*, Vol. 12, 1904, pp. 13–14.

"Protokół z I Walnego Zgromadzenia członków Towarzystwa Techników Naftowych odbytego dnia 11 czerwca 1893 w Jaśle" ["Minutes of the first General Meeting of members of the Society of Petroleum Technicians held on 11 June 1893 in Jasło"]. *Nafta*, Vol. 1, 1893, p. 2.

"Próba kotłów na Wystawie Krajowej" ["Boiler Test at the National Exhibition]". *Nafta*, Vol. 2, 1894, p. 133.

"Przemysł naftowy i woskowy. Szkic na 50 letni jubileusz" ["The Oil and Wax Industry. A Sketch for a 50-Year Jubilee"]. *Nafta*, Vol. 11, 1903, pp. 109–112.

"Przepisy górniczo-policyjne dla kopalń oleju ziemnego w Galicji z 24.06.1911 roku ogłoszone 18.08.1911 w Dzienniku Ustaw i Rozporządzeń Krajowych Nr 101" ["Mining and Police Regulations for Galician Oil Mines of 24.06.1911 Announced on 18.08.1911 in the Journal of Laws and National Regulations No. 101"]. *Nafta*, R. 1911, pp. 273–280.

"Spostrzeżenia nad 'Spostrzeżeniami nad systemem Raky'ego" ["Observations on 'Observations on the Raky System"]. *Nafta*, Vol. 9, 1901, pp. 58–61;

"Sprawozdanie z działalności Krajowego Towarzystwa Naftowego za rok 1908" ["Report on the Activities of the National Petroleum Society for the Year 1908"]. *Nafta*, Vol. 17, 1909, p. 33.

"Stanisław Jurski". *Nafta*, Vol. 13, 1905, p. 156.

"Statystyka przemysłu naftowego w Galicji w roku 1900" ["Statistics on the Oil Industry in Galicia in 1900"]. *Nafta*, Vol. 9, 1901, p. 67.

"System rotacyjny wiertniczy Parkera" ["Parker's Rotary Drilling System"]. *Nafta*, Vol. 21, 1913, pp. 168–170.

"Śp. dr inż. Olszewski Stanisław" ["The late Dr. Eng. Stanisław Olszewski"], *Przemysł Naftowy*, Vol. 14, 1939, pp. 47–48.

"Śp. inż. Władysław Dunka ds Sajo" ["The late Eng. Władysław Dunka ds Sajo"]. *Przemysł Naftowy*, Vol. 7, 1932, pp. 50–51.

"The First International Crude Oil Congress in Paris". *Naphta*, Vol. 8, 1900, pp. 297–299.

"Udział Polaków w II międzynarodowym kongresie naftowym w czerwcu 1905 r." ["Participation of Poles in the Second International Oil Congress in June 1905"]. *Nafta*, Vol. 13, 1905, pp. 209–212.

"Układy z Petroleą" ["Agreements with Petrolea"]. *Nafta*, Vol. 14, 1906, pp. 149–152.

"Układ z towarzystwem Standard Oil Company" ["An Arrangement with the National Oil Company"]. *Nafta*, Vol. 17, 1909, pp. 161–163.

"Ustawodawstwo naftowe wobec powszechnej ustawy górniczej". ["Oil Legislation on the General Mining Act"] *Nafta*, Vol. 12, 1904, pp. 226–230.

"W sprawie szkoły górniczej w Galicji" ["Concerning the Mining School in Galicia."] *Górnik*, Vol. 3, 1884, pp. 10–12, 17–18.

"W sprawie wyższych studiów górniczych w kraju"["On Higher Education in Mining in the Country"]. *Nafta*, Vol. 20, 1912, pp. 116–117.

"Walne Zgromadzenie Krajowego Towarzystwa Naftowego" ["General Meeting of the National Petroleum Society"]. *Nafta*, Vol. 21, 1913, p. 179.

"Wiadomości bieżące" ["Breaking News"]. *Górnik*, Vol. 2, 1883, pp. 5–6, 42, 43–44, 67–69, 71–77, 79–80, 137, 166–167.

"Wielmożnemu Ignacemu Łukasiewiczowi" ["To the Honourable Ignacy Łukasiewicz"]. *Wiek Nafty* no 3, 2003, pp. 2–3.

"Wiercenie rotacyjne w Galicji 'Parker-Karpath-Rotary' - świdry 'Sharp and Hughes" ["Rotary Drilling in Galicia, with the Parker-Karpath-Rotary – Sharp and Hughes Auger"]. *Ropa*, Vol. 3, 1913, pp. 218–221.

"Wniesienie do *Sejmu* ustawy naftowej" ["Submitting the Petroleum Act to the Sejm."] *Nafta*, 1907, pp. 33–37.

"Wnioski Krajowego Towarzystwa Naftowego na Państwową Radę Kolejową" ["Applications of the National Petroleum Society to the National Railway Council]". *Nafta*, Vol. 21, 1913, p. 333.

"Wróblowice, z księgi wspomnień B. Sulity". ["Wróblowice, From the Memoirs of B. Sulita"] *Kłosy* no. 804, 1880, p. 340.

"Wydobywanie ropy naftowej w kopalni w Bóbrce" ["Extraction of Crude Oil in the Bóbrka Mine"], *Wiek Nafty* no. 3, 2004, p. 21.

"Wystawa Paryska w roku 1900" ["The Paris Exhibition of 1900"]. *Nafta*, Vol. 6, 1898, R. 1898, p. 22.

"Z działu naftowego na przyszłorocznej wystawie krajowej" ["From the Oil Department at Next Year's National Exhibition."]. *Nafta*, Vol. 1, 1893, p. 74.

"Z historii Krajowego Towarzystwa Naftowego" ["From the History of the National Petroleum Society."]. *Nafta*, Vol. 12, 1904, pp. 375–377.

"Zmiana ustawy krajowej" ["Amendment of the National Law."]. *Nafta*, 1906, pp. 23–27.

"Zygmunt Bielski (Saryusz)". *Przemysł Naftowy*, Vol. 11, 1936, pp. 20–21.

Anczyc, Władysław. "Ignacy Łukasiewicz". *Kłosy,* No. 876, 1882, p. 229.

Angerman, Klaudiusz. "Krótkowidze" ["Short-sightedness"]. *Nafta*, Vol. 9, 1901, p. 28.

Bałda, Waldemar. "Klęczany – naftowa historia" ["Klęczany – an oil history"]. *Wiek Nafty,* No. 1, 2012, p. 13.

Bartoszewicz, Stefan. "Historia i stan ekonomiczny przemysłu naftowego w Galicyi" ["History and economic state of the oil industry in Galicia"]. *Nafta,* No. 19, 1906, pp. 294–295.

Bartoszewicz, Stefan. "Nowa organizacja naszego przemysłu" ["A New Organization for Our Industry"]. Nafta, Vol. 14, 1906, pp. 277–280.

Bartoszewicz, Stefan. "Embarras de richesse". *Nafta*, Vol. 15, 1907, pp. 237–239.

Bartoszewicz, Stefan. "Opalanie ropą" ["Heating with crude oil"]. *Nafta*, Vol. 15, 1907, p. 329.

Bartoszewicz, Stefan. "Pogląd na dzieje przemysłu naftowego i środki jego sanacji" ["View of the History of the Oil Industry and the Means of Its Remediation"]. *Nafta*, Vol. 16, 1908, pp. 367–377.

Bartoszewicz, Stefan. "Projekt większej assocjacji w naszym przemyśle" ["A Project for Greater Association in Our Industry"]. *Nafta*, Vol. 20, 1912, pp. 97–98.

Bartoszewicz, Stefan. "Przemysł naftowy i działalność Krajowego Towarzystwa Naftowego w latach 1902–1914" ["The Oil Industry and Activities of the National Petroleum Society in the Years 1902–1914"]. *Przemysł Naftowy*, Vol. 4, 1929, pp. 304–308.

Barzyk, Adam. "Wielkie jubileusze Ignacego Łukasiewicza" ["Great Anniversaries of Ignacy Łukasiewicz."]. *Wiek Nafty,* No. 1, 2002, pp. 3–4.

Berezowski, Edward, "*Józef Chodźko – geodeta i geograf XIX wieku* (1800–1881)" ["Józef Chodźko - surveyor and geographer of the 19th century (1800–1881)"]. *Przegląd Geodezyjny,* nr 10, 1970, pp. 430–431.

Biechoński, Wojciech. "Pierwsze lata działalności Krajowego T-wa Naftowego. Referat wygłoszony na Kongresie Naftowym w Przemyślu w r. 1882" ["The First years of Activity of the National Oil Association. A paper delivered at the Oil Congress in Przemyśl in 1882"], *Przemysł Naftowy*, June, 1929, pp. 300–301.

Biechoński, Wojciech. "Pierwsze lata działalności Krajowego Towarzystwa Naftowego" ["The First Years of Activity of the National Petroleum Society"]. *Przemysł Naftowy*, Vol. 4, 1929, pp. 300–303.

Bielski, Zygmunt. "Żuraw wiertniczy Express" ["Express drilling rig"]. *Nafta*, Vol. 12, 1904, pp.129–134, 178–179.

Bie1ski Zygmunt, "Wybór metody wiercenia. Referat na Zjeździe Naftowym we Lwowie 26 czerwca 1927 r." ["Choice of Drilling Maethod. Presentation at the Oil Congress in Lviv on June 26, 1927"]. *Przemysł Naftowy*, Vol. 2, No. 21, 1927, pp. 385–391.

Bielski, Zygmunt. "Ignacy Łukasiewicz, wynalazca nafty świetlnej" ["Ignacy Łukasiewicz, Discoverer of Lighting Kerosene."]. *Przemysł Naftowy,* No. 23, 1932, p. 591.

Bujak, Franciszek. "Przemysł naftowy w Galicji i obecne jego przesilenie" ["The Oil industry in Galicia and its current solstice"]. *Ekonomista*, Vol. 9, 1909, pp. 111–156.

Buniatzade, Zija Ali. *Witold Zglenicki a problem wydobycia ropy naftowej z dna morskiego* ["Witold Zglenicki and the problem of crude oil extraction from the seabed"]. "Notatki Płockie". 1978, 4/97, p. 31.

Caro, Leopold. "Projekt reformy ustawodawstwa naftowego" ["Draft Reform of Oil Legislation"]. *Nafta*,R. 1901, pp. 75–76, pp. 86–89.

Chłędowski, Kazimierz. "Źródła nafty w Bóbrce w Jasielskiem" ["Sources of Kerosene in Bóbrka in the Jasło Region"]. *Tygodnik Ilustrowany,* No. 325, 1865, p. 252.

Chłędowski, Kazimierz. "Nafciarze" ["Oilmen"]. *Dodatek miesięczny do Gazety Lwowskiej,* No. 2, 1872, p. 75.

Chodubski, Andrzej. *Paweł Potocki, 1879–1932) - pionier eksploatacji ropy naftowej spod dna morskiego* ["Paweł Potocki (1879–1932) – A Pioneer of Crude Oil Exploitation from under the Seabed"], "Kwartalnik Historii Nauki i Techniki", nr 4, 1988, pp. 981–994.

Chodubski, Andrzej. *Udział Polaków w poznaniu Morza Kaspijskiego* ["The Participation of Poles in Exploring the Caspian Sea"]. "Nautologia", nr 4, 1988, pp. 3–10.

Chodubski, Andrzej. *Działalność cywilizacyjna Polaków na Kaukazie* ["The Civilizational Activity of Poles in the Caucasus"], "Cywilizacje w czasie i przesterzeni", nr 1, 1995, pp. 65–77.

Chojnacki, Jakub. *Komunikat z dnia 30 listopada 1978 roku o ustaleniu daty i miejsca urodzenia Witolda Zglenickiego* – "Polskiego Nobla". ["Communique from November 30, 1978 about the date and place of birth of Witold Zglenicki - "the Polish Nobel"], "Notatki Płockie", nr 4/97, 1978, p. 36.

Chojnacki, Jakub. *O utrwalanie pamięci wybitnego Polaka inżyniera górnika – geologa Witolda Zglenickiego* ["On preserving the memory of the outstanding Polish mining engineer - geologist Witold Zglenicki"], "Notatki Płockie", nr 4, 98, 1978, pp. 23–30.

Cząstka, Jan J. "Doradcy Ignacego Łukasiewicza" ["Ignacy Łukasiewicz's Advisors"]. *Wiadomości Naftowe*, Vol. 3, 1973, pp. 271–272.

Cząstka, Jan Józef. "Kopalnia ropy naftowej w Bóbrce, (w 125-letnią rocznicę jej założenia)" ["Bóbrka Crude Oil Mine (on the 125th anniversary of its foundation)"]. *Wiek Nafty,* No. 1, 2004, pp. 23–24.

Cząstka, Jan J. "Ludzie kopalni ropy naftowej w Bóbrce" ["People at the Bóbrka crude oil mine"], *Wiek Nafty*, No. 4, 2004, pp. 12–13.

Cząstka, Jan J. "U źródeł powstania polskiego przemysłu naftowego. (Z okazji 120-rocznicy powstania tego przemysłu)" ["At the Source of the Emergence of the Polish Oil Industry. (On the Occasion of the 120th Anniversary of the Establishment of this Industry)"]. *Nafta*, Vol. 29, 1973, pp. 337–344.

Cząstka, Jan J. "Wacław Wolski". *Nafta*, Vol. 11, 1955, pp. 152–153.

Cząstka, Jan J. "Z dziejów górnictwa naftowego w Bieszczadach" ["History of Oil Mining in the Bieszczady Mountains"]. *Nafta*, Vol. 17, 1961, pp. 171–174.

Cząstka, Jan J. "Henryk Walter". *Nafta*, Vol. 17, 1961, pp. 316–319.

Czerwiński, Stanisław, "Taran p. Wolskiego" ["Mr. Wolski's *Taran*"]. *Nafta*, Vol. 11, 1903, pp. 141, 145–146

Długosz, Władysław. "Wspomnienia z Borysławia" ["Memoirs of Borysław"]. *Przemysł Naftowy*, Vol. 4, No. 9, 1929, pp. 329–332.

Fauck, Albert. "Ein Drittel Jahrhundert der Entwicklung der Tiefbohrtechnik in Galzien". *Naphta*, Vol. 8, 1900, pp. 99–102.

Fauck, Albert. "Użycie płuczki na nowych terenach" ["Use of the Flusher in New Areas"]. *Ropa*, Vol. 1, 1911, pp. 421–423.

Fedorowicz, Jan. "System kanadyjsko-polski w Nowej Zelandii" ["The Canadian Polish System in New Zealand"]. *Ropa*, Vol. 3, 1913, pp. 401–404.

Ferdynand, Jastrzębski. "Fragmenty z pamiętników starych nafciarzy" ["Fragments from the Memoirs of Former Oil Workers"]. *Przemysł Naftowy*, Vol. 8, 1933, pp. 337–342, 370–372, 396–398.

Franaszek, Piotr. "Zatrudnienie w przemyśle naftowym Galicji, do roku 1914" ["Employment in the Galician oil industry (until 1914)"]. *Studia Historyczne*, Vol. 33, No. 1, 1990, pp. 37-46.

Gorayski, August. "Ignacy Łukasiewicz". *Górnik*, No. 2, 1882, pp. 166–167.

Gorayski, August. "Przemysł naftowy" ["The Oil Industry"]. *Przegląd Polski*, Vol. 37, 1903, pp. 519–532.

Gorayski, August. "Siła stowarzyszenia" ["The Strength of the Association"]. *Przegląd* Naftowy, June, 1929, p. 298.

Grzybowski, Józef. "O potrzebie stacji geologicznej w Borysławiu" ["On the need for a geological station in Boryslaw"]. *Ropa*, Vol. 2, 1912, pp. 73, 104, 242–244.

J. P. "Ś. P. Ignacy Łukasiewicz" ["Late Ignacy Lukasiewicz"]. *Strażnica Polska*, No. 21, 1882, p. 169.

Jahoda, Karol. "Wkład Czytelni Polskich Akademików Górniczych w Leoben do rozwoju polskiego przemysłu naftowego" ["Contribution of the Reading Room of Polish Mining Academics in Leoben to the Development of the Polish Oil Industry]". *Nafta*, Vol. 32, 1976, pp. 428–431.

Jaros, Jerzy. *Polacy w Leningradzkim Instytucie Górniczym* ["Poles in the Leningrad Mining Institute"], Kwartalnik Historii Nauki i techniki, nr 3, 1972, p. 505.

Jędrkiewicz, Zygmunt. "Widoki rozwoju hutnictwa żelaznego w Galicji" ["Views of the Development of the Iron and Steel Industry in Galicia"]. *Prace Biura Statystyki Przemysłowej*, No. 2, 1913, p. 77.

Kachlik, Kazimierz. "Ignacy Łukasiewicz – od nafty świetlnej do rafinerii nafty" ["Ignacy Łukasiewicz - From Kerosene Lighting to Oil Refining"]. *Nafta*, No. 7–8, 1982, p. 82.

Kamiński, Zdzisław. "Wywłaszczenie Cracowskiego Zagłębia Węglowego" ["Expropriation of the Cracow Coal Basin"]. *Przegląd Górniczo-Hutniczy*, Vol., 9, No. 10, 1912, p. 359.

Klebert, Antoni. "Rozwój żurawia kanadyjskiego wiercenia w Galicji" ["The Development of the Canadian Drilling Rig in Galicia"]. *Przegląd*

Techniczny Naftowy, dodatek do czasopisma *Nafta*, R. 1907, pp. 17–20, 25–27, 29–31, 37–41, 45–46.

Kulczykowski, Mariusz. "'Deindustrializacja' w procesach uprzemysłowienia Galicji w XIX wieku. Problemy badawcze" ["Deindustrialisation" in the industrialisation of Galicia in the 19th century. Research problems."]. *Zeszyty Naukowe Uniwersytetu Jagiellońskiego. Prace Historyczne*, No. 78, 1986, pp. 75-87.

Liebelt, Stanisław. "Nasz Mazur na obczyźnie. Szkice z Sumatry, Łebak" ["Our Masuria abroad. Sketches from Sumatra, an Oil Carrier"]. *Nafta*, Vol. 14, 1906, p. 133.

Liebfeld. *Działalność inżyniera Witolda Zglenickiego w Baku w latach 1890- 1904 i jego zapis testamentowy dla Kasy im. Mianowskiego*, ["The activity of engineer Witold Zglenicki in Baku in the years 1890–1904 and his testamentary record for the Mianowski Fund"]. Kwartalnik Historii Nauki i Techniki, nr 2, 1977, p. 314.

Löwenherz, Oskar. "Ignacy Łukasiewicz jako organizator" ["Ignacy Łukasiewicz as an Organizer."] *Przemysł Naftowy*, No. 20, 1932, p. 483.

Löwenherz, Oskar. "Ignacy Łukasiewicz odkrywca nafty i twórca przemysłu naftowego" ["Ignacy Łukasiewicz, Discoverer of Kerosene and Creator of the Oil Industry"]. *Nafta,* No. 8–9, 1932, p. 206;

Łopatkiewicz, Zdzisław. "Seweryn Stawiarski. Materiały do biografii" ["Seweryn Stawiarski. Biographical materials"]. *Wiek Nafty,* No. 4, 2006, p. 49.

Łotysz, Sławomir. "Galicyjska nafta i wosk ziemny na Expo 1873. Relacja Agatona Gillera z Wiednia" ["Overview of the state of the Galician oil and earth wax mines in Galicia for Expo 1873. Report by Agathon Giller from Vienn."]. *Wiek Nafty,* No. 1, 2008, p. 4;

Margold, Stefan. "W rocznicę śmierci śp. Stanisława Szczepanowskiego" ["On the anniversary of the death of the late Stanisław Szczepanowski"]. *Dwutygodnik Naftowy*, Vol. 32, 1925, pp. 1–2.

Mierzecki, Roman. "Przemysł naftowy w Polsce w XIX i XX wieku" ["The Oil Industry in Poland in the 19th and 20th Centuries."]. *Analecta*, Vol. 8/2, 16, No. 55–71, 1999, pp. 55–71.

Mierzecki, Roman. "Stosowanie ropy naftowej w Galicji w XIX w." ["The Use of Oil in Galicia in the 19th century"]. *Wiek Nafty,* No. 3, 1999, p. 7.

Molenda, Danuta. "Polski ołów na rynkach Europy środkowej w XIII-XVII wieku" ["Polish lead in the markets of Central Europe in the 13th-17th centuries"]. *Studia i Materiały z Historii Kultury Materialnej,* Vol. 69, 2001, p. 14.

Müldner, Rudolf. "O naszym przemyśle naftowym" ["On our Oil Industry"]. *Czas,* No. 140, 1877, p. 1.

Munia, Jacek. "40 lat pracy Muzeum Przemysłu Naftowego i Gazowniczego im. Ignacego Łukasiewicza w Bóbrce w zachowaniu kultury technicznej i tradycji polskiego przemysłu naftowego i gazowniczego" ["40 years of work of The Ignacy Łukasiewicz Oil and Gas Industry Museum in Bóbrka in Maintaining the Technical Culture and Traditions of the Polish Oil and Gas Industry"]. *Wiek Nafty,* No. 2, 2001, pp. 16–19.

Olejarz, Barbara. "80-lecie odsłonięcia pomnika Ignacego Łukasiewicza" ["The 80th Anniversary of the Unveiling of the Monument to Ignacy Łukasiewicz"]. *Wiek Nafty,* No. 4, 2012, p. 5.

Olejarz, Barbara. "Wody mineralne z Bóbrki. Zapomniana historia zakładu wodoleczniczego" ["Mineral Waters from Bóbrka. The Forgotten History of the Hydrotherapy Facility"]. *Wiek Nafty,* No. 3, 2014, pp. 14–15.

Olszewski, Stanisław. "Uwagi do rządowego projektu prawa górniczego" ["Comments on the Government's Draft Mining Bill"]. *Górnik*, Vol. 3, 1884, pp. 103–109.

Olszewski, Stanisław. "Galicyjski przemysł naftowy i wosku ziemnego na wystawie krajowej we Lwowie" ["The Galician Oil and Earth Wax Industry at the National Exhibition in Lviv"]. *Sprawozdania Krajowego Towarzystwa Naftowego w Galicji*, Vol. 1, 1894, pp. 3–6.

Olszewski, Stanisław. "Kryzys naftowy w Galicji i zgromadzenie producentów naftowych we Lwowie" ["The Oil Crisis in Galicia and the Assembly of Oil Producers in Lviv"]. *Sprawozdania Krajowego Towarzystwa Naftowego we Lwowie*, Vol. 2, 1895, pp. 1–4.

Olszewski, Stanisław. "Rozwój przemysłu naftowego w Rumunii oraz w innych krajach" ["The Development of the Oil Industry in Romania and Other Countries"]. *Nafta*, Vol. 4, 1896, p. 128.

Olszewski, Stanisław. "Zapotrzebowanie nafty w Austro-Węgrzech" ["Oil Requirements in Austria-Hungary"]. *Nafta*, Vol. 6, 1898, p. 45.

Olszewski, Stanisław. "Przemysł naftowy w Galicji w latach 1884–1901" ["The Oil Industry in Galicia in the Years 1884–1901"]. *Nafta*, Vol. 10, 1902, pp. 133–137.

Olszewski, Stanisław. "Galicyjski przemysł naftowy jego dotychczasowe położenie i widoki na przyszłość" ["The Galician Oil Industry's Current Location and Future Prospects"]. *Nafta*, Vol. 18, 1910, p. 349.

Olszewski, Stanisław. "Ważne momenty w działalności Krajowego Towarzystwa Naftowego w czasie od 15 sierpnia 1881 r. do maja 1900 r." ["Important moments in the activity of the National Petroleum Society from 15 August 1881 to May 1900"]. *Przemysł Naftowy*, Vol. 4, 1929, pp. 303–304.

Olszewski, Stanisław. "Z dawnych lat przemysłu naftowego" ["Olden Times in the Oil Industry"]. *Przemysł Naftowy*, Vol. 4, No. 13, 1929, pp. 406–409.

Olszewski, Stanisław. "Ważniejsze momenty z działalności Krajowego Towarzystwa Naftowego w czasie od sierpnia 1881 do maja 1900" ["The Most Important Moments in the Activity of the National Oil Association in the Period from August 1881 to May 1900"]. *Przemysł Naftowy*, June, 1929, p. 303.

P. O., "Przemysł naftowy na wystawie rolno-przemysłowej w Przemyślu" ["The Oil Industry at the Agro-Industrial Exhibition in Przemyśl"]. *Górnik*, Vol. 1, 1882, pp. 267–274, 283–289.

Petit, Wiktor. "Kombinowany przyrząd wiertniczy systemu Petita" ["Petit System Combined Drilling Tool"]. *Ropa*, Vol. 4, 1914, pp. 278–282, 297–300.

Petit, Wiktor. "Nowy system wiertniczy" ["A New Drilling System"]. *Ropa*, Vol. 1, 1911, pp. 278–284.

Pietraszek, Edward. "Zagłębie Cracowskie w latach 1796–1848. Zarys społeczno-gospodarczy" ["The Cracow Basin in the years 1796–1848. A Socio-Economic Outline"]. *Kwartalnik Historii Kultury Materialnej*, No. 4, 1961, pp. 118–119, 746–753.

Plutyński, Antoni. "Inżynier Wacław Wolski" ["Engineer Wacław Wolski."] *Kwartalnik Historii Nauki i Techniki*, Vol. 11, No. 4, 1966, pp. 397–416.

Pruszkowski, Wacław, Horwath, Józef. "Hydraulisch betriebener Tiefbohrer". *Naphta*, Vol. 8, 1900, pp. 65–67.

Sozański, Józef. "Historyczny przegląd ustawodawstwa w górnictwie naftowym" ["Historical Overview of Legislation in Oil Mining"]. *Wiek Nafty,* No. 3, 1999, pp. 8–10.

Sulimirski, Wit. "O rozwoju kopalnictwa naftowego w Galicji" ["On the Development of Oil Mining in Galicia"]. *Ropa*, Vol. 3, No. 5, 1913, pp. 479–485, 502–507;

Syroczyński, Leon. "Sprawozdanie o egzaminie uczniów praktycznej szkoły kanadyjskiego wiercenia w Ropiance, odbytym dnia 22 grudnia 1885 r." ["Report on the Canadian Drilling School Practical Examination in Ropianka, 22 December 1885."]. *Górnik*, Vol. 5, 1886, pp. 14–15.

Szczuchniak, Andrzej. "Był żywym ogniem podwójnego światła – z dziejów pomników Ignacego Łukasiewicza" ["He Was a Living Flame of Double Light - from the History of the Monuments to Ignacy Łukasiewicz"]. *Wiek Nafty,* No. 4, 2013, pp. 24–25.

Szczur, Jerzy. "Karol Klobassa Zrencki senior 1823–1886". *Wiek Nafty,* No 4, 2008, p. 39.

Szczypiorski, Andrzej. *Bajeczna opowieść o złocie i miłości* ["A Fairytale about Gold and Love"], "Polityka", nr 27, 5th July 1969.

Szeliga-Żychliński, Franciszek. "Moje wspomnienia" ["My Recollections"]. *Przemysł Naftowy*, Vol. 6, 1931, p. 537.

Wachal, Władysław. "Stan techniki przeróbki ropy naftowej w okresie działalności Ignacego Łukasiewicza" ["State of the art of crude oil processing technology in the period of Ignacy Łukasiewicz's activit"]. *Wiek Nafty,* No 3, 2001, p. 5.

Wais, Tadeusz. "Kopalnia ropy naftowej w Bóbrce. 160 lat działalności" ["Bóbrka Oil Mine. 160 years of operation"]. *Wiek Nafty,* No. 4, 2014, p. 6.

Wolski, Marian. "Teofil Tytus Tomasz Trzecieski". *Wiek Nafty,* No. 3, 2009, p. 42.

Wo1ski, Wacław. "O nowych systemach wiertniczych" "On New Drilling Systems". *Nafta*, Vol. 9, 1901, pp. 1–4, 17–21.

Wolski, Wacław. "O taranie wiertniczym" ["On the Drilling *Taran*"], *Nafta*, Vol. 10, 1902, pp. 120–126, 137–142.

Wójcik, Zbigniew. *O niektórych polskich geologach i gleboznawcach kształcacych się bądź pracujacych na wyższych uczelniach Rosji* ["On some Polish geologists and soil scientists studying or working at Russian universities"], "Studia i Materiały z Dziejów Nauki Polskiej" 1976, seria C. z. 21, pp. 64–67.

Zeitleben, Jan. "Nowa Polonia w dziewiczych lasach Południowej Ameryki" ["New Polonia in the Virgin Forests of South America"]. *Nafta*, Vol. 3, 1895, pp. 5–8, 15–16, 22–24, 35, 30–35.

Zuber, Rudolf. "Nowa strefa ropna w Galicji" ["A New Oil Zone in Galicia."] *Górnik*, Vol. 3, 1884, pp. 29–30.

Zuber, Rudolf. "Antoni Błażowski". *Nafta*, Vol. 2, 1894, pp. 34–35, 51–53.

Zuber, Rudolf. "Dział naftowy na Wystawie Krajowej" ["The Oil Department at National Exhibition"]. *Nafta*, Vol. 2, 1894, p. 51.

Zuzak, Józef. "Zarys dziejów najstarszej w świecie kopalni ropy naftowej w Bóbrce" ["History of the world's oldest oil mine in Bóbrka"]. *Wiek Nafty,* No. 2, 1994, p. 15.

Articles in books

"Statut Krajowy Galicyjski z 1861 r." ["Galician National Statute of 1861"], In: *Sejm Krajowy galicyjski 1861–1914. Źródła*, Vol. 2, ed. Stanisław Grodziski. Warsaw: Wydawnictwo Sejmowe, 1993, pp. 5–6.

Bełza Władysław, "Kopalnia w Bóbrce" ["The Mine in Bóbrka"] In: *Iwonicz i okolice*, ed. Władysław Bełza. Lviv: Gubrynowicz i Schmidt, 1885, p. 110.

Brzozowski Stanisław, "Ignacy Łukasiewicz", In: *Historia polskiego przemysłu naftowego,* Vol. 1, ed. Ryszard Wolwowicz. Brzozów-Cracow: Muzeum regionalne PTTK im. Adama Fastnachta w Brzozowie, 1994, pp. 16–17.

Brzozowski Stanisław, "Mac Garvey William (1843–1914)", In: *Historia polskiego przemysłu naftowego* ["History of the Polish Oil Industry"], ed. Maciej Boryń, Andrzej Chodubski, Bartosz Duraj. Toruń: Wydawnictwo Adam Marszałek, 2014, p. 411.

Buniatzade Zija Ali, "Witold Zglenicki i problem morskiego wydobycia ropy naftowej w basenie Morza Kaspijskiego" *Kaspijskiego* [Witold Zglenicki and the Problem of Oil Extraction in the Caspian Sea], In: *Historia rosyjsko – polskich kontaktów w dziedzinie geologii i geografii*, ed. Antoni Łaszkiewicz, Henryk Barycz. Warsaw: Zakładad Narodowy im. Ossolińskich, 1972, pp. 32–34.

Chodubski Andrzej, "O zasługach inżyniera geologa Witolda Zglenickiego 'polskiego Nobla" [On the merits of geological engineer Witold Zglenicki, the "Polish Nobel"], In: *Historia polskiego przemysłu naftowego*, ed. Maciej Boryń, Andrzej Chodubski, Bartosz Duraj. Toruń: Wydawnictwo Adam Marszałek, 2014, p. 15.

Cząstka Jan Józef, "Dzieje przemysłu naftowego w Krośnieńskiem" ["History of the Oil Industry in Krośnieńsko"] In: *Krosno. Studia z dziejów miasta i regionu. Tom II (1918–1970)*, ed. Józef Garbacik. Cracow: Państwowe Wydawnictwo Naukowe, 1973, pp. 8–9.

Cząstka Jan Józef, "Ignacy Łukasiewicz twórca polskiego przemysłu naftowego" ["Ignacy Lukasiewicz the creator of the polish oil industry"] In: *Muzeum – skansen przemysłu naftowego im. Ignacego Łukasiewicza w Bóbrce*. Cracow: Przedsiębiorstwo Doświadczalne Naftochem, 1974, pp. 2–3.

Długoborski Wacław, "Górnictwo i hutnictwo do 1918 r." ["Mining and metallurgy until 1918"], In: *Uprzemysłowienie ziem polskich w XIX i XX wieku. Studia i materiały*, ed. Irena Pietrzak-Pawłowska. Cracow-Wroclaw: Ossolineum, 1970, pp. 118–121, 132–135.

Dybiec Julian, "Galicja na drodze do wielkiej przemiany" ["Galicia on the Road to Great Change"], In: *Cracow i Galicja wobec przemian cywilizacyjnych (1866–1914). Studia i szkice*, ed. Krzysztof Fiołek, Marian Stala. Cracow: Universitas, 2011, p. 32.

Franaszek Piotr, "Fabryka Maszyn i Narzędzi Wiertniczych w Gliniku Mariampolskim (do 1918 roku)" ["Factory of Drilling Tools and Machines in Glinik Mariampolski (until 1918)]" In: *Z dziejów przemysłu przed 1945 rokiem*, ed. Jędrzej Chumiński i Marek Zawadka. Wrocław: Wydawnictwo Gajt, 2012, pp. 161–175.

Franaszek Piotr, "Zagłębie Cracowskie" ["The Cracow Basin"], In: *Historia polskich okręgów i regionów przemysłowych*, Vol. 1, ed. Łukasz Dwilewicz, Wojciech Morawski. Warsaw: Polskie Towarzystwo Historii Gospodarczej, 2015, pp. 44–72.

Franaszek Piotr, "Eksploatacja ropy w Galicja a zagrożenia dla środowiska naturalnego" ["Oil Production in Galicia and Environmental Dangers"],

In: *Od regaliów po dobro narodowe. Ochrona i wykorzystanie zasobów środowiska narodowego na ziemiach polskich – aspekt historyczny*, ed. Tomasz Głowiński, Marek Zawadka. Wrocław: Wydawnictwo Gajt, 2016, pp. 39–54.

Franaszek Piotr, "Przemysł Zagłębia Cracowskiego w XIX i pierwszej połowie XX wieku (do 1945 roku)" ["The Industry of the Cracow Basin in the 19th and First Half of the 20th Century (until 1945)"], In: *Jaworzno interdyscyplinarnie. Przemysł pogranicza śląsko-małopolskiego na przestrzeni wieków*, ed. Adrian Rams. Jaworzno-Częstochowa: Muzeum Miasta Jaworzna, 2016, pp. 13–28.

Franaszek Piotr, "Walka wyborcza do galicyjskiego Sejmu Krajowego" ["Electoral struggle for the Galician National Sejm"], In: *Systemy reprezentacji i parlamentaryzm w Europie Środkowej w rozwoju historycznym*, ed. Antoni Barciak. Katowice-Zabrze: PAN, 2016, pp. 213–231.

Jaros Jerzy, "Górnictwo węgla kamiennego" ["Hard Coal Mining"] In: *Zarys dziejów górnictwa na ziemiach polskich*, Vol. 2, ed. Jan Pazdur. Katowice: Wydawnictwo Górniczo-Hutnicze, 1961, pp. 65–67.

Kaliński Janusz, "Bankowość austriacka na ziemiach polskich" ["Austrian Banking on Polish Soil"], In: *Między stabilizacją a ekspansją. System finansowy w służbie modernizacji (z warsztatów badawczych historyków gospodarczych)*, ed. Jerzy Łazor, Wojciech Morawski. Wrocław: Wydawnictwo Gajt, 2014, pp. 270-271.

Kapała Tomasz, "Zawodowe szkolnictwo naftowe" ["Vocational oil education"], In: *Historia polskiego przemysłu naftowego* ["History of the Polish Oil Industry"], ed. Maciej Boryń, Andrzej Chodubski, Bartosz Duraj. Toruń: Wydawnictwo Adam Marszałek, 2014, pp. 303–304.

Lalik Tadeusz, "Propinacja" ["Propination"] In: *Encyklopedia historii gospodarczej Polski do 1945 roku*, Vol. 2. Warsaw: Wiedza Powszechna, 1981, pp. 138–139.

Molenda Danuta, "Zastosowanie ołowiu na ziemiach polskich od XIV do XVII wieku" ["Use of lead on Polish soil from the 14th to the 17th centuries"] In: *Metale nieżelazne na ziemiach polskich od XIV do XVIII wieku*. Cracow-Wrocław: Ossolineum, 1987, pp. 13–14;

Morawski Szczęsny, "Ignacy Łukasiewicz i ropa galicyjska (zeznanie własne)" ["Ignacy Łukasiewicz and Galician oil (own testimony)"] In: *Światek Boży i życie na nim*, ed. Szczęsny Morawski. Rzeszów: Księgarnia Jana A. Pelara, 1871, p. 56.

Morawski Szczęsny, "Ropa w Bóbrce" ["Crude Oil in Bobrka"], In: *Światek Boży i życie na nim*, ed. Szczęsny Morawski. Rzeszow: Księgarnia Jana A. Pelara, 1871, p. 64.

Morawski Wojciech, *Słownik historyczny bankowości polskiej do 1939 roku roku* ["Historical Dictionary of Polish Banking until 1939"], Warsaw: Muza SA, 1998, p. 89.

Orłowski Stanisław, "Trzebinia i region w latach 1815–1918" ["Trzebinia and the District in the Years 1815–1918"], In: *Trzebinia. Zarys dziejów miasta i regionu*, ed. Feliks Kiryk. Cracow: Wydawnictwo i Drukarnia Secesja, 1994, pp. 210–218.

Piesowicz Kazimierz, "Cynk" ["Zinc"], In: *Encyklopedia gospodarcza Polski do 1945 roku,* Vol. 1, ed. A. Mączaka. Warsaw: Wiedza Powszechna, 1981, p. 111.

Szwykowski Z., *Zarys historii Kasy im. Mianowskiego* ["Outline of the history of the Mianowski Fund"], *w: Nauka polska i jej potrzeby, organizacja i rozwój*, Vol. 15. Warsaw: Fundacja Popierania Nauki, 1932, p. 135.

Wolwowicz Ryszard, "Maszyny, urządzenia i narzędzia dla górnictwa naftowego" ["Machinery, Equipment and Tools for Oil Mining"], In: *Historia polskiego przemysłu naftowego*, ed. Maciej Boryń, Andrzej Chodubski, Bartosz Duraj. Toruń: Wydawnictwo Adam Marszałek, 2014, p. 357.

Wójcikiewicz Zdzisław, "Górnik – pierwsze czasopismo naftowe" ["Górnik – the first oil journal"], In: *Krosno Studia z dziejów miasta i regionu,* Vol. III, ed. Stanisław Cynarski. Rzeszów: Krajowa Agencja Wydawnicza, 1995, pp. 119–120.

Zglenicki Tadeusz, *Polski Nobel* ["Polish Nobel"] In: Księga Pamiątkowa Zjazdu Małachowiaków 1180–1957/58. ed. Jakub Chojnacki. Płock: Zakład Graficzny Dom Słowa Polskiego, 1959, pp. 184–186.

Zyga Aleksander, "Pisarstwo polskie o początkach przemysłu naftowego w Galicji" ["Polish Writings on the Beginnings of the Oil Industry in Galicia"] In: *Krosno. Studia z dziejów miasta i regionu*, Vol. III, ed. Stanisław Cynarski. Rzeszów: Krajowa Agencja Wydawnicza, 1995, p. 181.

Internet sources

"75 lat temu pierwszą lampę naftową dla świata zapalił Polak, Był nim Ign. Łukasiewicz – pierwszy destylator nafty – twórca pierwszej kopalni nafty na świecie – twórca przemysłu naftowego i destylacyjnego" [*"75 years ago the first kerosene lamp for the world was lit by a Pole, Ign. Łukasiewicz – the first kerosene distiller – creator of the first kerosene mine in the world – creator of the oil and distillation industry"]*, Ilustrowany Kuryer Codzienny, 1927, nr 342. http://www.instytutpe.pl/inne/165-lat-temu-zaplonela-po-raz-pierwszy-lampa-ignacego-lukasiewicza/ (10 October 2018)

Kozicka – Kołaczkowska, Anna. "Ropa z głębi morza" ["Oil from the Depths of the Sea"], www.instytutpe.pl/wp-content/uploads/2017/09/Historia-sektora-energii-nr-1-2018-Witold-Zglenicki.pdf, http://www.instytutpe.pl/wp-content/uploads/2017/09/Historia-sektora-energii-nr-2-2018-Witold-Zglenicki-cz.-II.pdf (10 October 2018)

Słownik Geograficzny [Geographical dictionary], Vol. 6 http://dir.icm.edu.pl/Slownik_geograficzny/Tom_VI/763 (10 October 2018)

Słownik Geograficzny [Geographical dictionary], Vol. 11 http://dir.icm.edu.pl/Slownik_geograficzny/Tom_XI/532 (10 October 2018)

Staszic, Stanisław. "O Ziemiorodztwie Karpatów i innych gór i równiń Polski" ["On the Carpathian Mountains and other mountains and plains of Poland"] https://jbc.bj.uj.edu.pl/dlibra/doccontent?id=267 (10 October 2018)

Uchwała Senatu Rzeczypospolitej Polskiej z dnia 19 lutego 2003 roku w sprawie uczczenia wybitnych osiągnięć Ignacego Łukasiewicza, pioniera światowego przemysłu naftowego i górniczego [Resolution of the Senate of the Republic of Poland of February 19, 2003 on the commemoration of the outstanding achievements of Ignacy Łukasiewicz, pioneer of the global oil and mining industry]. http://ww2.senat.pl/k5/dok/uch/034/321uch.htm. (10 October 2018)

Materiały źródłowe

"Grupa IX. Wystawa przemysłu naftowego i wosku ziemnego urządzona przez członków Krajowego Towarzystwa Naftowego w Galicji" ["Group IX. Exhibition of the Oil and Earth Wax Industry Organized by Members of the National Petroleum Society of Galicia"], in: *Katalog Powszechnej Wystawy Krajowej we Lwowie 1894* (Lwów: Piller i Spółka, 1894), p. 28.

"Kasy oszczędnościowe i pożyczkowe" *["Savings and loan funds"] Dodatek miesięczny do Gazety Lwowskiej* no. 6 (1872), p. 138.

Adamczewski, P. "Zapis testamentowy Witolda Zglenickiego dla Kasy im. Józefa Mianowskiego, Studia z Dziejów Rosji i Europy Środkowo Wschodniej" ["The Testamentary Record of Witold Zglenicki for the Józef Mianowski Foundation, Studies in the History of Russia and Central and Eastern Europe"], PAN, XLVIII, p. 190.

Archive of the Jagiellonian University, Cracow, WL 231, Letter from F. Sawiczewski to the Dean of the Medical Faculty, UJ of 16 June 1851.

Bałajew, "Crude Oil in the Land of Fire", Baku 1969, p. 118.

Bogactwo Bóbrki ["The Riches of Bóbrka"], Muzeum Przemysłu Naftowego i Gazowniczego im. Ignacego Łukasiewicza w Bóbrce, Bóbrka 2009, p. 21.

Burlikowski, Jan. "Kto założył browar 'Okocim" ["Who Founded the 'Okocim' Brewery"], In: *Kronika miasta Brzeska*, Vol. 4, p. 93.

Burlikowski Jan, "Założenie browaru 'Okocim" ["The Foundation of the 'Okocim' Brewery"], In: *Kronika miasta Brzeska*, Vol. 1, p. 30.

Central State Historical Archive of Ukraine in Lviv (Центральний Державний Історичний Архів м. Львова – dalej: CDIAL), Fond 152, description 2, case 5678, p. 7-7v. Archiwum Uniwersyetetu Jagiellońskiego, Cracow, WL 216, List I. Łukasiewicza do Ministra Wyznań i Oświaty hr. Leona Thun z 17 maja 1851 r.

Hap, Wiesław. "Życie i działalność Ignacego Łukasiewicza na tle początków światowego i jasielskiego przemysłu naftowego" ["The Life and Activity of Ignacy Łukasiewicz against the Backdrop of the beginning of the world's and Jasło's oil industry"] In: *Rocznik Jasielski* vol. V (2003), p. 14.

Horoszkiewicz Julian, "Kasper Cięglewicz. Żywot, prace i cierpienia,"["Kasper Cięglewicz. Life, work and suffering"], In: *Gazeta Narodowa* no. 232 (1886), p. 1, no 233 (1886), p. 1.

"Ignacy Łukasiewicz i ropa galicyjska (Zeznanie własne)" ["Ignacy Łukasiewicz and Galician oil (own testimony)"] In: *1971–1974 Ku czci Ignacego Łukasiewicza*, Komitet Organizacyjny Obchodów Roku Jubileuszowego Ignacego Łukasiewicza przy Stowarzyszeniu Ziemi Krośnieńskiej, red. Zbigniew Ringier, Zdzisław Sroka.

"History of Azerbaijan", Baku 1960, vol. 2, pp. 204–206.

J. G., "Pawilon dla nafty i wosku ziemnego" ["Pawilon dla nafty i wosku ziemnego"], In: *Wystawa Krajowa Rolnicza i Przemysłowa. Organ Komitetu Wystawy* no. 2 (1877), pp. 1–2.

Jabłońska, Teresa, *"Lampa naftowa" ["The Kerosene Lamp"]*, Warsaw: 2012, *Blask lampy naftowej*, Muzeum Piastów Śląskich, Opole 2010, *W kręgu światła lampy naftowej [In the Light of the Kerosene Lamp]*, ed. Jan Gancarski, Krosno Muzeum Podkarpackie, Chorkówka, Muzeum Przemysłu Naftowego i Gazowniczego im Ignacego Łukasiewicza w Bóbrce, 2001

"John Rockefeller zrobił interes na… Łukasiewiczu" [*John Rockefeller did business on… Łukasiewicz"]*, Ilustrowany Kuryer Codzienny, 1927, nr 344.

Lisiczkin, "Distinguished activists of national science and technology related to oil.", Moscow 1967, p. 227.

Lisiczkin, "Distinguished activists of national science and technology related to crude oil", Moscow 1967, p. 227.

Łozińska Janina, "Pamięci Ignacego Łukasiewicza" ["Memories of Ignacy Łukasiewicz"], In: *Gazeta Narodowa* no. 11 (1882), p. 3.

Merunowicz Teofil, "Ignacy Łukasiewicz," In: *Gazeta Narodowa* no. 7 (1882), p. 1.

Oesterreichisches Patentamt, Wiedeń, p. 40, item 400:"*Zeh Johann, Magisster der Pharmacie in Lemberg, Erfindung die natűrliche Bergnaphte auf chemische Wege so zu làutern, dass sie dadurch zu technischen Zwecken unmittelbar vervendbar werde. Privilegium Urkunde 2 December,1853. Priv. Dauer 2 Jahre, geheim.*"

Połonskij, "Absheron Threshold", Moscow 1966, pp. 31–40.

Spis petycji po dzień 16 czerwca 1880 do Sejmu krajowego wniesionych, no. 128.

Stenographic report on the hearings of the Galician National Parliament (cyt. SS), 2 sitting I session IV period of the Galician National Parliament of 9 August 1877, pp. 53–54, 59.

SS, 12 sitting III session IV period of the Galician National Parliament of 1 July 1880, p. 345.

SS, 14 sitting III session IV period of the Galician National Parliament of 5 July 1880, p. 377.

SS, 15 sitting II session IV period of the Galician National Parliament of 9 October 1878, protokół 4 posiedzenia III session IV period of the Galician National Parliament of 9 czerwca 1880,

SS, 19 sitting II session IV period of the Galician National Parliament of 14 October 1878, pp. 597–601.

SS, 2 sitting II session IV period of the Galician National Parliament of 13 September 1878, p. 23.

SS, 21 sitting IV session IV period of the Galician National Parliament of 17 October 1881, p. 113.

SS, 27 sitting III session IV period of the Galician National Parliament of 22 July 1880, p. 931;

SS, 3 sitting I session IV period of the Galician National Parliament of 10 August 1877, p. 72.

SS, 7 sitting II session IV period of the Galician National Parliament of 25 September 1878.

SS, 8 sitting II session IV period of the Galician National Parliament of 27 September 1878, p. 166.

SS, 9 sitting IV session IV period of the Galician National Parliament of 30 September 1881, p. 154.

T. M., "W sprawie kongresu naftowego" ["On the Crude Oil Congress"], In: *Wystawa Krajowa Rolnicza i Przemysłowa. Organ Komitetu Wystawy* no. 2 (1877), pp. 2–3;

Tomanek Ludwik, Ilustrowany Kuryer Codzienny ["Illustrated Daily Courier"], 1927, nr 342.

W. Konstantinowicz Zglenickij. *Information at the meeting of the members of the Imperial Branch of the Russian Technical Society in Baku about the device for measuring curvature*. Papers of the Imperial Branch of the Russian Technical Society in Baku, 1893, September-October, p. 73.

Zeh, Jan. "Pierwsze objawy przemysłu naftowego w Galicji," In: *Przemysł naftowy* z 10 July 1935, [First Appearances of the Oil Industry in Galicia] of 10 July 1935, in: *Przemysł naftowy* [The Oil Industry] pp. 391–393.

Zglenicki Tadeusz, "Polski Nobel", ["The Polish Nobel"], In: Księga Pamiątkowa Zjazdu Małachowiaków 1180–1957/58, Płock 1959, pp. 184–186.

Zglenickij, Witold. *Sources of crude oil in the Kingdom of Poland. Nephtaquellen in Polen, Records of the Mineralogical Society in St. Petersburg*, series II, vol. 15, 1880, pp. 25–30.

Zglenickij, Witold. *About towns on the Absheron Peninsula and outside its borders, where plots for oil exploration should be created, on the basis of temporary rules approved by His Highness on May 14, 1900*, Crude Oil Business, no. 15 of 20.08, 1900, p. 837.

Źródła wschodnie w Archiwum Koronnym do dziejów orientalizacji smaku artystycznego w Polsce XVI – XVIII w., [Eastern Sources in the Crown Archive for the History of Orientalisation of Artistic Taste in Poland in the 16th-18th centuries], Warsaw 1946.

Index of Names

A
Abramowicz, Włodzimierz 187
Aeschylus 16, 261
Alth, Alojzy 104
Anczyc, Stanisław 201
Anczyc, Władysław Ludwik 52, 56–58, 65, 91, 92, 100–103, 108, 127, 130–132, 136, 145, 146, 149, 150, 152, 263, 266, 271
Angerman, Klaudiusz 224, 239, 254, 261, 271
Antoniewicz, Walentyna 134
Antoniewicz, Woysym 30, 32
Apolonia nee Świetlik 27
August III 46
Austen, Jane 18, 33

B
Badeni, Kazimierz 228
Bartkiewicz, Magdalena 26, 27, 261
Bartoszewicz, Stefan 100, 109, 215, 226, 230, 231, 232, 237, 240, 241, 243, 247, 249, 252, 256, 257, 268, 271, 272
Battaglia, Roger 202
Bełza, Władysław 91, 93, 111, 128, 145, 278
Benckendorff, Count 176, 180, 186
Benis, Artur 202, 207, 261
Bergheim, John Simeon 152, 216, 219, 222, 227, 236, 248, 249, 254, 255
Biechoński, Wojciech 109, 115, 116, 123, 240, 241, 272
Bielski, Zygmunt 91, 93, 106, 112, 119, 127, 155, 243, 244, 271, 272
Błażowski, Antoni 241, 243, 251, 267, 278
Bobrowska, Helena 99
Bobrowski, Napoleon 93, 99
Boczkowski, Father 49
Bogusławski, Wojciech 43
Bratkowski, Adam 69, 70, 71, 73, 214
Brzozowski, Piotr 253
Butowt-Andrzejkowicz 16
Byron, George 18, 28

C
Chałubiński, Tytus 187
Chekhov, Anton 175
Chłędowski, Kazimierz 82, 92, 100, 112, 126, 130, 262, 272
Chodźka, Józef 179
Chodźko-Borejko, Aleksander 14
Chopin, Fryderyk 18, 28
Choroszewski, Wincenty 170
Cięglewicz, Kasper Kufstein 137, 283
Ciołek, Erazm 12, 287
Coghen, Konrad 113
Copernicus, Nicolaus 12, 155
Curie-Skłodowska, Maria 33, 155, 157
Czarnecki, Jan 37, 44
Czartoryska, Jerzowa 128
Czerwiakowski, Ignacy 57
Czerwiński, Stanisław 221, 249, 273

D
D'Este, Ferdynand 69
Davies, Norman 12, 13, 21, 24, 263
de Lance, Karol 214
Dembowski, Edward 37, 40, 41, 48
Dittmar, Rudolf 25, 73, 74
Długosz, Jan 17

Długosz, Władysław 215, 219, 220, 240, 243, 244, 267, 269, 273
Doms, Robert 216
Drake, T. Edwin 75, 217
Dunka, de Sajo Władysław 242–244, 270
Dzieduszycki, Włodzimierz 202
Dzwonkowski, Edward 143

E
Edison 18
Erbreich, Leonard 206

F
Fabiański, Julian 243, 247
Fauck, Albert 95, 96, 107, 215, 220–222, 247, 248, 252, 253, 255, 273, 274
Fedorowicz, Mikołaj 104, 109, 235, 254
Fibich, Władysław 151, 245, 254
Fleming, Aleksander 55
Fraenkel, Samuel Anton 173
Franz Joseph I of Austria 56
Fredro, Aleksander 44

G
Galileo 12
Gąsiorowski, Kazimierz 243–245, 268
Gebauer, Aleksander 135, 136
General Legedisch 38
Gerlach 37
Glazor, Stanisław 104, 242
Goethe 18
Goetz-Okocimski, Jan Albin 203
Goetz-Okocimski, Jan Ewengelista 203
Gołuchowski, Agenor 198
Gołuchowski, Wojciech 154
Gombrowicz 21
Gorayski, August 115, 117, 123, 146, 149, 150, 201, 217, 230, 240, 274
Grabowski, Julian 107, 116

H
Habsburg, Ferdinand 56
Hacker Józef
Hecker, Józef 64, 67
Heindl, J. 72
Hirschler, Maciej 143
Hitler, Adolf 194
Horytiak, Jan 110, 137, 216
Howarth, Józef 221

I
Suszycki, Zenon 152, 241, 243, 245

J
Jabłecki, Tomasz 110, 216
Jabłonowski, Stanisław 79
Jabłoński, Adolf 97, 137, 138, 143, 144, 152, 243, 245, 251, 253, 263
Jabłoński, Bonawentura 137
Janiszewski, Stanisław 221
Jarry, Alfred 21, 263
Jaruzelski, Wojciech 192
Jastrzębski, Ferdynand 103, 142, 244, 274
Jefferson, Thomas 26
Jodłowski, Andrzej 121
Jurski, Stanisław 215, 243, 251, 253, 270

K
Kanty, Edmund Jan 201
Kanty, Leon 201
Kącki, Kacper 137
Klobassa, Helena
Klobassa, Karol 82, 84, 90, 91, 97–99, 102–104, 107, 119, 127, 135, 136, 148, 160, 215, 277
Klobassa, Stanisław 152
Klobassa, Wiktor 149, 152
Kluczyński, Józef 137
Kluk, Jan Krzysztof 17
Koschina, Edward 38

Kościuszko, Tadeusz 9, 11, 26–31, 33, 39, 46, 266
Krasiński, Zygmunt 45, 46, 264
Kraszewski, Józef Ignacy 37, 143, 144
Krynicka, Karolina 43, 49, 51

L
Łacki, Kasper 110
Laskowski, Felicjan 86, 87, 90, 92, 93, 103, 109, 214, 264
Lenin, Iljicz 9, 163, 170, 190, 191, 913, 274
Leo, Juliusz 201
Liebig, Justus 64
Linde, Samuel Bogumił 17, 265
Lisiecki, Józef 174
Łodziński, Felicjan 217, 235, 241, 251
Łozińska, Janina 149, 283
Longchamps de Berier, Jan 242
Longchamps de Berier, Mieczysław 242
Louis XIV 22
Löwenherz, Oskar 115, 117, 152–155, 275
Lubomirska, Izabela of the Czartoryski 206, 254
Lubomirski, Andrzej 202
Łukasiewicz, Aleksander 30–32
Łukasiewicz, Apolonia 27, 33, 34, 42, 43, 47
Łukasiewicz, Emilia 30, 32, 35, 43, 47–49, 84, 137, 141
Łukasiewicz, Franciszek 30
Łukasiewicz, Ignacy 9, 11, 13, 17, 19, 20, 23, 24, 26–38, 40–48, 51–63, 65–74, 76, 79, 81–97, 99–112, 114–122, 124–161, 163, 164, 167, 170, 174, 194, 195, 198, 213–215, 217, 227, 229, 230, 235, 240, 241, 244, 247, 248, 251, 256, 257, 259–261, 263–266, 269, 271–280, 282–284
Łukasiewicz, Józef 26, 28, 30–33, 48
Łukasiewicz, Maria 30, 43
Łukasiewicz, Władysław Anczyc 56–58, 91, 100, 101, 131, 136, 150

M
MacGarvey, William Harry 10, 152, 195, 216, 217, 219, 222, 227, 236, 240, 247–250, 252, 254, 255
MacIntosh, George 236, 250
Małachowski, Stanisław 167
Maślanka, Marcin 221, 245, 264
Mendelejew, Dymitr 263
Merunowicz, Teofil 115, 119, 122, 126, 130, 131, 133, 150, 202, 284
Metternich von, Clemens 38, 42, 46, 56
Mianowski, Henryk 202
Miączyński, Konstanty 145
Mickiewicz, Adam 16, 24, 25, 27, 35, 46, 47
Mikolasch, Piotr 51–53, 55, 56, 58, 60–67, 69–71, 73, 76, 82, 213, 214
Mikucki, Leon 215, 222, 240, 243, 244, 268
Miński, Kazimierz 245
Mrożek, Sławomir 21

N
Napoleon, Bonaparte 196
Nobel, Alfred 177, 178, 190
Norwid, Cyprian Kamil 11, 163, 265
Noth, Juliusz 95, 104

O
Olszewski, Stanisław 117, 118, 201, 220, 224, 225, 229, 230, 233, 238, 241, 243, 244, 253, 270, 276, 277

Orzeszkowa, Eliza 169, 265
Osiecki, Apolinar 104, 138
Otto von, Bismarck 111

P
Pawlewski, Bronisław 201, 246
Pawlikowski, Józef 29, 46, 265
Pelletier, Joseph 213
Penderecki, Krzysztof 21
Perkins, Cyrus 250
Perkins, Jacob 250
Pik, Maria 137
Pik, Wojciech 137
Pik-Mirandola, Franciszek 72
Pilat, Stanisław 233
Pilat, Tadeusz 201
Plater, Emilia 35
Plater, Grażyna 35
Poniatowski, Stanisław 57
Potocki, Adam 206
Potocki, Alfred 41
Potocki, Paweł 190
Prometeusz 28, 263

R
Radziwiłł, Michał 172, 211
Raszka, Jan 154
Rau, Henryk 206
Rau, Wilhelm 206
Ressig, Emil 62
Rockefeller, John Davidson 25, 105, 170, 229, 233, 283
Roeske, Wojciech 53, 54, 74
Rogawski, Walery 80, 84, 87, 133
Rollinson 47
Romanowicz, Tadeusz 202
Rothschild 180, 185, 186, 192, 193
Rutowski, Tadeusz 202
Rzączyński, Gabriel 17

S
Sapieżyna, Jadwiga 128
Sawiczewski, Florian 59, 282
Schmerling, Anton von 198
Schreiner, Abraham 62, 73, 74, 153
Scott, Elgin 250
Setkowicz, Paweł 242
Silliman, Beniamin 74, 75
Siterski, Walery 221
Skrzyńska, Kajetanowska 49
Skrzyński, Adam 227, 254
Smarzewski, Seweryn 104
Smoleński, Władysław 188, 192, 193
Smolka, Franciszek 150
Sobieski, Jan 21, 207
Stacherska, Alojza 43
Stacherska, Emilia 35, 47, 48, 135, 141, 151
Stacherska, Józefa Honorata Maria 84, 134, 137, 268
Staszic, Stanisław 9, 17, 155, 163, 171–173, 265, 282
Stawiarski, Seweryn 151, 275
Stebnicki, Ignacy 179
Stierman, Lejba 62
Stocker, Wiliam 104, 107, 109
Stokowski, Apolinary 150
Stokowski, Feliks 104
Stopczyński 98
Strauss, Johann 18
Suszycki, Zenon 138, 152, 241, 243, 245
Swoboda, Antoni 34, 48
Syroczyński, Leon 215, 223, 240, 245, 247, 267, 267, 268, 277
Szajnocha, Władysław 102, 112, 113, 120, 266
Szaynok, Władysław 243
Szczepanowski, Stanisław 200, 201, 215, 218, 235, 236, 240, 243, 244, 248, 250, 254, 266, 275
Szczepański, Aleksander 202, 233, 266
Szczęsny, Morawski 63, 82, 99, 102, 280
Szczęsny, Potocki 113

Szela, Jakub 39, 40, 266
Shakespeare, William 28
Szpaczek 100
Szumsk, Kazimierz 245

T
Tałasiewicz, Jan 37, 44, 45, 48
Tarłowski, Aleksander 41, 42, 44, 50
Timoftiewicz, Julian 218
Tomanek, Ludwik 25, 26, 61, 73, 109, 117, 122, 128, 131–134, 137, 140, 145, 151, 153, 154, 266, 284
Trzecieska, Anna 90
Trzecieski, Tytus 40, 81, 82, 84, 90–92, 98, 99, 103, 104, 119, 152, 160, 214, 215, 240, 241, 278

V
da Vinci, Leonardo 12, 264
Vitello 11, 12, 262

W
Wachtel, Henryk 244
Wagemann, Gustav 105, 254
Walter, Filip 213
Walter, Loeben Henryk 95, 99, 102, 104, 107, 123, 216, 243, 244, 273
Wesołowska, Helena 248
Wiesiołowski, Franciszek 37, 40, 44, 45, 47, 48
Windakiewicz, Edward 100–102, 105, 109, 110, 113, 114, 266
Winkelried 29
Winogradov, Maria 178, 188, 192
Witkacy 21
Włodkowic, Paweł 167
Wolski, Wacław 215, 217, 221, 222, 235, 240, 241, 243, 244, 252, 273, 277, 278
Wrotnowski, Andrzej 202
Wrzosek, Gabriel 17
Wyrzykowski, Aleksander 170
Wyspiański, Stanisław 23, 24, 267

Z
Zabłocki, Feliks 150
Załoziecki, Roman 246, 252
Załuska, Olimpia 174
Zamojski, Franciszek 243
Zdanowicz, Ludwik 243
Zeh, Jan 62, 63, 65–68, 71, 73, 76, 82, 93, 214, 284, 285
Zeitleben, Jan 250, 251, 278
Zejszner, Ludwik 57, 66, 71
Zeus 16
Zglenicki, Bolesław 167–169, 174, 188
Zglenicki, Konstanty 167
Zglenicki, Witold 9, 18, 163–174, 176–178, 180–182, 184–186, 188–190, 192, 194, 225, 260, 262, 272, 273, 275, 279, 282, 285
Zieliński, Apolinary 88
Zieliński, Eugeniusz 88
Zieniewicz, Porofir 88
Zima, Franciszek 109
Znamierowski, Józef 109
Zuber, Rudolf 101, 224, 241, 251, 254, 267, 278

Geographical Index

A

Absheron Peninsula 10, 171, 172, 175, 179, 180, 281
Africa 12
Agulhas 12
Amiradzan 190
Argentina 251
Artpatakan 16
Aterpatakan 16
Ateszgiah 14, 262
Atlantic 12
Austrian 9, 10, 21, 29–31, 33, 38–40, 44–46, 49–51, 56, 62–64, 76, 83, 87, 95, 98, 105, 109, 111, 120, 124, 127, 131, 137, 161, 167, 172, 195–202, 204, 207–209, 211–213, 219, 222, 228, 229, 231–234, 237, 239, 243, 244, 248, 252, 262, 263, 276, 280
Azerbejdżan 179, 262
Azja 11

B

Baku 9, 14, 16, 163–165, 170, 175–192, 194, 220, 225, 275, 282, 283, 285
Bardejov 134, 142
Benkowa Wisznia 44
Berlin 69, 217
Bochnia 35, 204
Bohorodczan 65
Borysław 5, 57, 58, 61, 64, 69, 98, 109, 116, 120, 121, 149, 191, 209, 210, 212, 214–217, 219, 220, 222, 224, 228, 232–234, 241, 244, 245, 256, 269, 270
Bóbrka 5, 9–11, 15, 16, 32, 68, 71, 77–81, 85–89, 91–95, 97–110, 114, 115, 123, 124, 144, 148, 149, 151–153, 155–157, 159, 162, 168, 172, 191, 192, 194, 196, 198, 200, 202, 204, 206, 208, 210–214, 216, 218, 220, 222, 224–226, 228, 232, 234, 236, 238, 240, 242–244, 246, 247, 248, 250, 252, 253, 255, 256, 258, 267–269, 272, 274, 278
Bratforth 231
Brody 193
Brzostek 9, 83, 84, 86
Bukareszt 233, 253, 264
Bukowina 206

C

Canada 5, 191, 212, 213, 244
Chmielów 169
Chorkówka 5, 15, 39, 75, 89, 90, 95, 98–101, 103–105, 107, 108, 110, 113, 115, 116, 123–125, 128, 129, 131–135, 137, 139, 141, 142, 144, 147, 151, 155, 223, 250, 255, 256, 279
Chorzelow 26
Copenhagen 171
Czajkowa 26
Czechowice 225, 229

D

Dąbrowa 55
Dębe 164, 165, 170, 188
Drohobycz 57–59, 63, 70, 209, 216, 221, 224, 227–229, 245
Dukla 9, 34, 38, 91, 117, 211, 241
Dynow 26

E

Erkelenz 218
Estonia 60

Europe 7–9, 13–15, 17–19, 22, 29, 47, 50, 51, 53, 65, 69, 72, 94, 95, 107, 161, 174, 185, 196, 201, 205, 208, 209, 213, 218, 259, 271, 278

F
Filipowice 202
Franzesbad 130, 138

G
Galicja 96, 195, 205, 223, 258, 275
Gdańsk 258, 260
Glinik Mariampolski 151, 232, 233, 250, 251, 275
Gorlice 9, 34, 36, 57, 71, 72, 75–77, 79, 80, 82, 83, 89, 100, 111, 112, 114, 128, 151, 152, 154, 163, 210, 211, 214, 223, 232, 236, 244, 245
Góry Luszowskie 202
Greece 7
Greenwich 10
Grojec 201

H
Hamburg 65, 68, 105
Harklow 120, 249
Huntington 244

I
Iwonicz 87, 94, 250, 274

J
Jarosław 124
Jasło 9, 34, 44, 71, 75, 78, 80, 82, 83, 86, 89, 91, 96, 99, 111, 114, 144, 145, 163, 214, 216, 224, 236, 238, 245, 250, 260, 265, 268, 279
Jaworzno 201, 202, 276
Jeleń 203
Jerusalem 8

K
Kałusz 200, 207
Karachchuri 182
Karlsbad 130, 138
Karlsruhe 239
Kaukaz 172, 175, 183, 260, 269
Kielce 167, 170, 183, 230
Klęczany 84, 86, 110, 250, 267
Kobylanka 214
Kołaczyce 147
Kombornia 115
Korczyna 123, 142
Kraków 8
Krosno 9, 10, 15, 22, 57, 65, 66, 68, 71, 77, 78, 90, 96, 100, 103, 109, 114–118, 122, 125, 133, 140, 142–145, 147, 148, 150–152, 163, 197, 214, 217, 223, 241, 255, 256, 260, 265, 275, 277, 279
Krynica 130, 138
Krzeszowice 201

L
Leśniówka 89
Leżajsk 31, 34, 38, 39, 123
Libiąż 201–203
Libusza 214, 223
Limanowa 112, 225, 229
Lipinki 214
London 8, 9, 15, 36, 56, 171, 173, 182, 197, 239
Lwów 262
Łańcut 30, 31, 37–40, 44, 46, 58

M
Maczki 207
Małopolska 202, 276
Męcin 105
Middle East 15
Mielec 25, 26
Moczydło 202
Munich 72, 139
Moscow 40, 166, 170, 171, 179, 180, 259, 279, 280

Mroczkow 167–169
Myślenice 207

N
New Delhi 8
New Zealand 247, 270
Niżna Łąka 109
Nordhausen 218
Norway 7
Nowosielce Gniewoszowe 100
Nowy Sącz 84

O
Okocim 199, 200, 262, 279, 284
Ontario 244
Oświęcim 203, 205

P
Pacific 8
Paris 15, 23, 29, 43, 51, 56, 65, 69, 96, 133, 140, 182, 239, 248, 266, 267
Peczeniżyn 120, 250
Pennsylvania 5, 70, 71, 92, 176, 191, 213, 216, 243, 247
Petersburg 5, 15, 159, 161, 164–167, 169, 170, 172, 178, 179, 281
Pilzno 35
Płock 151, 160, 162–164, 170, 173, 176, 184, 185, 256, 258, 268, 269, 277, 281
Podbuża 121
Podgórzyn 100
Polanka 36, 71, 77, 78, 87–89, 105, 210, 223
Polanka near Krosno 77
Poland 5, 7–9, 13, 15, 17–20, 22–25, 27, 29, 35, 36, 42, 53, 57, 58, 65, 68, 69, 87, 93, 96, 107, 108, 116, 125, 149, 151, 152, 154, 156, 162, 163, 167, 169, 174, 176, 182, 184, 185, 187–189, 191, 192, 196, 198–201, 208, 211, 212, 214, 222, 224, 229, 230, 231, 236, 243, 247, 258–262, 265, 271, 278, 281
Potok 145, 214
Poznań 29, 33,124, 125, 140, 151, 219, 251, 261, 264
Prague 62, 63, 239
Prószków, (Silesia) 139
Prussia 17, 36, 88, 191, 201, 203, 204, 208
Przemyśl 40, 95, 98–100, 105, 111, 139, 203, 228, 237, 249, 264, 268, 273
Pstążnica 169

R
Rabka 94
Ropa 100
Ropianka near Dukli 91, 100, 103, 148, 241, 273
Russia 17, 24, 25, 52, 60, 108, 154, 161, 165, 167, 169, 170, 171, 173, 181, 183, 185, 186–191, 204, 207, 208, 216, 245, 247, 257, 278
Rudki 49
Rumunia 225, 276
Riga 9, 38, 148, 163, 174, 177
Rymanów 128, 138, 148
Rzeszów 34, 50, 280, 281

S
Sambor 62, 65–67, 71, 227
Sanbuczyn 179
Sanok 104, 148, 204, 210, 216, 236, 255
Schodnica 120, 124, 228, 235, 249, 254
Sękowa 218
Siary 109
Siersza 205, 206, 211, 265
Słoboda Rungurska 124, 235, 248
Słupsk 95, 215, 247
Solina 104
Sołotwina 65

Stalingrad 194
Stanisławów 61, 221, 227, 232, 245, 264
Stary Sambor 227
Stebnik 231
Strasbourg 207
Sucha 208
Suchedniów 171–174
Sumatra 251, 275
Surachany 189, 190
Swoszowice near Cracow 209
Szczecin 117
Switzerland 29, 69, 169
Śląsk 19

T
Tarnów 32, 37, 104, 109, 267
Tatra 57, 143
Tenczynek 205, 206, 267
Titusville 75, 217
Trzebinia 205, 206, 209, 228, 229, 281
Tustanowice 230, 237, 238, 249

U
Uherce Mineralne 104
Ulaszowice 84, 90–92, 227
USA 64, 180, 217, 220, 235, 247, 251

V
Vatican 22

W
Warsaw 9, 11, 17, 19, 23, 24, 26, 28, 29, 35, 37, 40, 47, 53, 57–59, 61, 79, 99, 101, 109, 111, 129, 131, 138, 155, 163–166, 168–171, 173, 174, 178–180, 187–189, 193, 196, 198, 201, 204, 205, 207, 209, 210, 243, 261–267, 278–281, 283
West Point 29
Węgierska Górka near Żywiec 208
Wiedeń 107, 275, 284
Wieliczka 57, 102, 204
Wiener Neustadt 95
Wietrznica 113
Wola Kiełpińska near Zegrze 188
Wrocanka 113
Wyoming 248

Z
Zachwiejow 30
Zaczerń 49, 50
Zaduszniki 30, 31, 59, 134
Zagłębie Cracowskie 206, 207, 261, 277
Zakopane 208
Zielona Góra 155
Zręcin 82, 99, 111, 114, 119, 127, 135, 136, 148, 151, 152
Zwierzyniec 206
Zwięczyca 38, 49
Żarki 205
Żeglce 127
Żmigród 119, 124
Żołyń 35
Żywiec 203, 208

This text is from Peter Lang publishing house:
This book is dedicated to the life and work of Ignacy Łukasiewicz, Polish pharmacist whose world-renowned achievements include construction of the world's first oil refinery and invention of the modern kerosene lamp. The authors also portray the history of the Galician oil industry and set it in a context of political, social and technological changes taking place in the 19th-century Central and Eastern Europe.

"The work adds substantially to existing scholarship in English. As the author of the only English-language academic monograph devoted to a general history of the Galician oil industry, I can attest that this manuscript adds significant and important information, details, depth of investigation that is not provided in my book or any other book. It therefore makes a novel contribution that will be very valuable to anyone looking for a truly detailed account of Ignacy Łukasiewicz's contribution within the context of the Galician oil industry in general."
Alison Frank Johnson
Professor of History and of Germanic Languages and Literatures
Harvard University, Center for European Studies

In colourful language, the authors sketch the profiles of two outstanding Poles, pioneers of the oil industry – Ignacy Łukasiewicz, MSc. in Pharmacy, and mining engineer and geologist Witold Zglenicki, called the Polish Nobel. They were not scholars from university faculties, but explorers who brought great added value to the development of civilisation worldwide. Łukasiewicz, active in Austrian Galicia, extracted energy from "rock oil", which in the following decades radically changed the rhythm of human life, while Zglenicki, a generation younger, whose fate drew him to Baku in the Russian empire made available to mankind, through the construction of a drilling platform, the rich reserves of crude oil on the seabed. The book presents the values that both Łukasiewicz and Zglenicki professed and were faithful to – patriotism, and respect for culture, science and religion. It shows people full of passion, giants among workers, visionaries, and sober realists, courageous, excellent organizers, and promoters of progress. Nor does it overlook many of the problems they had to face before they were successful. It takes the reader on a journey to the 19th-century world where the pioneers of the oil industry lived, i.e. to the lands of south-eastern Poland, which at that time functioned as the northern province of the Habsburg Empire far from Vienna, as well as to Baku, where the oil industry developed dynamically and the Nobel brothers, among others, ran their extensive oil business.

The book also shows the universality and timelessness of the values recognized by the Polish pioneers of the oil industry, which constitute a good model of conduct for contemporary entrepreneurs. This scientific work is an interesting and captivating read. It can be used not only by scientists and students, but also by everyone who is interested in industrial cultural heritage and also those who are or intend to be entrepreneurs.

Prof. Krzysztof Broński
Head of the Department of Economic and Social History
Cracow University of Economics

www.ingramcontent.com/pod-product-compliance
Ingram Content Group UK Ltd.
Pitfield, Milton Keynes, MK11 3LW, UK
UKHW021421091026
15624UKWH00010B/316

9 783631 792919